A Mechanistic Exploration of Natural Compounds for Neuronal Health

Sadaf Jahan • Arif Jamal Siddiqui
Andleeb Khan
Editors

A Mechanistic Exploration of Natural Compounds for Neuronal Health

Springer

Editors
Sadaf Jahan
Department of Medical Laboratory Sciences
College of Applied Medical Sciences
Majmaah University
Al-Majmaah, Saudi Arabia

Arif Jamal Siddiqui
Department of Biology
College of Science, University of Hail
Hail, Saudi Arabia

Andleeb Khan
Department of Biosciences
Faculty of Science, Integral University
Lucknow, Uttar Pradesh, India

ISBN 978-981-95-0124-3 ISBN 978-981-95-0125-0 (eBook)
https://doi.org/10.1007/978-981-95-0125-0

This Springer imprint is published by the registered company Springer Nature Singapore Pte Ltd.
The registered company address is: 152 Beach Road, #21-01/04 Gateway East, Singapore 189721, Singapore

Preface

The proposed book, *Unlocking Neuronal Health: A Mechanistic Exploration of Natural Compounds*, will delve into the fascinating realm of natural compounds and their mechanistic effects on restoring and promoting neuronal health. This book provides a comprehensive overview of the underlying mechanisms through which natural compounds exert their neuroprotective properties, offering a deeper understanding of their therapeutic potential for neurodegenerative disorders. By exploring the intricate interactions between natural compounds and neuronal pathways, this book seeks to bridge the gap between scientific research and practical applications, facilitating the development of novel therapeutic strategies. Chapters included in the book provide detailed information regarding the neurodegenerative disorders, neuronal complications, and reasons behind the culprit. Moreover, many chapters shed light on the mechanistic approach and involvement of natural compounds as therapeutic strategies, thus helpful in the restoration of neuronal health. Ultimately, this book will contribute to the advancement of knowledge in the field and inspire further research and development in this promising area of neuroscience.

Neurodegenerative disorders, such as Alzheimer's disease, Parkinson's disease, and Huntington's disease, pose significant challenges to individuals, families, and healthcare systems worldwide. While conventional treatments offer limited symptomatic relief, the search for more effective and disease-modifying therapies continues. The use of natural compounds derived from plants in the treatment of neurodegenerative disorders is an area of growing interest. These compounds possess diverse chemical structures and biological activities that can target multiple underlying mechanisms involved in neurodegeneration. Through a mechanistic approach, natural compounds offer several potential ways to mitigate the progression of neurodegenerative disorders and promote neuronal health, like anti-inflammatory, anti-oxidative, neurotransmitters, etc.

We would like to express our gratitude to all authors who participated to complete the book. Without their support, this project seems to be incomplete. We also appreciate the role of Springer staff who constantly supported us to complete the project.

There are numerous people who deserve thanks. I, Sadaf Jahan, would like to begin by thanking the Almighty and my late grandma, Mrs. Akhtari Begam, for their unceasing support. My parents, Mr. Furkan Ali and Mrs. Farhat Sultana, my brothers, Mr. Shoeb Ali and Mr. Suhail Ali, and all family members have been a major source of inspiration for me throughout my life. My husband Muhammad Alam, my son Azlaan Muhammad Umar, and my daughter Huda Jahan Alam are great blessing and support to complete this book successfully.

I, Arif Siddiqui, express my special thanks to my parents Mr. Hakimuddin Siddiqui and Mrs. Shahnaz Begum, wife Ruhi Parveen, son Ariz Siddiqui, daughter Ayzal Siddiqui, and all family members for their everlasting love, enthusiasm for science, and encouragement to pursue every task successfully.

I, Andleeb Khan, would like to express my thanks to the Almighty for giving us this vision. My sincere thanks to my parents, Mr. Jameel Uddin Khan and Late Mrs. Anis Un Nisa. My sister, Qudsiya Fatima, has always guided me in all the wax and wane of life. I would like to thank my husband Dr. Syed Wajahat Ali and my kids Syed Muhammad Ali and Muhammad Shayaan Syed for their seamless support while compiling this book.

Al-Majmaah, Saudi Arabia — Sadaf Jahan
Hail, Saudi Arabia — Arif Jamal Siddiqui
Lucknow, Uttar Pradesh, India — Andleeb Khan

Contents

About the Editors

Sadaf Jahan is currently working as Associate Professor in the Department of Medical Laboratory Sciences, College of Applied Medical Sciences, Majmah University, Saudi Arabia. She has earlier served as Assistant Professor (Microbiology) in the Department of Microbiology, Chinmaya Degree College, H.N.B. Garhwal University, Haridwar, India (2018–2019). Previously, she also worked as a National Post-Doctoral Fellowship through DST-Science and Engineering Research Board (SERB) at CSIR-Central Drug Research Institute, Lucknow, India, between 2016 and 2018. In 2009, she qualified for a National Level Exam (UGC CSIR NET-JRF) in Life Sciences and pursued a PhD in Biological Sciences in 2010 at CSIR-Indian Institute of Toxicology Research (IITR), Lucknow, India, through the Academy of Scientific and Innovative Research (*AcSIR*), New Delhi, India, and completed her PhD in 2015. Her research experience and interest are broadly focused on developing and validating in vitro model systems to study developmental neurotoxicity of environmental chemicals and drugs using chemically induced differentiating neural cells, derived from human umbilical cord blood stem cells. She is an editorial board member of *Biotechnology and Bioinformatics*. She has served as referee for a number of international journal publishers, including MDPI (Biomolecules, Pharmaceuticals, JCM), Frontiers in Genetics, Frontiers in Pharmacology; SAGE Open Medical Case Reports; International placenta stem cell society (IPLASS); Oxidative Medicine and Cellular Longevity; Scientific Reports; PLoS One; Toxicology International; etc. She has more than 10 years of research experience in neurosciences and stem cells.

She has also published more than 100 research articles in the peer-reviewed international journals and authored or co-authored many book chapters. She is a member of many international scientific societies and organizations such as International Placenta Stem Cell Society (IPLASS), International Society for Neurochemistry (ISN), and American Society for Neurochemistry, and Life Member of Indian Academy of Neuroscience (IAN), India, and Society of Free Radical Research (SFRR), India. She has reviewed more than 100 manuscripts for reputed journals.

She is running several projects as PI (principal investigator) or Co-PI obtained from the Deanship of Research, Majmaah University, as well as other Saudi universities. Since 2020 to present, she has also been a member of Quality Assurance

Committee at the Department of Medical Laboratory Sciences, Majmaah University, Saudi Arabia.

Arif Jamal Siddiqui is currently working as Associate Professor of Biology Department, University of Hail, Saudi Arabia. He is an Associate Editor in Chief of Infection and Drug Resistance (Taylor and Francis), Associate Editor of three Biotech (Springer Nature), Academic Editor of Journal of Immunology Research, PeerJ, Brain Science, Stem Cells International, BioMed Research International, Canadian Journal of Infectious Diseases and Medical Microbiology, Advisory board member of Heliyon and other journal involve as an Editorial board member. Previously, he has worked as a Post-Doctoral Fellowship in Texas Tech University Health Sciences Center, Lubbock between 2015 and 2018. In 2011, he was awarded a scholarship by the Indian Council of Medical Research (ICMR), New Delhi, India, for Higher Education Council to pursue a PhD in Parasitology and Immunology at the CSIR-Central Drug Research Institute (CDRI), Lucknow, India, and completed his PhD in 2015. His research experience and interest are to study host-parasite interaction, role of stem cells in neurodisorder diseases, immune response (specific T, B, MQ, dendritic cells and specific memory T, B related study) during parasite interaction, and drug and vaccine development against parasites infection. Currently, he is working on parasite diversity, novel role of stem cells in various neuro-related diseases, natural bioactive compound role in cancer, immune behaviors and responses, and diagnosis and vaccine development for *Schistosoma mansoni*, *Schistosoma japonicum*, and *Schistosoma haematobium* (trematode worm) and *Plasmodium* parasites (malaria). He has more than 12 years of research experience in the field of studying clinically and economically important human pathogens. He has served as referee for a number of International journals, including *Pathogens* (MDPI), *Journal of Biomolecular Structure and Dynamics* (Taylor and Francis), *Current Drug Targets* (Bentham Science), *Biology* (MDPI), *Vaccine* (MDPI), *Animals* (MDPI), *Pathogens* (MDPI), *Antioxidants* (MDPI), *Molecules* (MDPI), *Veterinary Sciences* (MDPI), *Antibiotics* (MDPI), *Microorganism* (MDPI), *Life* (MDPI), *Frontiers in Nutrition* (Frontier), *Infection and Drug Resistance* (Dove Press), *Parasites and Vectors* (BioMed Central), *Immuno Targets and Therapy* (Dove Press), *BMJ Open*, *PLOS One*, *Environmental Science and Pollution Research* (Springer), *International Journal of Nanomedicine* (Dove Press), *Infectious Disorder—Drug Targets*, and many more. He has also published more than 100 research articles in peer-reviewed international journals and authored or co-authored many book chapters. He is running several projects as PI (principal investigator) or Co-PI obtained from the Deanship of Research, University of Hail, as well as other Saudi universities. He is also a member of the Indian Science Congress, India.

Andleeb Khan PhD, is an Associate Professor at the Department of Biosciences, Faculty of Science, Integral University, Lucknow, India. She has completed her PhD from Jamia Hamdard, New Delhi, India, in Toxicology with a specialization in natural and active constituents from plants for prevention of Alzheimer's disease. She has teaching and research experience of more than 12 years in neuroscience and neurotoxicology. Dr. Khan has taught many subjects like Toxicology, Pharmacology, Biochemistry, Molecular Biology, Neurology, Molecular Pharmacology, Pharmacogenomics, Biochemistry, Natural Chemistry, Drug and Cosmetics, and Alternative Medicine. Her accomplishments in academics in the Master course are outstanding, and she was able to get the Gold medal, the highest achievement for the topper of this course. She has published more than 100 research publications (citation: 4983, H-index: 38), has been involved in 10 research grants, and contributed many book chapters with international publishers. She is the editor of books with international publishers. She is also an Editorial board member and reviewer of various international journals and is the principal investigator of some significant projects on neurodegenerative diseases and their prevention. She has received many awards, appreciation, and recognition for her services to science. She holds life memberships in various international organizations and is an active participant in national and international scientific events. Currently, Dr. Khan is engaged in studying the molecular mechanism of Alzheimer's disease, ischemic stroke, and Parkinson's disease, and their prevention by various active principles from natural sources and nanoformulations.

Abbreviations

6-OHDA	6-hydroxydopamine
AA	amino acids
Ach	acetylcholine
AD	Alzheimer's disease
AIF	apoptosis-inducing factor
ALS	amyotrophic lateral sclerosis
AO	antioxidant
APP	amyloid precursor protein
ARE	antioxidant response element
Aβ	amyloid beta
Aβ	beta-amyloid peptide
BBB	blood-brain barrier
CAT	catalase
CNS	central nervous system
COX	cyclooxygenase
CPS	caspase
CREB	cAMP-response element binding
CYP	cytochrome P450
DOPAC	3,4-dihydroxyphenylacetic acid
DP	dopamine
DR	death receptor
ER	endoplasmic reticulum
ERK	extracellular signal-regulated kinase
ETC	electron transport chains
FR	free radical
GAP-43	growth-associated protein 43
GCLc	glutamate cysteine ligase catalytic subunit
GCLM	glutamate cysteine ligase modifier subunit
GSH	glutathione
GSH/GSSG	glutathione disulphide
GSH-Px	glutathione peroxidase
GSK-3	glycogen synthase kinase
GSR	glutathione reductase

GST	glutathione S-transferase
H_2O_2	hydrogen peroxide
HD	Huntington disease
HNE	4-hydroxynonenal
HO•	hydroxyl radicals
HO-1	haem oxygenase-1
IL-1β	interleukin 1-beta
IL-6	interleukin-6
JNK	c-Jun N-terminal kinase
LOX	lipoxygenase
MAO	monoamine oxidase
MAPKs	mitogen-activated protein kinases
MAPs	microtubule-associated proteins
MDA	malondialdehyde
MMP	matrix metalloproteinases/mitochondrial membrane potential
MMP, ΔΨm	mitochondria membrane potential
MPO	myeloperoxidase
MT	mitochondrial
NADPH	nicotinamide adenine dinucleotide phosphate hydrogen
NFTs	neurofibrillary tangles
NF-κB	nuclear factor-kappaB
NLRP3	nucleotide-binding oligomerization domain-like receptor
NMDA	N-methyl-D-aspartate
NO	nitric oxide
NO_2 •	nitrogen dioxide radical
NOS	nitric oxide synthase
NOX	NADPH oxidase
NR	NMDA receptor
Nrf2	nuclear factor erythroid 2-related factor 2
NS	nitrosative stress
ONOO−	peroxynitrite anions
OS	oxidative stress
OXPHOS	oxidative phosphorylation
PARP	poly ADP-ribose polymerase
PD	Parkinson's disease
PI3K	phosphatidylinositol 3-kinase
PLC	phospholipase C gamma
RC	respiratory chain
RNS	reactive nitrogen species
ROS	reactive oxygen species
SDH	succinate dehydrogenase
SN	substantia nigra
SOD	superoxide dismutase

SYN	synuclein
TNF-α	tumour necrosis factor alpha
Trx	thioredoxin
UPS	ubiquitin proteasome system
VDAC1	voltage-dependent anion channel-1
XO	xanthine oxidase

Mechanisms of Neuronal Damage and Dysfunction

1

Uzair Ahmad Ansari

Abstract

Neurodegenerative disorders represent a significant burden on global health and are characterized by progressive deterioration of neurons, which leads to cognitive and motor dysfunction. Understanding the mechanisms underlying neuronal demise is crucial for developing effective therapeutic strategies. Neuronal damage often stems from the interplay of multiple factors, including oxidative stress, excitotoxicity, protein misfolding, mitochondrial dysfunction, neuroinflammation, and impaired neurotrophic support. Several key signaling pathways involved in the regulation of these factors are NADPH oxidase, ionotropic glutamate receptors such as N-methyl-D-aspartate (NMDA) receptors and α-amino-3-hydroxy-5-methyl-4-isoxazolepropionic acid (AMPA) receptors, nuclear factor kappa B (NF-κB), mitogen-activated protein kinase (MAPK) pathways, receptor tyrosine kinases such as Trk receptors and the downstream phosphoinositide 3-kinase (PI3K)/Akt, and unfolded protein response (UPR). Understanding the intricate crosstalk between these pathways is essential to develop targeted therapeutic interventions aimed at halting or slowing neurodegeneration. Emerging therapeutic strategies include antioxidant therapies, modulators of excitotoxicity, proteostasis regulators, mitochondrial enhancers, anti-inflammatory agents, and neurotrophic factors. This chapter explores the complex pathways underlying neuronal damage and dysfunction, with a focus on the critical molecular and cellular mechanisms involved. A deeper understanding of these processes holds promise for the development of innovative therapeutic strategies aimed at mitigating neurodegeneration and improving patient outcomes

U. A. Ansari (✉)
CSIR-Central Drug Research Institute, Lucknow, India

S. Jahan et al. (eds.), *A Mechanistic Exploration of Natural Compounds for Neuronal Health*, https://doi.org/10.1007/978-981-95-0125-0_1

Keywords

Neurodegeneration · Oxidative stress · Mitochondrial dysfunction · Protein aggregation · Neuroinflammation · Excitotoxicity

1.1 Introduction

Our brain is an intricate web of billions of neurons, specialized cells that communicate with each other to process and transmit information throughout the nervous system. These neurons form complex networks that regulate everything from basic reflexes to higher cognitive functions like thinking, memory, and emotions (Herculano-Houzel 2009; Telesford et al. 2011). However, like all cells in the body, neurons can suffer damage and dysfunction, leading to a host of neurological disorders that can profoundly affect the quality of life (World Health Organization 2006). Neuronal damage and dysfunction can arise from various causes, each affecting the delicate balance and intricate functions of the nervous system in unique ways. Neurodegenerative diseases, such as Parkinson's disease (PD), amyotrophic lateral sclerosis (ALS), and Alzheimer's disease (AD), progressively destroy neurons, leading to symptoms like loss of memory, impaired movement, and muscle weakness (Erkkinen et al. 2018; Yacoubian 2017). The exact mechanisms behind these diseases are still being unraveled, but they often involve the buildup of abnormal proteins that disrupt cellular processes and eventually kill neurons (Maiti et al. 2017).

In this exploration of neuronal damage and dysfunction, we delve into the various causes, mechanisms, and potential treatments that form the forefront of neurological research. By gaining a deeper understanding of how our neurons become damaged and how we might counteract these processes, we move closer to mitigating the devastating impacts of these conditions and improving the lives of those who suffer from them.

1.2 Cellular and Molecular Mechanisms of Neuronal Damage

The cellular and molecular levels of neuronal injury must be thoroughly explored to comprehend why and how these damages occur in the brain. The neurons of the brain and nervous system function as tiny messengers, transmitting and receiving messages that support our ability to move, think, feel, and do an array of other functions (Evans-Martin 2022). However, malfunctions can cause damage to neurons. Several pathways might cause this harm at the cellular level. One of them is excitotoxicity, in which the neurons are harmed by an excess of excitation caused by certain chemicals (Mehta et al. 2013). Another factor is oxidative stress, which is triggered by noxious substances called free radicals and is comparable to the rusting of cells (Guo et al. 2013). Neural damage can also result from excessive or protracted inflammation, which is the body's reaction to an injury or illness (Aktas et al. 2007). The complexity increases much more at the molecular level. The energy source of cells, the mitochondria, can malfunction and cause cell death. If calcium levels get too high, it can become dysregulated and wreak chaos. Calcium is an essential chemical for neuron signaling (Calvo-Rodriguez and Bacskai 2021). Under some circumstances, neurons can undergo apoptosis, a process that results in programmed cell death and ultimately leads to their mortality. Furthermore, misfolded or aggregated proteins can build up within neurons, impairing their functionality and eventually causing death (Ochneva et al. 2022).

It is important to fully understand these cellular and molecular pathways of neuronal damage in order to prevent or cure ailments such as traumatic brain injury, neurodegenerative diseases, and stroke. Researchers can create more effective strategies to safeguard neurons and sustain the health of our brains by understanding what goes wrong at the tiniest scales.

1.2.1 Excitotoxicity

Cell death brought on by harmful effects of excitatory amino acids is known as excitotoxicity. This process has been specifically connected to hyperactivation of the glutamate receptor, which can have negative effects such as generating free radicals and disrupting calcium homeostasis (Chen and Lipton 2006). The demise of dopaminergic neurons in the substantia nigra (SN) results in significant reduction of dopamine levels in the striatum, which is a hallmark of excitotoxicity in PD (Rama and García 2016; Regoni et al. 2024). This process is intimately linked to oxidative stress, which plays a critical role in the initiation and progression of neuronal damage in PD. Excitotoxicity, more precisely, the overactivation of NMDA receptors has been connected to the pathogenic mechanisms of AD-related chronic neurodegeneration (Yu et al. 2023). The disturbances in L-glutamate-using systems point to a crucial role in the onset of neurodegeneration. Overactivation of NMDA receptors has been connected to degeneration caused by Aβ in AD. The authors proposed that excitotoxicity occurs when both synaptic *N*-methyl-D-aspartate receptors (sNMDARs) and extrasynaptic *N*-methyl-D-aspartate receptors (eNMDARs) are activated simultaneously (Yu et al. 2023). Any change in equilibrium that would decrease sNMDAR or increase eNMDAR signaling is probably going to be harmful to the survival of neurons. Hence, it is necessary to maintain the appropriate amount of NMDAR signaling both inside and outside of the synapse. This level should be sufficient to sustain neuronal viability and activity, but not so high as to result in excitotoxic neurodegeneration. Mild NMDAR activity increase, which is frequently observed in AD patients and animal models, does not have an immediate lethal effect. Nonetheless, persistent NMDAR activation and Ca^{2+} dysregulation that goes through for months or years may be harmful for gradually developing events, such as deleterious excitotoxicity, in the course of AD/AD-related dementia (ADRD) development (Ma 2023). Specifically, excitotoxicity is driven by Ca^{2+} influx through eNMDARs, along with downstream signaling regulated by transient receptor potential cation channel subfamily M member (TRPM). (Hong et al. 2023). The NMDAR component GluN3A, on the other hand, functions as a "gatekeeper" for NMDAR activity and provides neuroprotection against both acute and long-term excitotoxicity (Bauminger and Gaisler-Salomon 2022).

Excitotoxicity may contribute to the localized vulnerability of neurons and is consistent with the neuropathology observed in AD. It was suggested that reduced glutamate absorption from the synaptic cleft, which is connected with the compromised EAAT2 activity in presynaptic astrocytes, is the cause of increased extracellular glutamate concentrations that mediate Aβ-induced excitotoxicity (Passlick et al. 2021). In addition to the excitotoxicity, oxidative stress is a crucial component in the progression of AD. The presence of oxidative stress complicates the neurotoxic effects of glutamate and Aβ further and may accelerate the degenerative processes.

1.2.2 Oxidative Stress

All eukaryotic life forms require oxygen to function properly. Since it has a high redox potential and can readily take electrons from reduced substrates, it plays an important role in survival. Depending on their metabolic requirements, various tissues have varying oxygen requirements. Since the brain makes up just around 2% of the body weight, it consumes more than 20% of the oxygen in the body (Siwicka-Gieroba et al. 2022). The two main kinds of brain cells—neurons and astrocytes, are primarily accountable for this tremendous O_2 and glucose consumption. Even while oxygen is necessary for all living things, excessive oxygen levels may be harmful, even neurotoxic. The molecular makeup of oxygen determines its toxicity and chemical activity. Specifically, decreased level of reactive oxygen species (ROS) exists in different forms, such as hydrogen peroxide (H_2O_2), hydroxyl radical (OH•), and superoxide ($O_2^{\bullet-}$). The term "ROS" is used nowadays to refer to both oxygen radicals and non-radicals (O_3, H_2O_2, $O_2^{\bullet-}$) that can be quickly transformed into free radicals. The hydroxyl radical (OH•) is one of the most reactive ROS. Different ROS have varying levels of reactivity. ROS have a high level of reactive activity, which causes them to chemically interact with biological components and alter how cells operate or even destroy them (Simpson and Oliver 2020).

A high level of ROS damages cells, a condition known as oxidative stress. It is brought about by a major disruption in equilibrium between the generation of ROS and the level of antioxidants. Since these reactive species alter the way biomolecules operate, they play a role in the progress of neurodegeneration. Targeting several substrates inside the cell, ROS can result in DNA oxidation, lipid peroxidation, protein oxidation, and RNA oxidation. The brain's high concentration of polyunsaturated fatty acids (PUFAs), particularly arachidonic acid and docosahexaenoic acid, makes it susceptible to oxidation, leading to the formation of malondialdehyde (MDA) and 4-hydroxynonenal (HNE) (Jaganjac et al. 2021).

The brains of AD patients exhibit signs of ROS-mediated damage; the brain and CSF fluid of AD patients have higher concentrations of malondialdehyde and 4-hydroxynonenal in comparison to control groups (Dakterzada et al. 2023). In addition, AD samples have higher levels of hydroxylated guanine than their age-matched controls (Graille et al. 2020). These findings from the human brain are consistent with those from *in vivo* models of AD, wherein the cortical and hippocampal regions showed increased levels of protein and lipid peroxidation prior to the development of plaque or neurofibrillary tangle (NFT) pathology (Pádua et al. 2024). The brain of PD exhibits a decrease in PUFA content in the substantia nigra, accompanied by an increase in lipid peroxidation indicators like MDA and HNE (Barrera et al. 2018). Protein carbonyls, a byproduct of protein oxidative degradation, are seen in the PD brain when compared with the healthy group (Fabelo et al. 2011). Additionally, there is some evidence that suggests the reactive nitrogen species in the brain of PD may have a role in the nitration and nitrosylation of several proteins. Apart from elevated 8-hydroxydeoxyguanosine levels in PD brain, there is a proven rise in frequent loss of mitochondrial DNA among the healthy dopaminergic neurons in the SN of PD. These results show that oxidative stress may cause the deleterious effect on mtDNA (Puspita et al. 2017).

Numerous neurodegenerative conditions, such as Huntington's disease, AD, PSP, PD, ALS, Friedreich's ataxia, neurodegeneration with buildup of iron in the brain region, and optic atrophy, are associated with hampered mitochondrial activity. There has been significant interest in mitochondrial damage and the production of ROS within the context of PD, evidenced by studies conducted in human brain samples from both toxin-induced and genetic disease models. It has been shown that PD patient's lymphocytes, platelets, and substantia nigra have decreased mitochondrial complex I activity. Since, previously, it was shown that the neurotoxic 1-methyl-4-phenyl-1,2,3,6-tetrahydropyridine (MPTP) is responsible for the PD-like symptoms in animal models, it has been widely utilized as an *in vivo* model of PD. Research has shown that the active metabolite of MPTP, 1-methyl-4-phenylpyridinium (MPP^+), can obstruct electron transport by showing affinity to the same location to that of rotenone, the known complex I disruptor, resulting in the creation of a lesser amount of ATP. The observation that rotenone and MPP+ generate superoxide anions in submitochondrial particles further reinforces the hypothesis that MPP+ primarily functions as a mitochondrial toxin. Since antioxidants may efficiently block MPP^+ and rotenone, it is more likely that oxidative stress, rather than metabolic alterations, is the origin of their neurotoxic effects. Intoxicated (MPP^+, rotenone) mice models of PD show reduced ROS generation when the modest uncoupling of mitochondria is combined with the elevated expression of uncoupling protein 2 (UCP2). While UCP2 increased expression reduces MPTP-mediated cell death in the region of nigral dopamine, UCP2 deficiency likewise enhances the susceptibility of dopamine neurons to MPTP. The discovery of many PD-related genes (PINK1, DJ-1, Parkin) that are highly correlated with mitochondrial function supports the idea that oxidative stress caused by mitochondrial malfunction is a major factor in the pathophysiology of Parkinson's disease. Oxidative stress is caused by DJ-1 loss of function, and DJ-1 protects neurons by means of its antioxidant mechanism in mitochondria. PD is linked with the mutations in the PINK1 gene. The mitochondrial kinase PINK1 is implicated in defective respiration.

Previous studies have shown that PINK1 deficiency inhibits complex 1, reduces substrate availability, and increases ROS generation in mitochondria similar to rotenone. Due to a reduction in mitochondrial calcium efflux, PINK1 reduced expression also causes mitochondrial calcium excess, which makes it difficult for the body to cope with cytosolic calcium difficulties. The generation of ROS and mitochondrial Ca^{2+} work together to open the mitochondrial permeability transition pore (PTP); this change permits pro-apoptotic chemicals to be translocated from the mitochondria into the cytosol and ultimately results in cell death. Models of PD that have been linked to PINK1 gene mutations and toxicant-induced (MPP^{+}) models have demonstrated the early activation of the PTP. More recently, PINK1-coordinated activity with parkin to control the autophagic clearance of damaged mitochondria has become apparent. The failure of mitophagy caused by the deleterious effect on either the parkin or PINK1 gene leads to the buildup of defective mitochondria and increased generation of ROS, which exacerbates the oxidative stress on the neuron.

Another hallmark of neurodegenerative diseases is abnormal protein aggregation, and it seems that ROS generation and mitochondrial malfunction may have an impact on the aggregation of α-synuclein, a protein that builds up in every PD brain (Feng et al. 2021). Autosomal-dominant PD in a family type is brought on by mutations in the α-synuclein gene. When mutant α-synuclein is expressed in neurons or mice models, mitochondrial malfunction and increased generation of reactive oxygen species are seen (Jiang et al. 2007). Nevertheless, α-synuclein oligomerization and toxicity also depend on protein oxidation brought on by the generation of ROS in the mitochondria. Recently, it has been depicted that α-synuclein binds to mitochondria and causes mitochondrial fragmentation by inhibiting membrane fusion. α-synuclein binds to mitochondria and leads to the fragmentation of mitochondria. Interestingly, this phenotype may be avoided by co-expressing the mitochondrial PD proteins PINK1, parkin, and DJ-1.

In animal models of AD, cybrid cells and mitochondria extracted from the hippocampal region and platelets of AD patients showed a reduction in complex IV functioning (Swerdlow 2023). The buildup of β-amyloid (βA) precedes plaque pathology and is related to oxidative stress, dysfunctioning in mitochondria, and energy depletion. In Alzheimer's disease (AD), disruptions in calcium balance have been noted, with β-amyloid (βA) causing elevated cytoplasmic calcium levels and mitochondrial calcium overload. This imbalance leads to increased reactive oxygen species (ROS) production and induces the opening of the permeability transition pore (PTP), contributing to neuronal damage (Boyenle et al. 2022). In primary astrocytes and isolated mitochondria, βA can cause PTP to open. Moreover, βA may directly bind to cyclophilin D, a component of the PTP, resulting in the formation of a complex within the mitochondria. This complex exhibits a decreased threshold for opening in the presence of mitochondrial PTP (mPTP) inducers (Shevtsova et al. 2021).

1.2.3 ROS Production via NADPH Oxidase

There is strong evidence linking NADPH oxidase and oxidative stress to AD, despite the fact that other alternative pathways of neurodegeneration have been postulated. In the early stages of postmortem AD brain, there is overexpression of NOX1 and NOX3, while NOX2 activation was found in the brains of AD patients (Ma et al. 2017). Animal models have likewise verified the NADPH oxidase critical role in AD. Therefore, NOX2 deficiency led to better results in an AD mouse model. Animals lacking NOX2 were crossbred with Tg2576 mice, which carry the Swedish mutation of APP leading to the accumulation of Aβ fragments. In this model, the harmful effects of βA deposition were prevented due to the absence of efficient NOX2 activity. Both young and old Tg2576/NOX2-deficient animals showed improvements in their behavioural abnormalities, and neuronal oxidative stress was eliminated (Sorce and Krause 2009). It was also suggested that NADPH oxidase has a role in AD at the cellular level. NADPH oxidase was directly activated by βA in primary cultures of rat microglial cells and human phagocytes. Microglial NOX is activated by βA via the B-class scavenger receptor CD36 (Gandhi and Abramov 2012). In order for active NADPH oxidase to operate normally, an ion/anion channel must be opened in order to compensate for charge transfer by moving protons across the membrane (DeCoursey et al. 2003). Through the blockage of NADPH oxidase charge-compensating mechanism, suppression of the CLIC1 channel in βA-activated microglia prevented the formation of superoxide and shielded the cells (Milton et al. 2008). Because βA-mediated induction of NADPH oxidase occurs in conjunction with large NO production and peroxynitrate formation, it may cause harm to nearby cells (Brown 2007). When NADPH oxidase is activated, oxidative stress is produced. This depolarizes the membrane of mitochondria, and together with calcium, it opens the mPTP and modifies the membrane structure by activating phospholipase C (Mironova and Pavlov 2021). Neighbouring neurons get exposed to the signal of oxidative stress, which are more susceptible than astrocytes. Oxidative stress was observed in rotenone and MPTP-mediated toxin model of PD. These models show that NADPH oxidase (NOX2) is activated in microglia (Chéret et al. 2008). Moreover, pharmacological inhibition of NADPH oxidase has been shown to protect mesencephalic dopaminergic (N27) cells from MPP+-induced dopaminergic loss (Anantharam et al. 2007). Genetic models of Parkinson's disease (PD) also demonstrate elevated oxidative stress levels. In one such model, the loss of PINK1 function in midbrain neurons is linked to increased ROS production by NADPH oxidase (Fiadeiro et al. 2024). It's exciting to observe that a high cytosolic calcium concentration activates NADPH oxidase, which increases superoxide generation. The plasmalemmal glucose transporter is inhibited by ROS generation from NADPH oxidase, which leads to a dysregulation of mitochondrial metabolism.

In the AD mouse model related with the presenilin 2 gene, oxypurinol was able to lower ROS generation and safeguard neurons (Abe et al. 2004; Angelova 2021). Furthermore, βA has the ability to stimulate H_2O_2 synthesis in the cytoplasm of neocortical neurons. In lethal PD models induced by paranonylphenol and MPP(+), the rat striatum exhibited a significant reduction in OH production when treated with the XO inhibitor allopurinol, suggesting that XO may play a role in the oxidative stress associated with PD (Obata et al. 2001).

1.2.4 Mitochondrial Dysfunction

The term "mitochondrial dynamics" refers to the balance between fission and fusion processes that maintain mitochondrial function and integrity. These continuous cycles of fission and fusion are essential for cellular homeostasis. Mitochondria play a crucial role in energy production by generating adenosine triphosphate (ATP) through oxidative phosphorylation, providing the necessary energy for neuronal function. The ATP biogenesis may be impacted by excessive mitochondrial fragmentation or malfunction, which can result in mitochondrial dysfunction and be fatal to neuronal capacity to survive and thrive. Excessive mitochondrial fission causes brain damage because mitochondrial activities are crucial. Research has shown that the activity of mitochondrial fission proteins is related to hallmark features in the brains of individuals with ND, including the accumulation of βA plaques, hyperphosphorylated tau, and α-synuclein (Taghizadieh et al. 2023). Defects in the mitophagy pathway, indicated by PINK1/parkin abnormalities, reflect the connection between mitochondrial alteration and the pathology of these disorders. Deficits in PINK/Parkin not only affect quality control but also have the potential to cause neuroinflammation, which accelerates or is involved in the progression of ND (Goudarzi et al. 2021).

The mutated α-synuclein was found to be localized to mitochondria where it combines with spectrin and translocates Drp1, which causes mitochondria to degrade (Devi and Anandatheerthavarada 2010). Misfolded α-synuclein accumulates erroneously and is found within Lewy bodies, which are an intracerebral hallmark of PD patients (Mahul-Mellier et al. 2020). Moreover, traumatic brain injury, aging, and oxidative stress all contribute to a greater risk of developing PD. The active metabolite MPTP is called MPP+. MPP+ preferentially penetrates nigrostriatal dopamine neurons and severely impairs the respiratory chain, which increases oxidative stress. The MPP+ has been shown to cause excessive mitochondrial fission by causing Parkin to become S-nitrosylated and Drp1 Ser616 to get phosphorylated (Montagna et al. 2020). Yang et al. have demonstrated that fly models with mutated PINK1 and Parkin exhibit key characteristics of PD, including motor disability, death of dopaminergic neurons, mitochondrial dysfunction, and reduced longevity. (Yang et al. 2006). Disruption of the pathway of PINK/Parkin suggests that the effects of AD and PD may overlap (Xie et al. 2020). It is known that mitophagy contributes to anti-neuroinflammation, and it has been proposed that it counteracts neuronal injury through quality control of mitochondria (Lautrup et al. 2019). A study on the role of mitophagy in inborn immunity suggests that PINK1 and Parkin could be classified as inhibitors of inflammation. PINK1-Parkin-induced mitophagy can reduce inflammation, as well as inhibit innate immune responses (Rakovic et al. 2019). It's interesting to notice that Parkin, which has been activated by AMPK may have tumor-suppressive properties. The AMPK and Parkin axis prevents the development of the RIPK1-RIPK3 necrosome and suppresses inflammatory carcinogenesis by encouraging the polyubiquitination of RIPK3 (Lee et al. 2019). Research on Parkin have showed that, at least partially, mitochondrial quality control mediates the connection between PD pathogenesis and PINK1- and Parkin-induced mitophagy (Ge et al. 2020).

Mitochondrial shrinkage occurs during the initial phase of mitochondrial fission, triggered by the interaction between the endoplasmic reticulum (ER) and the outer mitochondrial membrane (OMM) at sites where the ER and mitochondria come into contact (ER-mitochondria contact sites, or EMCS) (Friedman et al. 2011). FIS1, MFF, and MiDs are examples of mitochondrial-bound proteins that function as "membrane-anchored adaptors" (Koirala et al. 2013). These proteins support the Drp1 recruitment to EMCS in its oligomeric forms. Subsequently, Drp1 attaches to mitochondria, creates a ring-shaped structure that strengthens the preexisting neck of constriction in the mitochondria and ultimately triggers fission. The significant functions that EMCS plays in mitochondrial dynamics and associated disorders have been revealed by ongoing research in this field. The EMCS links intracellular calcium management to mitochondrial stability, which is essential for proper mitochondrial function (Yang et al. 2023). Phosphorylation of Drp1 is the most well-known example of post-translational modification, which is essential for GTPase functioning. Mitochondrial fission is caused by cdk1/cyclin B kinase phosphorylating serine 616 on the GED of Drp1 during mitosis, which promotes its oligomerization (Guo et al. 2018). A fascinating new body of work suggests that it is possible for PINK1 to phosphorylate Drp1 on S616. This promotes the fission process, which separates the mitochondrion into the damaged and healthy parts. The injured mitochondria will be further broken down by PINK1/parkin-mediated mitophagy following the fission process.

It is significant to remember that Mfn1 and Mfn2 most likely have separate functions in the mitochondrial fusion process, and that inner mitochondrial membrane (IMM) fusion happens after outer mitochondrial membrane (OMM) fusion. In particular, Mfn2 is necessary for OPA1-induced IMM fusion and does not depend much on Mfn1 (Maruyama et al. 2024). However, while Mfn2 has been the subject of previous study, it is unknown how Mfn1 and OPA1 could interact. Since mitochondrial fission and fusion together constitute a cohesive dynamic process comprising several pathways, a change in one of these links is likely to have a broad effect on mitochondrial dynamics, potentially leading to neural malfunction. This means that while treating mitochondrial dynamics, we must pay more attention to the chain reaction and downstream repercussions.

At synaptic terminals, mitochondria perform several vital tasks and act as local energy sources. One characteristic of some age-related neurodegenerative cases is the absence of synaptic mitochondria. Neural synapses depend on viable mitochondria to operate; thus, the movement and elimination of damaged mitochondria as well as the appropriate distribution of normal ones are essential. At the distal end, newly produced organelles have to move down the axon to restock the mitochondria (Cheng et al. 2022). The clearing and repair of mitochondria depend on retrograde transport. According to recent studies, mitophagy and mitochondrial retrograde transport are coordinated (Han et al. 2020a, b). This study focused on three proteins: the dynein-SNAPIN motor-adaptor complex, BNIP3L/Nix, and RHEB (Ras homolog abundant in brain). In non-neuronal cells, RHEB has been shown to target mitochondria for autophagy. According to this work, neuronal axons with RHEB-targeted mitochondria have dominant retrograde movement and high motility. It also states that the membrane protein BNIP3L/Nix makes it easier for RHEB to associate with injured mitochondria. Phagophores then ingest these combinations to create mitophagosomes, which only travel retrogradely in axons. For mitophagic clearance, degradative lysosomes are known to be mostly found in the soma of neurons (Cai and Ganesan 2022). Increasing SNAPIN levels in AD neurons enhances the downstream transporting motility of mito-amphisomes. These mito-amphisomes then fuse with LEs loaded with the dynein and SNAPIN transport complex, reducing mitophagic occurrence and ameliorating axon mitochondrial abnormalities (Han et al. 2020a, b). This work offers novel mechanistic insights into the mechanisms underlying the coordination between mitophagy and synaptic mitochondria's retrograde transport, as well as how impairments in retrograde transport exacerbate mitochondrial abnormalities and synaptic degeneration linked to AD.

Huntington's disease (HD) is a neurodegenerative disorder inherited in an autosomal-dominant manner, resulting from genetic mutations in the huntingtin protein (Htt), specifically an expanded CAG triplet repeat in the polyglutamine (polyQ) region. Research has indicated that mitochondrial dysfunction plays a significant role in the progression of HD. Along with a wide range of other symptoms, a progressive decline in control over thought, movement, and emotional expression is a hallmark of HD. It has been shown that peripheral blood cells from HD patients exhibit mitochondrial degeneration (Askeland et al. 2018). They demonstrate that as the quantity of Drp1 increased the levels of Mfns and reduced Opa1. According to a recent study, Drp1 hyperactivation can be inhibited by using P110, which prevents the Drp1-Fis1 interaction, which is sufficient to reduce neuropathology and behavioural abnormalities linked to HD (Rios et al. 2023). This might aid in the creation of a viable therapy for HD treatment. A recent research found that via interacting with Drp1 on mitochondria, the mutated Htt could increase its GTPase functioning (Sawant et al. 2021). Moreover, different research observed that the mutated Htt might be involved in the elevated expression of NO, which may be a plausible mechanism of this interaction (Teleanu et al. 2022). When Drp1 interacts with NO, it is S-nitrosylated, which causes excessive mitochondrial fission and damage to neurons (Lee and Kim 2018). Interestingly, AD has also been reported to exhibit this SNO-Drp1. Furthermore, this study demonstrates that mutant Htt can undergo S-nitrosylation, allowing it to transmit NO to Drp1, and causes the Drp1 to be S-nitrosylated. A research revealed that Drp1 S-nitrosylation facilitates the Drp1 and mutated Htt to interact (Song et al. 2011; Sesaki et al. 2014). Alterations in dynamics of mitochondria in neuronal cells produced by the mutated Htt were consistently reduced by the introduction of the non-nitrosylatable altered Drp1 (C644A).

1.2.5 Apoptosis, Autophagy, and Necrosis

Neurodegeneration involves the widespread death of neurons through various mechanisms, including autophagic cell death, apoptosis, and necroptosis. Neuronal death in neurodegenerative diseases often results from a combination of various death pathways, with different pathways predominating in different conditions. The extent of interaction between these pathways depends on the cell's energy status and the severity of the stimuli. Understanding these complex mechanisms is crucial for designing effective therapeutic strategies.

Necrosis occurs when cells cannot withstand stress that exceeds their threshold for tolerance. Necrosis-induced pathological neuronal death frequently occurs alongside neurodegeneration. Highly susceptible neurons suffer selective neuronal necrosis as a result of ischemic brain damage (Taoufik and Probert 2008). It was stated that extreme loss of energy and the activated Ca^{2+} reliant breakdown of proteins cause neurons to die (Yamashima 2004). Gradients of ion, necessary for the survival of neurons, disrupted when energy is lost. Following this sequence of events, glutamate builds up at synaptic clefts (Grewer and Rauen 2005). Death of downstream target neuronal cell at the synapse by necrosis results from increased glutamate buildup in the synaptic clefts (Al-Nasser et al. 2022). Metabolic disruption, which results in an excessive release of ROS and necrosis, is a typical abnormality of neurodegeneration (Simpson and Oliver 2020; Bhatt et al. 2021). According to studies by Saito et al. (1993) and Adamec et al. (2000), the release of calpain and cathepsin D and E can result in death by necrosis in AD and temporary frontal brain ischemia, respectively (Yamashima 2013; Di Domenico et al. 2016). Aspartyl proteases ASP-3 and ASP-4, which are homologs of human cathepsin proteases in *C. elegans*, have also been connected to cell death by necrosis (Yamashima 2004). Cathepsin spatial localization is important for neurodegeneration. Cathepsins are localized in lysosomes in young rat brain cortices but are shown to be cytosolically localized in old rat brain cortices (Stoka et al. 2016). More research indicates that death of cells by necrosis could be exhibiting a well-synchronized process rather than a chaotic architecture. Therefore, necroptosis is the programmed type of cell mortality, characterized by cell inflammation and malfunctioning mitochondria. Necroptosis is mediated in large part by the receptor-interacting serine/threonine protein kinase1 (RIPK1) and RIPK3 (Vandenabeele et al. 2010; Ni et al. 2019). RIPK1 phosphorylates RIPK3, which is its downstream target. RIPK1 and RIPK3 can bind together due to the RIP homotypic interacting domains. These interaction motifs contain the consensus strands VQVG (RIPK3) and IQIG (RIPK1), respectively. Wang et al. state that when mixed lineage kinase domain-like (MLKL) is phosphorylated, RIPK3 can cause phosphorylation and activation of it (Wang et al. 2014). The necrosome is known to be the central molecular complex of RIPK1 and RIPK3. They also cause the lysosome and plasma membrane to permeate, which releases reactive oxygen species and lysosomal content. This causes cells to lyse by necrosis and ultimately leads to cell death (Seo et al. 2021).

One of the most important intracellular breakdown processes for the clearance of abnormal proteins that build up inside the cell is autophagy. Beyond the autophagy core function of disintegrating damaged organelles and protein deposition, it is also linked to cell cycle dysregulation, senescence, genomic instability, cellular energetics, metabolism, and changes in subpopulation of stem cells. Chaperone-mediated autophagy (CMA), known as macro-autophagy, is a type of selective autophagy that has been shown to be important in the process of neurodegeneration and to specialize in protein degradation (Auzmendi-Iriarte and Matheu 2021; Wang and Lu 2022). Numerous pathophysiological conditions, including dementia, cancer, and cardiovascular diseases, have been linked to the autophagic response. Damaged proteins and organelles build up inside the cell as a result of autophagy defects (Rana et al. 2021). Alpha-synuclein is the primary component of aggregation in PD, whereas mutant huntingtin is the main component in HD (Srinivasan et al. 2021). Since neurodegenerative conditions in mice are caused by suppression of basal levels of autophagy, increasing autophagy to get rid of macromolecule aggregates would make sense as a treatment strategy for neurodegenerative illnesses. In fact, using specific proteins like p62 and hsc70, autophagy may remove macromolecule aggregates or inclusion bodies with specificity (Johansen and Lamark 2011). Autophagy stimulation reduced the severity of disease in models of HD and ALS and elevated the removal of a-synuclein in PC12 cells model of PD (Walter et al. 2016; Zou et al. 2017). Huntingtin inclusions sequester Beclin 1 and trigger autophagy. According to Yue et al., mice deficient in either of the two essential components of the autophagy system, Atg5 or Atg7, had inclusions of ubiquitin, aberrant response, and neuronal cell death (Yue et al. 2009; Park et al. 2020).

The development of the central nervous system depends on apoptosis, which mediates exact and programmed normal neuronal death. Other important types of cellular apoptosis include ferroptosis and mitoptosis, which denote iron-dependent and mitochondria-dependent cell death, respectively. However, the pathologic mechanism of neurodegeneration is associated with a number of chronic disease states, premature apoptosis, or inadequate apoptosis control in PD, HD, and ALS (Hartmann et al. 2001; Bhatia and Sharma 2021; Hickey and Chesselet 2003), spinal muscular atrophy, and other neurological disorders (Ghavami et al. 2014). Activation of neuronal cytomembrane receptors triggers several processes, including caspase activation, increased intracellular calcium (Ca^{2+}) levels, and the production of reactive oxygen species (ROS). These factors play a key role in initiating programmed cell death (Nita and Grzybowski 2016). High ROS and Ca^{2+} concentrations are involved in supporting the intrinsic apoptotic pathway. Numerous signals, Bcl-2 proteins, elevated Ca^{2+} marker, and generation of ROS all trigger the mitochondrial route of apoptosis (Morris et al. 2021). There is a release of pro-death molecules into the cytoplasmic region when these signals build up at the mitochondria. Cytochrome-c (cyt-c), a pro-death protein found in the mitochondrial membrane's electron transport chain, is one example with particular significance for neurodegenerative diseases (Kalpage et al. 2020). After entering the cytoplasmic region, cyt-c binds to the complex of the apoptosome via nucleotide, procaspase, and Apaf-1. This combination then leads to neuronal death by promoting protein activation.

Cell death provides the essential elements of life and orchestrates life during brain development. Research has been conducted to understand the function of p53 in the process of apoptosis (Wang et al. 2023). Fascinatingly, current evidence also points to p53 involvement in autophagy. On the other hand, in an in-vitro model of toxicant-induced neurodegeneration caused by proteasomal suppression, pifithrin-induced p53 inhibition effectively counteracted the activation of autophagy and enhanced the accumulation of proteins (Yang et al. 2020). Additionally, the production of p53 and increased autophagic activity were promoted by administration with the ubiquitin antagonist lactacystin. Kainic acid (KA) treatment of primary striatal neurons resulted in increased expression of p53, p62, Beclin1, and light chain 3 (LC3). However, pifithrin-a treatment prevented these alterations, suggesting that p53 have a role in KA-mediated autophagy induction and death in neuronal cells caused by mitochondrial failure (Dong et al. 2012). In a model of excitotoxic neuronal death, rats were injected with the NMDA receptor agonist quinolinic acid (QA) into the striatum. Studies have demonstrated that QA resulted in a reduction of Bcl-2 levels and an increase in p53 production. P53, in turn, upregulated key regulators of apoptosis, including Beclin 1, PUMA, LC3, and Bax, while also contributing to further decreases in Bcl-2. Pre-treatment with the p53 inhibitor Pifithrin-alpha (PFT-α), the autophagy inhibitor 3-methyladenine (3-MA), or the NF-kB inhibitor SN50 significantly mitigated these changes (Wang et al. 2009). Remarkably, current research shows that supplementing with PFT-a induces Parkin-mediated mitophagy. Moreover, it has been demonstrated that PINK1 downregulation is necessary for nuclear p53-mediated autophagy suppression (Yin et al. 2021). The findings of these investigations suggest a new interaction between apoptosis and mitochondrial selective autophagy, suggesting that p53 plays a role in both apoptotic and autophagic pathways that cause cell death.

1.2.6 Protein Aggregation and Misfolding

Proteinopathy is a category of conditions in which some proteins get misfolded and do not maintain structural stability, hence compromising the body's capacity to function normally. These proteins frequently are unable to go through proper folding; in this abnormal folding condition, they may either lose their usual function or become harmful in some way (a toxic gain-of-function). Proteinopathies encompass a broad spectrum of neurodegenerative diseases, including multiple system atrophy, AD, PD, amyloidosis, and prion diseases, and ALS (Marsh 2019). In most proteinopathies, alterations in the 3D folding structure of a protein increase the likelihood that the protein will bind to itself, contributing to the development of disease. The protein is hard to clear once it is aggregated, and it may impair the affected organs' ability to function normally. The protein may occasionally lose its normal function as a result of misfolding (Sweeney et al. 2017). All proteins have the ability to misfold in specific situations since they all share the polypeptide backbone, a common structural element. But only a limited number of proteins are associated with proteopathic diseases, maybe because the susceptible proteins have peculiar structural features. Certain risk factors, which promote protein self-assembly, also increase the probability of developing proteinopathy. These include alterations destabilizing the protein main amino acid sequence, PTM, variations in pH or temperature, increased protein synthesis, or reduced protein clearance (Abbasbeigi 2020; Basak and Basak 2022).

Different physiological mechanisms, including deficiencies in chaperone proteins, reduced functioning of mitochondria, dysregulated autophagy processes, and disruption of the ER under stress, can lead to the accumulation of protein aggregates in cells. Since post-translational modification of proteins takes place in the ER, it is the most significant part of the eukaryotic cell. Chaperone proteins identify non-native polypeptide conformations and take role in the process of folding proteins. The ER interacts crucially with other organelles, especially mitochondria, to control proteostasis and degrade erroneous proteins in order to carry out many roles (van Vliet and Agostinis 2018). The primary connection between mitochondria and the endoplasmic reticulum (ER) is the mitochondria-associated membrane (MAM), which constitutes around 20% of the total membrane. MAM not only facilitates the physical transfer of signals between the two organelles but also enables chemical communication, such as the exchange of calcium (Ca^{2+}) and reactive oxygen species (ROS). This communication is crucial for understanding cell viability. MAMs play a role in various biological processes, including inflammation, autophagy, phospholipid synthesis, calcium regulation, mitochondrial protein homeostasis, and the cell's ability to adapt to harmful stimuli (Degechisa et al. 2022). The aforementioned mechanisms are intricately intertwined to preserve proteostasis, guarantee appropriate protein folding, and avert the formation of protein aggregates. The unfolded protein response (UPR) is the mechanism through which cells ensure proper protein folding and maintain cell survival under conditions of stress, such as the accumulation of misfolded proteins (Read and Schröder 2021). Upon identifying mutant proteins, UPR facilitates their breakdown via catabolic pathways, utilizing autophagic lysosomes and the ubiquitin-proteasome system. ER-associated degradation and biogenesis are examples of adaptive responses triggered by UPR-activated pathways. The stress sensors initiate apoptosis when UPR regulating systems are overloaded.

Decompensation of the above functions results in cellular death under extended periods of stress. Neurodegenerative processes are facilitated by ER malfunction. However, research has also shown that ER stress contributes to neuronal malfunction in mental diseases including bipolar disorder, schizophrenia, and depression. Three key kinases are responsible for initiating the stress response in the endoplasmic reticulum: inositol-requiring enzyme 1 (IRE1), protein kinase R-like endoplasmic reticulum kinase (PERK), and activating transcription factor 6 (ATF6). These kinases activate signaling pathways that manage cellular stress, helping in maintaining protein folding homeostasis (Ong et al. 2024). The buildup of misfolded proteins can trigger these kinases. Their activation triggers a series of metabolic events that lead to the promotion of proapoptotic pathways. The IRE1 pathway controls ER-induced apoptosis, whereas PERK is essential for a cell to survive under ER stress. Transmembrane kinase PERK phosphorylates eIF2α, which reduces overall protein translation and boosts the expression of ATF4. Consequently, this amplifies the gene expression of particular genes of interest. Increased transcription of genes producing ER chaperones results from the transmembrane transcription factor ATF6 being degraded by the protein proteases S1P and S2P as it enters the Golgi apparatus. In the meanwhile, the UPR-related genes transcription is activated by the transmembrane RNase IRE1 (Shacham et al. 2021; Siwecka et al. 2021).

The pathophysiology of neurodegenerative illnesses is commonly associated with impaired proteostasis, as seen by the buildup of certain proteins like tau, α-synuclein, and amyloid. Through proteolytic cleavage, beta-amyloid is created from its transmembrane protein precursor. The development of new synapses, neuroplasticity, and general neuron survival all depend on APP (Höhn et al. 2020). Amyloid plaques in AD include beta-amyloid. Unfortunately, there is no information available on a possible correlation between amyloid levels in the brain's cortical area and cognitive impairment. However, a number of animal experiments have shown that animals exhibit disrupted behaviour when amyloid is directly injected into brain areas.

The gradual buildup of beta-amyloid plaques is directly linked to cognitive decline and memory problems. According to clinical research, beta amyloid has been linked to depression and anxiety disorders (Krell-Roesch et al. 2021). Preclinical research has also produced supporting findings. The most popular test in behavioural investigations of mice used to identify the anxiogenic and anxiolytic characteristics of the drugs used is the elevated plus maze (EPM) test (Lin et al. 2021; Builders et al. 2023). Treatment with beta-amyloid was found to significantly decrease the open arm time (OAT) in the elevated plus maze (EPM), with a reduction of approximately 71.3% compared to the control group (Zare et al. 2015). The experiment used the forced swimming test, or FST, to measure depressive reactions in rats. The time of immobility is the depression index. As a result, a deeper depression condition is indicated by a larger immobility time, while a lesser depressive state is indicated by a shorter immobility time. After beta-amyloid injection, the immobility duration in the forced swim test (FST) was 5.4 times longer compared to the control group. Additionally, the swimming time in the Aβ group was significantly reduced, approximately 71.3% less than that of the control group. The role of beta-amyloid and other misfolded proteins in development of ND is depicted in (Fig. 1.1) (Dutta and Trapp 2011; Ni et al. 2022).

1.3 Crosstalk Between Pathways: Unravelling the Intricate Interactions

One of the primary routes implicated in several disorders of the central nervous system is the hippocampal signaling pathway, which has gained prominence recently (Sahu and Mondal 2020). Transcriptional repression-induced atypical cell death (TRIAD), a form of necrotic cell death, has been identified in Huntington's disease (HD) (Yamanishi et al. 2017). The Hippo pathway, which involves a series of kinases, ultimately regulates the transcriptional coactivator with PDZ-binding

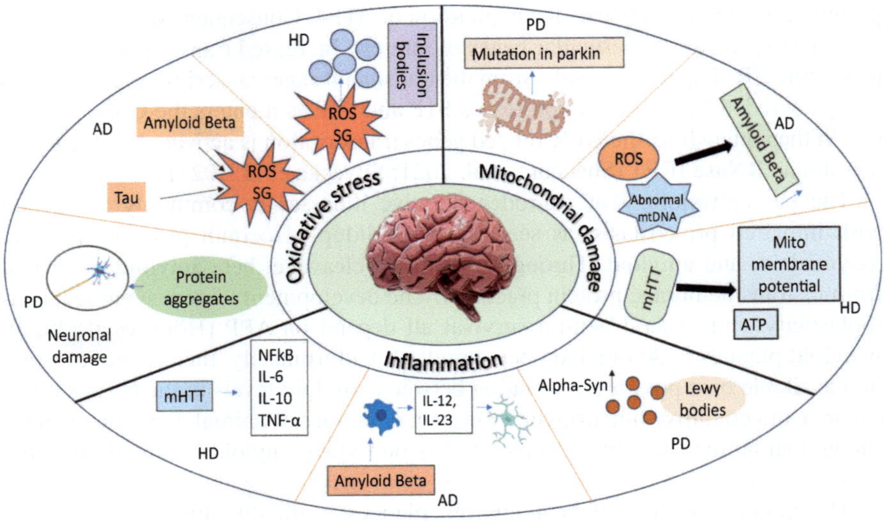

Fig. 1.1 Mechanisms of neurodegenerative diseases: Neurodegenerative diseases are characterized by several key cellular and molecular processes, including mitochondrial dysfunction, oxidative stress, and inflammation. In Alzheimer's disease (AD), mitochondrial dysfunction contributes to the accumulation of amyloid-beta (Aβ) by promoting mitochondrial DNA mutations and the generation of reactive oxygen species (ROS). This dysfunction also exacerbates the impairment of Parkin, which plays a role in Parkinson's disease (PD). In Huntington's disease (HD), mutant huntingtin (mHTT) disrupts mitochondrial ATP production and respiration. Oxidative stress is another critical factor across these diseases. In AD, oxidative stress is driven by Aβ-induced imbalance and the phosphorylation of tau (P-tau) proteins, contributing to neurodegeneration. In PD, increased oxidative stress hampers the ubiquitin-proteasome system (UPS), leading to further damage to dopaminergic neurons in the substantia nigra. In HD, ROS impairs neurotransmitter transmission by promoting protein misfolding and the formation of inclusion bodies. Neuroinflammation plays a pivotal role in the progression of these diseases. In AD, microglia regulate inflammation as part of the central nervous system's innate immune response. In PD, the accumulation of alpha-synuclein within Lewy bodies contributes to inflammation. In HD, inflammatory processes are driven by mHTT-induced cytokine release and the activation of the NF-kB signaling pathway

domain (TAZ) and controls the phosphorylation of Yes-associated protein (YAP). This phosphorylation event determines the cellular localization of YAP and TAZ, influencing various cellular processes (Zhang et al. 2018b). YAP was shown to be present in the Aβ particles, which inhibited the activation of YAP in the nuclear region of cells and in the neuronal cortex of AD in humans. TRIAD has also been linked to ALS; however, with this case, when neuronal degeneration takes place, YAP deltaC decreases (the pro-survival version of YAP) (Mao et al. 2016). In the later stages of symptoms, the phosphorylation ratio of total p73 to p73was enhanced, suggesting a major involvement of YAP-p73 dysfunctioning in the neuronal degeneration associated with ALS (Sahu and Mondal 2020). The length of whole YAP and p73, on the other hand, persisted till the last phase of symptoms, even though the levels of total p73 declined as the illness continued. Interestingly, YAP target gene expression was shown to be lower in the cortical region of neurons of ALS patients as well as a considerably lower amount of nuclear YAP when compared to the controls. Ahn et al. demonstrated that PD is associated with downregulation of Netrin1 (NTN1), a secretory laminin-associated protein that regulates nervous system formation, axon orientation, and proliferation (Ahn et al. 2020). NTN1 attaches to and leads to the phosphorylation of the UNC5B (known as netrin receptor), which causes UNC5B to become activated by apoptosis and cause the demise of dopaminergic neurons. As a result of the drop in Netrin1, MST1 is more activated, and the kinase communicate with UNC5B to phosphorylate it. This lowers YAP expression and increases cell mortality. According to these findings, neuronal apoptosis is decreased when UNC5B is inhibited or when UNC5B is phosphorylated by MST1 (Ahn et al. 2021). However, the growing field of research on this subject indicates that the Wnt pathway may potentially play a significant role in neurodegenerative illnesses.

Tau phosphorylation and neurofibrillary tangles are inhibited by Wnt/β-catenin signaling by blocking GSK-3βb in AD (Engel et al. 2006). Conversely, Aβ toxicity increases DKK1 expression, which in turn increases GSK-3β activity. DKK1, a Wnt inhibitor, prevents the Wnt/β-catenin pathway, which hyperphosphorylates the tau and leads to death of neurons (Ren et al. 2019). Similar data was also shown in PD. Mutations in the LRKK2 gene boost LRRK2 capacity to regulate GSK-3β, and much like in AD, this results in tau phosphorylation, which causes ND (Kofoed et al. 2020). Mutated LRRK2 are associated with neurodegeneration and a decrease in the activity of the canonical Wnt pathway (Galloway and Ichida 2014; Guo et al. 2018). Similar to AD, PD was also shown to have low β-catenin and high DKK1, together with elevated GSK-3β activity. These findings once again linked Wnt/β-catenin downregulation to neurodegeneration (Ramakrishna et al. 2023). Another similarity to AD is that high levels of DKK1 cause the neurodegeneration to be rescued when LiCl is administered (Marchetti 2018). Furthermore, it was shown how Wnt1 signaling contributes to the deterioration of dopamine-releasing neurons in the midbrain and the process of self-healing in mice with PD injuries in midbrain. It was proposed that astrocytes protect neurons from neurodegeneration and promote neurogenesis by producing Wnt1 ligand in the extracellular environment, as a result interacting with Fzd1 receptors, and initiating the Wnt regulatory pathway (L'episcopo et al. 2011).

1.4 Role of NLRP3 Inflammasomes in ND

The NLRP3 (NOD-like receptor family pyrin domain-containing 3) inflammasome is an essential element of the innate immune system, acting as a sensor for cellular stress and damage. Its activation leads to an inflammatory response that, while protective in the short term, can contribute to chronic inflammation and neurodegeneration when dysregulated. In order to activate the inflammasome, NLRP3 senses the accumulated Aβ. Fibrillary Aβ may also cause the secretion of IL-1β from microglial cell culture; however, this process is reliant on the NLRP3 inflammasome adapter (ASC) and sensor (NLRP3) (Hulse and Bhaskar 2022). In APP/PS1 mice (a double transgenic mice for AD model), ASC deficiency improves cognitive impairments and reduces brain Aβ accumulation. In the same mice model, the treatment of ASC specks causes Aβ deposition to spread. Mice model lacking in ASC do not exhibit the spread of Aβ deposits, and co-treatment of antibody against the ASC prevents the progression of Aβ-related hallmarks. Hence, AD development is caused by ASC specks produced by microglial cell undergoing pyroptosis, which seed and propagate Aβ oligomers and clumps (Li et al. 2023a, b). In APP/PS1 animals, NLRP3 or deficient caspase 1 expression results in decreased brain stimulation of IL-1β, and caspase-1 enhanced phagocytosis of Aβ by microglia, decreased Aβ load in the brain, and protection against long-term potentiation (LTP) decline, memory loss, and loss of the neuronal spine (Thawkar and Kaur 2019; Hanslik and Ulland 2020). Nevertheless, by 16 months of age, the decreased Aβ burden is apparent in APP/PS1 mice with deficiency in NLRP3 and not at four months (Heneka et al. 2013). Expression of caspase-1 functioning and IL-1β are markedly elevated in individuals with initial AD stage or moderate memory loss related to AD (Li et al. 2018), and ASC-linked Aβ is detected in the brains of patients having AD (Venegas et al. 2017). These findings imply that the activation of the NLRP3 inflammasome is an early pathogenic event in AD. Apart from its pivotal function in the Aβ pathogenesis of AD, NLRP3 inflammasome also has an impact on tau pathology (Stancu et al. 2022). Fibrillar Aβ with the homogenates of brain given intracerebrally in animals increased the phosphorylated tau as well as Tau22 aggregates in mice, a phenomenon that is inhibited by deficient ASC or NLRP3, indicating that inflammation associated with tau in AD and related *in vivo* models is worsened by Aβ-induced NLRP3 inflammasome activation (Van Zeller et al. 2021; Hulse and Bhaskar 2022).

A variety of human cell types have been demonstrated to respond to α-Syn aggregates *in vitro* by acting as an NLRP3 inflammasome inducer: primary monocytes7, THP-1-derived, macrophage-like cells67,68, and primary microglia. Mature IL-1β is processed in response to each of these reactions and secreted into the culture media. There is indication of *in vivo* stimulation of the NLRP3 inflammasome in connection to α-syn in brain tissue damaged by Parkinson's disease patients; in particular, caspase-1 has been linked with α-syn in Lewy bodies with a spatial arrangement pattern (Li et al. 2021). This spatial arrangement could be a consequence of the way that caspase binds to the inflammasome scaffold and suggestive of the function of the inflammasome in the creation of Lewy bodies. The stimulation of the NLRP3 inflammasome, especially by α-syn, has been suggested to take part in the development and promotion of progressive PD pathology (Soraci et al. 2023). A molecular connection has been found between α-syn and canonically triggered caspase-1. Specifically, the active caspase breaks down α-syn at Asp121 to shortened version α-syn 1–121, a shorter form that is seen in Lewy bodies *in vivo* and is more likely to form aggregate. The same study findings suggested that caspase-1 activity may eventually encourage α-syn accumulation and development of Lewy body in neuronal cells as it found caspase-1 in correlation with Lewy bodies with α-syn in PD patients brain (Ma et al. 2018; Pike et al. 2022). The correlation between α-syn activation and caspase-1 activity might perhaps explain the findings that caspase-1, although being attracted to the scaffold of inflammasome and possibly triggered, was not required for primary human microglia to process IL-1β in response to α-syn activation. Other caspases or proteases would probably take up the job of processing IL-1β in the event of activated NLRP3 inflammasome if caspase-1 is redirected to process of α-syn-activation in human microglial cells (Fig. 1.2) (Pike et al. 2022; Hulse and Bhaskar 2022).

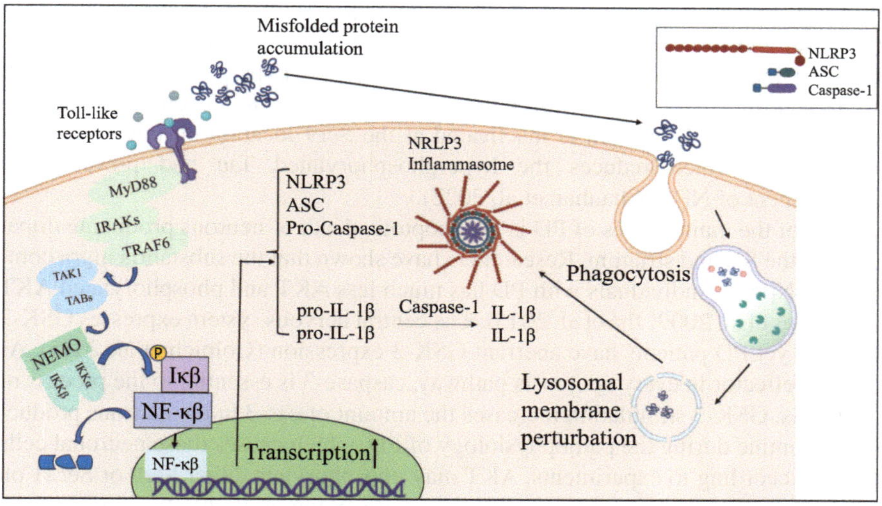

Fig. 1.2 Protein aggregates associated with Alzheimer's disease (AD) and Parkinson's disease (PD), including Aβ, tau, and α-synuclein, act as triggers for activating the NLRP3 inflammasome and forming ASC specks. These aggregates interact with toll-like and scavenger receptors on microglial cells, initiating a MyD88-dependent signaling cascade that leads to the translocation of NF-κB to the nucleus. Once there, NF-κB promotes the production of key inflammasome components and pro-inflammatory cytokines such as pro-IL-1β and pro-IL-18. Aggregates are engulfed through phagocytosis and transported to the lysosome, where they compromise the lysosomal membrane, triggering inflammasome activation. The NLRP3 inflammasome assembles by forming oligomers, facilitating the recruitment of ASC via PYD domain interactions. This complex then attracts pro-caspase-1 through CARD domain interactions, leading to its autolytic cleavage and the generation of active caspase-1. Once activated, caspase-1 processes pro-IL-1β and pro-IL-18 into their mature, functional cytokines (IL-1β and IL-18). Key signaling molecules involved include MyD88 (myeloid differentiation primary response gene 88), IRAK (IL-1 receptor-associated kinase), TRAF6 (TNF receptor-associated factor 6), TAK1 (transforming growth factor-β-activated kinase 1), TAB (TAK1-binding proteins), NEMO (NF-κB essential modulator), IKKα/β (inhibitor of NF-κB kinase), IκB (inhibitor of NF-κB), and NF-κB (nuclear factor-kappaB)

1.5 Roles of PI3K/AKT Pathway in ND

PI3K is first activated by RTK, which then initiates the PI3K/AKT pathway. The plasma membrane receptor RTK is bound by a number of cytokines or growth factors, including platelet-derived growth factor (PDGF) and fibroblast growth factor (FGF), in response to activation by external ligands. Tyrosine sequences in the cellular regions of the receptor are cross-phosphorylated and receptor dimerization results from this interaction (Heras-Sandoval et al. 2014). Activation of the PI3K/AKT signaling might help to slow the course of AD as it was shown that the PI3K/AKT signaling pathway is responsible for the production of two unique pathogenic confirmations in AD (Heras-Sandoval et al. 2014; Kitagishi et al. 2014). The P3IK/AKT signal pathway can be activated to shield neurons from the neurotoxic effects of Aβ. Numerous targets located downstream of this route are intimately linked to the onset and progression of the illness. For instance, there is a clear correlation between increased Aβ synthesis and deposition, tau hyperphosphorylation, and NFT (neurofibrillary tangles) formation and increased GSK-3β activity. The phosphorylated GSK-3β protein is inactivated at the Ser9 location during AKT phosphorylation, which reduces the hyperphosphorylated Tau and prevents the development of NFT (Chauhan et al. 2022).

One of the main causes of PD is the apoptotic death of neurons producing dopamine in the SN and striatum. Researchers have shown that the substantia nigra compacta (SNpc) of individuals with PD has much less AKT and phosphorylated AKT (Timmons et al. 2009; Jha et al. 2015). The central nervous system expresses GSK-3 broadly, yet PD patients have aberrant GSK-3 expression (Golpich et al. 2015). As a major effector in every apoptosis pathway, caspase-3 is essential to the process of apoptosis. GSK-3 stimulation increases the amount of cas-3 in the neurons producing dopamine during the pathophysiology of PD, which causes these neuronal cells to die. According to experiments, AKT may phosphorylate either Ser9 or Ser21 on GSK-3β or GSK-3α to decrease GSK-3 activity (Song et al. 2010). Thus, by preventing apoptosis, the PI3K/AKT pathway activation promotes development as well as survival of dopaminergic neuronal cell. Additionally, PTEN deletion promotes the fast development of dopaminergic (DA) neurons and contributes to neuroprotection (Li et al. 2024).

The beginning of Parkinson's disease (PD) is also influenced by oxidative stress, which causes excitatory toxicity, lipid peroxidation, intracellular calcium excess, and DNA damage, all of which cause neuronal cell death and apoptosis. The PI3K/AKT pathway modifies downstream molecular targets, including mTOR, FoxO3a, and GSK-3, to affect oxidative stress. Reduced mTOR activity might result in neurodegeneration. The stress-responsive protein regulation during development as well as responses to damaged DNA 1 (REDD1) is enhanced and has the capacity to regulate mTOR activity in dopaminergic neurons in PD patients (Miller et al. 2021). The PI3K/AKT/FoxO3a pathway is inhibited when Parkin, a mutation-prone gene linked to Parkinson's disease, is expressed inappropriately. This leads to an imbalance in oxidative stress and, eventually, the development of Parkinson's disease (Fig. 1.3) (Goyal et al. 2023; Gong et al. 2018).

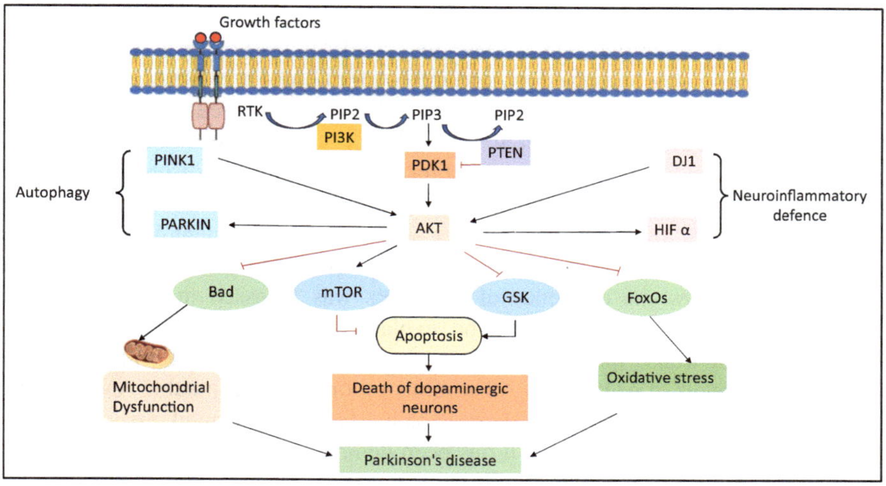

Fig. 1.3 The connection between the PI3K-AKT signaling pathway and Parkinson's disease (PD). This pathway is initiated by the activation of receptor tyrosine kinases (RTKs), which subsequently activates PI3K. PI3K converts PIP2 to PIP3, which then recruits PDK1 and AKT to the cell membrane. PDK1 phosphorylates AKT, activating it, allowing AKT to regulate multiple downstream processes. DJ1 enhances this pathway by offering protection against oxidative stress and inhibiting PTEN, while AKT, regulated by HIF-1α, promotes neuroinflammatory defense. The activation of the PI3K-AKT pathway plays a critical role in the survival and development of dopamine neurons by preventing apoptosis. This pathway inhibits apoptosis by modulating Bad (Bcl2-antagonist of cell death) and provides neuroprotection through PTEN knockout, which accelerates the growth of dopaminergic neurons. Oxidative stress, a significant factor in neuronal death and PD progression, is also regulated by the PI3K-AKT pathway, which influences molecules such as GSK-3, mTOR, and FoxO3a. Disruptions in this pathway, such as from abnormal Parkin gene expression, lead to imbalanced oxidative stress, contributing to dopaminergic neuron death, a hallmark of PD

1.6 AMPA Receptor in ND

Cognitive decline is one of the earliest symptoms of AD and has been linked to reduced synaptic communication caused by AMPA receptors in models of AD (Zhang and Bramham 2020) and synapse loss in AD human brain (Li and Selkoe 2020). According to several studies, adding Aβ to neurons causes synaptic loss, reduces LTP, increases LTD, and promotes AMPA receptor internalization to suppress its transmission (Ripoli et al. 2014; Li and Selkoe 2020; Babaei 2021). Drugs that modify AMPA receptors are being tested in clinical trials to treat AD-related cognitive loss because of the substantial impact of the disease on AMPA receptor transmission. AMPAKines are a family of pharmaceuticals that increase excitatory glutamatergic neurotransmission by acting as positive modulators of AMPA receptors (Li and Selkoe 2020). To improve glutamatergic synaptic transmission, they delay the onset of desensitization and deactivation of AMPA receptors (Arai and Kessler 2007; Kadriu et al. 2021).

The mRNA of AMPAR and protein synthesis is significantly downregulated in humans as well as animal AD models, and AMPAR trafficking is dysregulated. GluA1–3 subunit immune reactivity and mRNA for GluA1 and GluA2 were significantly reduced in the hippocampus areas of AD patients, an area that is particularly vulnerable to the disease (Italia et al. 2021; Wu et al. 2022). Additionally, AMPAR protein was much lower in brains of AD model with cognitive disorder during the early stage, and deficiency of GluA4 subunits was linked to a reduction in cognitive function (Qu et al. 2021). Further evidence supporting the deficient AMPARs in AD was found in genetic mice models or following Aβ-treated brain (Gu et al. 2009). Depending on how the existence of AMPARs was assessed in each study, the decrease in AMPAR amount may be explained by post-transcriptional, post-translational, or transcriptional regulatory mechanisms or by other physiological changes. In models of AD or after Aβ administration in tissue, notable AMPAR loss from synapses leading to synaptic depression has been documented, associated with a reduction in their total number. It has been demonstrated that Aβ overexpression causes endocytosis of synaptic AMPAR, decrease in dendritic spine, and synaptic downregulation in hippocampus slice cultures from AD mice models (Hsieh et al. 2006). Hippocampal neurons in double knock-in mice with mutant presenilin-1 and amyloid precursor protein genes displayed decreased AMPAR synaptic currents, including decrease in LTP, and long-term depression (LTD) (Auffret et al. 2009, 2010).

It was found that the mice had a notable reduction in working memory, which was correlated with a rise in Aβ accumulation. Furthermore, it was found that oligomeric Aβ, rather than the more soluble fibrillar form, prevented GluA-1 and GluA-4 containing AMPAR synaptic insertion and conditioning acquisition in an in vitro model of eyeblink classical conditioning (Zheng et al. 2010) Interestingly, it was discovered that the presence of oligomeric Aβ impeded pro-BDNF proteolytic conversion to the mature form needed for AMPAR trafficking. It has been consistently observed that exposure to Aβ causes a dramatic reduction in the surface expression of synaptic AMPARs, indicating the Aβ deleterious impact on synaptic AMPAR trafficking in AD (Hsieh et al. 2006; Parameshwaran et al. 2008).

Synaptic currents induced by AMPAR and cell surface expression were demonstrated to be significantly decreased in mouse genetic models of Parkinson's disease. Reduced whole-cell currents and reduced surface expression of AMPARs, especially GluA1, are the outcomes of loss of function mutations in *in vitro* hippocampal neuronal cells carrying the Park2 gene, which codes for the ubiquitination enzyme Parkin and is linked to inception juvenile and sporadic PD (Cortese et al. 2016; Zhu et al. 2018). Reduced expression of Homer1, a postsynaptic scaffolding protein required for AMPAR capture inside the PSD, appears to be at least partially responsible for this regulation. Reduced Parkin expression affects synaptic plasticity processes, such as LTD and LTP, as well as neuronal excitability. There is a theory that the non-motor impairments including cognitive dysfunction and dementia that are seen in PD patients might be caused by these findings in the hippocampal region. Remarkably, AD was associated with deficiency in Parkin (Lonskaya et al. 2013). Lastly, disruption of the endosomal protein sorting pathways is linked to PD;

this causes the reduction of transmission of neurotransmitters through the depletion of synaptic surface expression and diminished dendritic spines caused by inappropriate trafficking of AMPARs to the plasma membrane (Perrett et al. 2015).

Additionally, in mouse models of HD, disrupted trafficking and surface expression of AMPARs was shown (Zhang et al. 2018a; Keifer 2022). In studying multiple HD models, it was shown that deficiency in BDNF–TrkB signaling causes disruption of AMPAR synaptic surface diffusion in cultured hippocampal neurons by disrupting protein–protein interactions (stargazin–PSD95, in particular), which are necessary for PSD trafficking. Remarkably, the anti-depressant medication tianeptine, which functions by activating the BDNF–TrkB pathway, restored the compromised AMPAR mobilization. Furthermore, on hippocampal-dependent learning tasks, this therapy restored LTP and performance. A research was conducted whereby they employed TTX to block the neuronal functioning in the *in vitro* cortical neurons derived from both wild-type and transgenic mice expressing the human HD gene with CAG repeats (Keifer 2022; Smith-Dijak et al. 2019). This led to the expression of synaptic AMPARs and the creation of new synapses in wild-type cells, but it did not affect dendritic development or AMPAR trafficking in HD neurons. However, pridopidine, a medication that increases BDNF signaling, may cause AMPAR trafficking and synapse formation in the HD neurons. Significantly, both of these findings provide compelling evidence that raising BDNF levels in the brain may help to compensate for the behavioural function and plasticity deficits associated with HD. There are several underlying processes for severe AD and other brain diseases like PD and HD. The most common ones are diminished glutamatergic neurotransmission, which is connected to the loss of AMPARs and ultimately causes synapse retraction and neuronal degeneration.

1.7 Therapeutic Interventions in Neurodegenerative Diseases

Given that ER stress has a crucial role in the occurrence of neuronal cell death in ND, focusing on related pathways may provide hope for preventing the death of neurons. The PERK pathway has been researched of the three arms of UPR under ER stress situations, with respect to the neuroprotective benefits of ER stress inhibition. Simultaneously, salubrinal, an antagonist of ER stress molecule, is thoroughly studied in the pathophysiology and management of neurodegenerative diseases (Wang et al. 2020; Davis et al. 2022). Salubrinal increases the amounts of ER chaperones, including BiP, via activating UPR. Through disruption of death-associated signaling pathways, such as ATF4 or ASK1, it suppresses dephosphorylated eIF2α, thus decreasing neuronal cell mortality (Shi et al. 2023; Zhang et al. 2024). Increasing body of research indicates that in transgenic mice models, salubrinal greatly prevented α-synuclein-mediated dopaminergic (DA) cell mortality and decreased ER aggregation of α-synuclein (Colla et al. 2012). Additionally, by upregulating p-eIF2α and BiP, salubrinal prevented neuronal mortality and decreased the deposition of mutated huntingtin (mHTT) (Yuan et al. 2024). Furthermore, in models of *in vivo* of PD, the medication Adaptaquin prevents Tribbles pseudokinase 3 (Trib3) production by suppressing the activity of CHOP and ATF4 activity, most likely it is not dependent on eIF2α pathway (Demmings et al. 2021). Neuronal cell death may also be mitigated by inhibition of the IRE1 pathway. For example, by disintegrating IRE1 oligomers and reducing IRE1α RNase activity, Kinase inhibitor RNase attenuator 6 (KIRA6) prevents apoptotic death (Volgyi et al. 2015). Targeting and deleting ASK1 in mutated SOD1 genetic mice has been suggested to reduce motor neurons mortality because of ASK1 critical function in IRE1-induced UPR and induction of programmed cell death (Matus et al. 2013). Further data indicates that DA neurons are shielded from neurotoxin-mediated ER tress linked ND by overexpression of XBP1 (Shi et al. 2022). Additionally, it is known that another method of preventing neuronal cell death in ALS and HD models is to upregulate autophagy by targeting XBP1 (Vidal et al. 2012). It has been shown via studies that ATF6 ablation promotes the death of DA neuronal cells brought on by neurotoxins such as MPP+ and 6-hydroxydopamine (6-OHDA) (Xiang et al. 2017). This depicts that the third arm of UPR pathways may have an essential role in the neuronal death. ATF6, PERK, IRE1α, and CHOP have all shown comparable reductions in expression in Brefeldin A (BFA) mediated ER stress in IMR32 cell lines when treated with kaempferol, a plant-based extract which blocks ER stress-mediated ND (Abdullah and Ravanan 2018).

Unquestionably, the pathophysiology of neurodegenerative illnesses involves the neurotoxic consequences of oxidative stress. Numerous potential compounds with antioxidant qualities have been documented in this subject matter. Encoded by mtDNA, Humanin (HN) is a cytoprotective polypeptide with neuroprotective properties against cellular stressors including oxidative stress. HN reduces apoptosis by upregulating Bcl-2 and downregulating Bid and Bax, and it controls mitochondrial activity by targeting apoptotic proteins (Matsunaga et al. 2016; Zuccato et al. 2019). Furthermore, in vitro ALS models have demonstrated anti-apoptotic properties of some derivatives of amino acids (Zheng et al. 2005). In H_2O_2-mediated NSC-34 motor neurons, *N*-acetyl-DL-tryptophan (DL-NAT) and *N*-acetyl-L-tryptophan (L-NAT) prevented mortality in neurons. L-NAT prevents neuronal apoptosis by blocking the excretion of mitochondrial Cyt C/Smac/AIF and inhibiting caspase activation (Sirianni et al. 2015). In another study, it was shown that *N*-acetyl serotonin (NAS) had an inhibitory effect on apoptosis by activating neurotrophic signaling pathways TrkB/CREB/BDNF (Yoo et al. 2017). To lower the amount of ROS, NAS stimulates and promotes the production of antioxidant enzymes. In models of neurodegenerative diseases, it also modulates variables that are pro-apoptotic and anti-apoptotic and restores the potential of the mitochondrial membrane to prevent the death of neurons. Evidence suggests that mPTP might cause cell death by retaining more calcium in the mitochondria. CoQ10 shields neurons from oxidative stress and death by inhibiting the mitochondrial permeability transition pore (Frontiñán-Rubio et al. 2021).

Some studies showed that a possible treatment strategy to reduce inflammation and its subsequent problems in neurodegenerative illnesses is to target neuroinflammation. As a result, it can enhance patients' neuronal performance during both mental and physical activities. In the initial stages of ND, some particular microRNAs, including miR-155-5p, may play a significant role in controlling inflammatory cascades. Patients with AD and MS have been found to have overexpressed miR-155-5p in their CSF (Zingale et al. 2021). Recently, it was shown that rosmarinic acid (RA) attenuates neuronal death caused by inflammatory pathways via modulating miR-155-5p (Lv et al. 2020). Calpains, as non-caspase proteases, collaborate with important elements of neuronal cell death and take part in the process of neuronal cell death. Thus, neuroprotection might arise from targeting these proteases. Research has shown that calpain activity suppression and microglial neuroinflammation may both be reduced by alpha1-antitrypsin (A1AT) (Zhang et al. 2023). A focus on neuroinflammation has led to the recruitment of further innovative tactics. The natural antioxidant found in green tea, epigallocatechin-3-gallate (EGCG), was used in recent research to target inflammatory markers in the SN region of rats treated with lipopolysaccharide (LPS) (Payne et al. 2022). Through the suppression of neuroinflammation, it displayed neuroprotection. Since the NLRP3 inflammasome contributes significantly to the increase of neurotoxicity, blocking NLRP3 and its related pathways may be a useful strategy to reduce the cell death brought on by neuroinflammation. MCC950, a small-molecule NLRP3 blocker, significantly reduced neuroprotection, mitigated motor impairments, and accumulated α-synuclein aggregates in mouse PD models by inhibiting inflammasome activation (Huang et al. 2021). Notably, dopamine is a neurotransmitter that may bind to the dopamine D1 receptor. This interaction causes NLRP3 to be degraded and ubiquitinated, and it also limits NLRP3 activation by binding of cAMP to NLRP3 (Yan et al. 2015). The GW501516, an agonist of peroxisome proliferator-activated receptor beta/delta (PPAR-β/δ), was shown in certain studies to attenuate inflammatory effect caused by NLRP3 in PD mice. This medication has several drawbacks, such as the BBB resistance to the drugs ability to pass through and reach the brain parenchyma. Therefore, unless the issue is resolved, this chemical cannot be regarded as a potential therapy for Parkinsons' disease (PD) (Li et al. 2023a, b). Phytopharmaceutical extracts, which include a diverse diversity of active molecules, have the potential of multitarget therapeutic impact. Here is a list of the phytopharmaceutical extracts that have been utilized to treat neurodegenerative disease models both in vivo and in vitro. Hydrophobic polyphenolic chemicals known as curcumins are isolated from the rhizome of *Curcuma longa*, herbaceous tuberous plant that is widely distributed in South Asia, China, and India. Curcumin decreases inflammatory effect by preventing NF-κB from cellular signaling. Curcumin was found to have neuroprotective effects in early preclinical investigations on the motor skills and the activity patterns in genetic PD *Drosophila melanogaster* (Sarawi et al. 2021). In the paraquat-mediated PD model, curcumin has also been shown to ameliorate the motor symptoms of PD (Esquivel et al. 2020). Similar neuroprotective effects of curcumin have been shown in 6-OHDA PD animal models (He et al. 2022). The neuroprotective effect on the mice nigrostriatal pathways was facilitated by α7 nicotinic receptor. Regarding AD, many cell lines studies showed that curcumin binds to Aβ, prevents Aβ from aggregating, and destabilizes fibrils that have already formed (Orjuela et al. 2021). The therapeutic benefits of the free form when taken orally are compromised by its hydrophobic nature, poor bioavailability, fast metabolism, limited capacity to pass the blood–brain barrier, and rapid elimination. A tall, ancient Chinese tree that is now found all over the world, *Ginkgo biloba*, is a member of the Ginkgoaceae family. Their dried leaves are used to make ginkgo biloba extract (GbE), which is a yellow powder. In models of cerebral ischemia, several preclinical studies revealed that GbE has protective benefits. Remarkably, it was found that long-term pretreatments with this extract, which lasted 12 months prior to the damage caused by middle cerebral artery occlusion, also dramatically reduced the ischemia-mediated neuronal death in 24-month-old rats (Naveen and Bhattacharjee 2021; Singh et al. 2022). In particular, treated mice showed lower levels of oxidative stress and infarct volume in comparison to the normal. The results were linked to prevention of the ERK activity.

The root of the Panax genus, ginseng, is known for its polyphenols, flavonoids, vitamins, and ginsenosides. Scientific interest has mostly been directed toward ginsenoside Rg1, a triterpenoid saponin with neuroprotective qualities. Rb1 makes better motor impairments in MPTP mice PD model (Gong et al. 2022; Kabir et al. 2023). Furthermore, this saponin prevents learning and spatial memory problems linked to Parkinson's disease in MPTP rats. Recent high-throughput metabolomics research in AD animal models has identified Rg1 pharmaceutical targets which are 14 metabolites that are part of the linoleic acid, tryptophan, arachidonic acid, and sphingolipid metabolic pathways (Li et al. 2019). Preclinical trials employing the Rg1 for the therapy of memory loss in AD models were reported in a recent comprehensive review by (Liang et al. 2021). In the majority of AD models, this ginsenoside markedly enhanced cognitive performance; however, it did not affect SAMP8 transgenic mice or animals under long-term stress. This molecule's ability to preserve synapses and combine its anti-inflammatory and antioxidant characteristics appears to be important in the Rg1-mediated enhancement of cognitive function.

The flavonoid quercetin is most frequently found in higher plants as a glycoside in the forms of hyperin, rutin, and iso-quercetin. In AD, quercetin has demonstrated therapeutic efficacy in enhancing learning and memory (Wróbel-Biedrawa et al. 2022). According to the research findings, quercetin administration inhibited the enzymes secretase and AChE, which prevented acetylcholine from degrading and reduced the generation of Aβ, respectively (Orhan 2021). In the amygdala and hippocampus area, quercetin treatment reverses extracellular amyloidosis and decreases tauopathy, microgliosis, and astrogliosis (Kaushik et al. 2023). This preserves memory and emotional performance in age-triple genetic AD animals. Additionally, because of its antioxidant properties, quercetin inhibits the development of fibril Aβ proteins and blocks the inflammatory cascade and cell lysis (Chiang et al. 2023). Moreover, it has been demonstrated that quercetin inhibits APP maturation, which modifies Aβ production and aggregation (Khan et al. 2019). According Ay, Luo et al., isoquercetin decreased α-synuclein fibrillization and enhanced motor capabilities in PD. In transgenic mice models of PD, quercetin also guards against mitochondrial damage and the gradual degradation of dopaminergic neurons (Ay et al. 2017). PD is clearly associated with cholinergic deficit and memory loss in PD, and quercetin treatment was shown to significantly alleviate cognitive dysfunction brought on by 6-OHDA administration (Madiha et al. 2021). On the other hand, oral quercetin treatment has been demonstrated in research by Haleagrahara and colleagues to have a positive impact in the same paradigm by decreasing striatal dopamine loss and elevating oxidative stress markers (Haleagrahara et al. 2013). It has been demonstrated that quercetin decreases the dose-dependent breakdown of striatal dopamine, a finding that is connected with a notable rise in striatal dopamine levels and a marked decrease in the expression of lipid peroxidation markers (Kaushik et al. 2023). Oral quercetin treatment attenuates behavioural impairments, nigrostriatal degeneration, and striatal dopamine loss in a moderate but substantial way. The quercetin glycoside rutin was also investigated and found to partially improve motor impairments in the 6-OHDA rat model (Tahir et al. 2021). This effect was associated with an increase in brain GSH levels and a moderate yet significant elevation in striatal dopamine levels. Conversely, markers of lipid and protein oxidation were found to be reduced.

In Parkinson's disease (PD) animal models, the flavonoid baicalein, extracted from the roots of *Scutellaria baicalensis*, has demonstrated both anti-inflammatory and anti-pyroptotic properties. Experimental studies suggest that baicalein protects dopaminergic neurons by reducing cytokine levels that drive inflammation. Additionally, in the MPTP-induced mouse model of PD, baicalein targets Gasdermin D to suppress pyroptosis while also inhibiting NLRP3 inflammasome activation and caspase-1 activity (Zheng et al. 2023).

Nobiletin, a flavonoid extracted from citrus fruits, has gained growing interest for its potential neuroprotective effects in neurodegenerative diseases. Molecular docking studies have shown that nobiletin can bind to beta-site amyloid precursor protein cleaving enzyme 1 (BACE1), inhibiting its activation and subsequently reducing the production of Aβ protein (Jahan et al. 2022; Hung et al. 2023). By triggering neprilysin, nobiletin can facilitate the breakdown of Aβ (Wang et al. 2021). Furthermore, nobiletin inhibits inflammation through the pathways of TLR4/NF-kB/Nrf2 and JNK, as well as inhibiting the production of COX-2 and iNOS, as well as cell death, in an AD model generated by Aβ (Bunbupha et al. 2020; Hosokawa et al. 2021; Yadav et al. 2024). In AD models, nobiletin therapy decreased the formation of HMGB1 and pyroptosis-related proteins, including ASC, cleaved caspase-1, N-terminal portion of gasdermin D, and nucleotide-binding oligomerization domain-like receptor 3 (Chai et al. 2022). Nobiletin has been shown to enhance ChAT expression, considerably reduce AChE activity, and improve cognitive impairment (Matsuzaki and Ohizumi 2021). Nobiletin was able to alleviate learning and memory deficits caused by stroke by enhancing the production of Ca2+/calmodulin-mediated protein kinase II (CaMKII), glutamate receptor 1, and microtubule-associated protein 2, as well as by raising the levels of phosphorylation of ERK and CREB (Ramezani et al. 2023). Nobiletin reduced the hypoxia-induced oxidative stress response and enhanced haem oxygenase-1 (HO-1) and Nrf2 nuclear translocation production in a glial cell hypoxia model (Song et al. 2024). Nobiletin increases the expression of claudin-5, which decreases cerebral oedema and the permeability of the blood–brain barrier (Matsuzaki and Ohizumi 2021). *In vitro* study demonstrated that nobiletin showed protective role in the prevention of arsenic-mediated toxicity in iPSC-derived human NPCs (Jahan et al. 2022, 2024). Nobiletin is injected intravenously, and it quickly builds up in the infarcted region to prevent nerve damage by lowering neuronal apoptosis and neutrophil infiltration, boosting BDNF expression and activating the Akt/CREB pathway (Xu et al. 2021). Parkinson-like symptoms caused by 1-methyl-4-phenyl-1,2,3,6-tetrahydropyridine can be improved by nobiletin by increasing the amount of CaMKII autophosphorylation, increasing the functioning of tyrosine hydroxylase, and reducing the damage to dopamine neurons. Additionally, nobiletin decreases neurodegenerative lesions in Parkinson's disease, increases GDNF expression, and suppresses microglial activation (Jeong et al. 2015; Nakajima and Ohizumi 2019). Nobiletin is a strong contender for the creation of novel therapies for neurodegenerative illnesses due to its capacity to lower inflammation, oxidative stress, and amyloid-beta aggregation while enhancing neuroplasticity and normal neurotransmission.

1.8 Conclusion

Neurodegenerative disorders present a significant global health challenge, as they gradually impair neuronal function, leading to severe cognitive and motor deficits. This chapter has examined the intricate mechanisms underlying neuronal damage and dysfunction, highlighting the importance of understanding these processes to develop effective therapeutic strategies. Neuronal degeneration arises from a complex interplay of factors, including oxidative stress, excitotoxicity, protein misfolding, mitochondrial dysfunction, neuroinflammation, and compromised neurotrophic support. Several key signaling pathways are involved in these pathological processes, such as NADPH oxidase, ionotropic glutamate receptors (including NMDA and AMPA receptors), nuclear factor kappa B (NF-κB), mitogen-activated protein kinase (MAPK) pathways, receptor tyrosine kinases like Trk receptors, and the phosphoinositide 3-kinase (PI3K)/Akt pathway. Additionally, the unfolded protein response (UPR) is integral to managing cellular stress linked to protein misfolding. The intricate crosstalk between these pathways highlights the complexity of neurodegeneration. However, it also offers multiple targets for therapeutic intervention. Emerging treatments focus on antioxidants to combat oxidative stress, modulators to reduce excitotoxicity, regulators of proteostasis to manage protein misfolding, mitochondrial enhancers to improve energy production, anti-inflammatory agents to reduce neuroinflammation, and neurotrophic factors to support neuronal health and survival (Fig. 1.4).

Advancing our understanding of these pathways offers hope for the development of targeted therapies that may slow or even halt the progression of neurodegenerative diseases. By elucidating the cellular and biochemical mechanisms underlying neuronal degeneration, this chapter underscores potential trajectories for future investigative and therapeutic endeavors. The overarching objective is to enhance the well-being of individuals afflicted by these debilitating disorders, fostering hope through scientific innovation and cutting-edge therapeutic strategies.

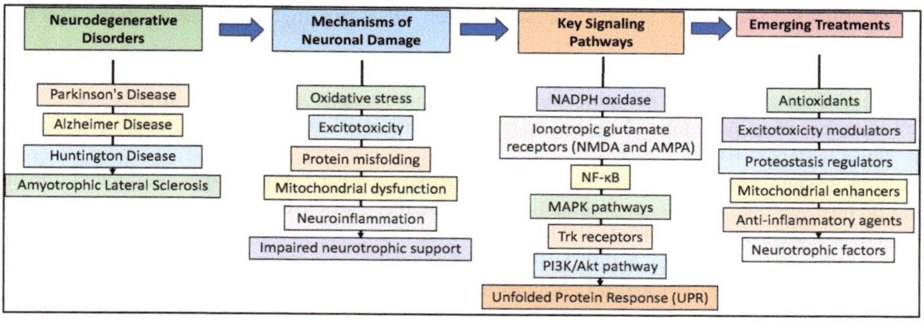

Fig. 1.4 The progression from neurodegenerative disorders to potential therapeutic advancements, highlighting key mechanisms and pathways involved in neuronal damage. It focuses on Parkinson's disease, Alzheimer's disease, Huntington's disease, and amyotrophic lateral sclerosis, all of which are linked to various neurodegenerative processes, including oxidative stress, excitotoxicity, protein misfolding, mitochondrial dysfunction, neuroinflammation, and reduced neurotrophic support. Additionally, the figure emphasizes critical signaling pathways contributing to these conditions, such as NADPH oxidase, ionotropic glutamate receptors (NMDA and AMPA), NF-κB, MAPK pathways, Trk receptors, PI3K/Akt signaling, and the unfolded protein response (UPR). Lastly, it presents emerging therapeutic strategies targeting these pathways, including antioxidants, excitotoxicity modulators, proteostasis regulators, mitochondrial function enhancers, anti-inflammatory compounds, and neurotrophic factors

References

Abbasbeigi S (2020) Misfolded structuresl a brief insight into protein aggregation criteria, which may lead to Proteopathy diseases. J Chem Rev 3(1):97–108

Abdullah A, Ravanan P (2018) Kaempferol mitigates endoplasmic reticulum stress induced cell death by targeting caspase 3/7. Sci Rep 8(1):2189

Abe Y, Hashimoto Y, Tomita Y, Terashita K, Aiso S, Tajima H, Niikura T, Matsuoka M, Nishimoto I (2004) Cytotoxic mechanisms by M239V presenilin 2, a little-analyzed Alzheimer's disease-causative mutant. J Neurosci Res 77(4):583–595

Ahn EH, Kang SS, Qi Q, Liu X, Ye K (2020) Netrin1 deficiency activates MST1 via UNC5B receptor, promoting dopaminergic apoptosis in Parkinson's disease. Proc Natl Acad Sci 117(39):24503–24513

Ahn EH, Kang SS, Ye K (2021) Netrin-1/receptors regulate the pathogenesis in Parkinson's diseases. Precis Med Future 5(2):50–61

Aktas O, Ullrich O, Infante-Duarte C, Nitsch R, Zipp F (2007) Neuronal damage in brain inflammation. Arch Neurol 64(2):185–189

Al-Nasser MN, Mellor IR, Carter WG (2022) Is L-Glutamate toxic to neurons and thereby contributes to neuronal loss and neurodegeneration? A systematic review. Brain Sci 12(5):577

Anantharam V, Kaul S, Song C, Kanthasamy A, Kanthasamy AG (2007) Pharmacological inhibition of neuronal NADPH oxidase protects against 1-methyl-4-phenylpyridinium (MPP+)-induced oxidative stress and apoptosis in mesencephalic dopaminergic neuronal cells. Neurotoxicology 28(5):988–997

Angelova PR (2021) Sources and triggers of oxidative damage in neurodegeneration. Free Radic Biol Med 173:52–63

Arai AC, Kessler M (2007) Pharmacology of ampakine modulators: from AMPA receptors to synapses and behavior. Curr Drug Targets 8(5):583–602

Askeland G, Dosoudilova Z, Rodinova M, Klempir J, Liskova I, Kuśnierczyk A, Bjørås M, Nesse G, Klungland A, Hansikova H (2018) Increased nuclear DNA damage precedes mitochondrial dysfunction in peripheral blood mononuclear cells from Huntington's disease patients. Sci Rep 8(1):9817

Auffret A, Gautheron V, Repici M, Kraftsik R, Mount HTJ, Mariani J, Rovira C (2009) Age-dependent impairment of spine morphology and synaptic plasticity in hippocampal CA1 neurons of a presenilin 1 transgenic mouse model of Alzheimer's disease. J Neurosci 29(32):10144–10152

Auffret A, Gautheron V, Mattson MP, Mariani J, Rovira C (2010) Progressive age-related impairment of the late long-term potentiation in Alzheimer's disease presenilin-1 mutant knock-in mice. J Alzheimer's Dis 19(3):1021–1033

Auzmendi-Iriarte J, Matheu A (2021) Impact of chaperone-mediated autophagy in brain aging: neurodegenerative diseases and glioblastoma. Front Aging Neurosci 12:630743

Ay M, Luo J, Langley M, Jin H, Anantharam V, Kanthasamy A, Kanthasamy AG (2017) Molecular mechanisms underlying protective effects of quercetin against mitochondrial dysfunction and progressive dopaminergic neurodegeneration in cell culture and MitoPark transgenic mouse models of Parkinson's disease. J Neurochem 141(5):766–782

Babaei P (2021) NMDA and AMPA receptors dysregulation in Alzheimer's disease. Eur J Pharmacol 908:174310

Barrera G, Pizzimenti S, Daga M, Dianzani C, Arcaro A, Cetrangolo GP, Giordano G, Cucci MA, Graf M, Gentile F (2018) Lipid peroxidation-derived aldehydes, 4-hydroxynonenal and malondialdehyde in aging-related disorders. Antioxidants 7(8):102

Basak A, Basak S (2022) Protein aggregation and self assembly in health and disease. Curr Proteomics 19(1):4–19

Bauminger H, Gaisler-Salomon I (2022) Beyond NMDA receptors: homeostasis at the glutamate tripartite synapse and its contributions to cognitive dysfunction in schizophrenia. Int J Mol Sci 23(15):8617

Bhatia V, Sharma S (2021) Role of mitochondrial dysfunction, oxidative stress and autophagy in progression of Alzheimer's disease. J Neurol Sci 421:117253

Bhatt S, Puli L, Patil CR (2021) Role of reactive oxygen species in the progression of Alzheimer's disease. Drug Discov Today 26(3):794–803

Boyenle ID, Oyedele AK, Ogunlana AT, Adeyemo AF, Oyelere FS, Akinola OB, Adelusi TI, Ehigie LO, Ehigie AF (2022) Targeting the mitochondrial permeability transition pore for drug discovery: challenges and opportunities. Mitochondrion 63:57–71

Brown GC (2007) Mechanisms of inflammatory neurodegeneration: iNOS and NADPH oxidase. Biochem Soc Trans 35(5):1119–1121

Builders MI, Oyefeso RA, Ede SO, Joseph SO, Ise PU (2023) Evaluation of anxiolytic activity of ethanolic extract of Thymus vulgaris leaves using elevated plus maze [EPM] test. J Adv Pharmacol 3(1):12–19

Bunbupha S, Apaijit K, Maneesai P, Prasarttong P, Pakdeechote P (2020) Nobiletin ameliorates high-fat diet-induced vascular and renal changes by reducing inflammation with modulating AdipoR1 and TGF-β1 expression in rats. Life Sciences 260:118398

Cai Q, Ganesan D (2022) Regulation of neuronal autophagy and the implications in neurodegenerative diseases. Neurobiol Dis 162:105582

Calvo-Rodriguez M, Bacskai BJ (2021) Mitochondria and calcium in Alzheimer's disease: from cell signaling to neuronal cell death. Trends Neurosci 44(2):136–151

Chai W, Zhang J, Xiang Z, Zhang H, Mei Z, Nie H, Xu R, Zhang P (2022) Potential of nobiletin against Alzheimer's disease through inhibiting neuroinflammation. Metab Brain Dis 37(4):1145–1154

Chauhan N, Paliwal S, Jain S, Verma K, Paliwal S, Sharma S (2022) GSK-3β and its inhibitors in Alzheimer's disease: a recent update. Mini Rev Med Chem 22(22):2881–2895

Chen H-SV, Lipton SA (2006) The chemical biology of clinically tolerated NMDA receptor antagonists. J Neurochem 97(6):1611–1626

Cheng X-T, Huang N, Sheng Z-H (2022) Programming axonal mitochondrial maintenance and bioenergetics in neurodegeneration and regeneration. Neuron 110(12):1899–1923

Chéret C, Gervais A, Lelli A, Colin C, Amar L, Ravassard P, Mallet J, Cumano A, Krause K-H, Mallat M (2008) Neurotoxic activation of microglia is promoted by a nox1-dependent NADPH oxidase. J Neurosci 28(46):12039–12051

Chiang M-C, Tsai T-Y, Wang C-J (2023) The potential benefits of quercetin for brain health: a review of anti-inflammatory and neuroprotective mechanisms. Int J Mol Sci 24(7):6328

Colla E, Coune P, Liu Y, Pletnikova O, Troncoso JC, Iwatsubo T, Schneider BL, Lee MK (2012) Endoplasmic reticulum stress is important for the manifestations of α-synucleinopathy in vivo. J Neurosci 32(10):3306–3320

Cortese GP, Zhu M, Williams D, Heath S, Waites CL (2016) Parkin deficiency reduces hippocampal glutamatergic neurotransmission by impairing AMPA receptor endocytosis. J Neurosci 36(48):12243–12258

Dakterzada F, Jové M, Cantero JL, Pamplona R, Pinoll-Ripoll G (2023) Plasma and cerebrospinal fluid nonenzymatic protein damage is sustained in Alzheimer's disease. Redox Biol 64:102772

Davis CK, Bathula S, Hsu M, Morris-Blanco KC, Chokkalla AK, Jeong S, Choi J, Subramanian S, Park JS, Fabry Z (2022) An antioxidant and anti-ER stress combo therapy decreases inflammation, secondary brain damage and promotes neurological recovery following traumatic brain injury in mice. J Neurosci 42(35):6810–6821

DeCoursey TE, Morgan D, Cherny VV (2003) The voltage dependence of NADPH oxidase reveals why phagocytes need proton channels. Nature 422(6931):531–534

Degechisa ST, Dabi YT, Gizaw ST (2022) The mitochondrial associated endoplasmic reticulum membranes: a platform for the pathogenesis of inflammation-mediated metabolic diseases. Immun Inflammation Dis 10(7):e647

Demmings MD, Tennyson EC, Petroff GN, Tarnowski-Garner HE, Cregan SP (2021) Activating transcription factor-4 promotes neuronal death induced by Parkinson's disease neurotoxins and α-synuclein aggregates. Cell Death Differ 28(5):1627–1643

Devi L, Anandatheerthavarada HK (2010) Mitochondrial trafficking of APP and alpha synuclein: relevance to mitochondrial dysfunction in Alzheimer's and Parkinson's diseases. Biochim Biophys Acta Mol basis Dis 1802(1):11–19

Di Domenico F, Tramutola A, Perluigi M (2016) Cathepsin D as a therapeutic target in Alzheimer's disease. Expert Opin Ther Targets 20(12):1393–1395

Dong XX, Wang YR, Qin S, Liang ZQ, Liu BH, Qin ZH, Wang Y (2012) p53 mediates autophagy activation and mitochondria dysfunction in kainic acid-induced excitotoxicity in primary striatal neurons. Neuroscience 207:52–64

Dutta R, Trapp BD (2011) Mechanisms of neuronal dysfunction and degeneration in multiple sclerosis. Prog Neurobiol 93(1):1–12

Engel T, Goñi-Oliver P, Lucas JJ, Avila J, Hernández F (2006) Chronic lithium administration to FTDP-17 tau and GSK-3β overexpressing mice prevents tau hyperphosphorylation and neurofibrillary tangle formation, but pre-formed neurofibrillary tangles do not revert. J Neurochem 99(6):1445–1455

Erkkinen MG, Kim M-O, Geschwind MD (2018) Clinical neurology and epidemiology of the major neurodegenerative diseases. Cold Spring Harb Perspect Biol 10(4):a033118

Esquivel AR, Douglas JC, Loughran RM, Rezendes TE, Reed KR, Cains THL, Emsley SA, Paddock WA, Videau P, Koyack MJ (2020) Assessing the influence of curcumin in sex-specific oxidative stress, survival and behavior in Drosophila melanogaster. J Exp Biol 223(22):jeb223867

Evans-Martin F (2022) The nervous system. Infobase Holdings, Inc

Fabelo N, Martín V, Santpere G, Marín R, Torrent L, Ferrer I, Díaz M (2011) Severe alterations in lipid composition of frontal cortex lipid rafts from Parkinson's disease and incidental Parkinson's disease. Mol Med 17:1107–1118

Feng S-T, Wang Z-Z, Yuan Y-H, Sun H-M, Chen N-H, Zhang Y (2021) Update on the association between alpha-synuclein and tau with mitochondrial dysfunction: implications for Parkinson's disease. Eur J Neurosci 53(9):2946–2959

Fiadeiro MB, Diogo JC, Silva AA, Kim Y-s, Cristovao AC (2024) NADPH oxidases in neurodegenerative disorders: mechanisms and therapeutic opportunities. Antioxid Redox Signal 41:522

Friedman JR, Lackner LL, West M, DiBenedetto JR, Nunnari J, Voeltz GK (2011) ER tubules mark sites of mitochondrial division. Science 334(6054):358–362

Frontiñán-Rubio J, Rabanal-Ruiz Y, Durán-Prado M, Alcain FJ (2021) The protective effect of ubiquinone against the amyloid peptide in endothelial cells is isoprenoid chain length-dependent. Antioxidants 10(11):1806

Galloway KE, Ichida JK (2014) Modeling neurodegenerative diseases and neurodevelopmental disorders with reprogrammed cells. In: Stem cells, tissue engineering and medicine, regenerative. World Scientific, Singapore, p 215

Gandhi S, Abramov AY (2012) Mechanism of oxidative stress in neurodegeneration. Oxid Med Cell Longevity 2012:1

Ge P, Dawson VL, Dawson TM (2020) PINK1 and Parkin mitochondrial quality control: a source of regional vulnerability in Parkinson's disease. Mol Neurodegener 15(1):20

Ghavami S, Shojaei S, Yeganeh B, Ande SR, Jangamreddy JR, Mehrpour M, Christoffersson J, Chaabane W, Moghadam AR, Kashani HH (2014) Autophagy and apoptosis dysfunction in neurodegenerative disorders. Prog Neurobiol 112:24–49

Golpich M, Amini E, Hemmati F, Ibrahim NM, Rahmani B, Mohamed Z, Raymond AA, Dargahi L, Ghasemi R, Ahmadiani A (2015) Glycogen synthase kinase-3 beta (GSK-3β) signaling: implications for Parkinson's disease. Pharmacol Res 97:16–26

Gong J, Zhang L, Zhang Q, Li X, Xia X-J, Liu Y-Y, Yang Q-S (2018) Lentiviral vector-mediated SHC3 silencing exacerbates oxidative stress injury in nigral dopamine neurons by regulating the PI3K-AKT-FoxO signaling pathway in rats with Parkinson's disease. Cell Physiol Biochem 49(3):971–984

Gong L, Yin J, Zhang Y, Huang R, Lou Y, Jiang H, Sun L, Jia J, Zeng X (2022) Neuroprotective mechanisms of ginsenoside Rb1 in central nervous system diseases. Front Pharmacol 13:914352

Goudarzi S, Hosseini A, Abdollahi M, Haghi-Aminjan H (2021) Insights into Parkin-mediated mitophagy in Alzheimer's disease: a systematic review. Front Aging Neurosci 13:674071

Goyal A, Agrawal A, Verma A, Dubey N (2023) The PI3K-AKT pathway: a plausible therapeutic target in Parkinson's disease. Exp Mol Pathol 129:104846

Graille M, Wild P, Sauvain J-J, Hemmendinger M, Canu G, Irina and Hopf, Nancy B. (2020) Urinary 8-OHdG as a biomarker for oxidative stress: a systematic literature review and meta-analysis. Int J Mol Sci 21(11):3743

Grewer C, Rauen T (2005) Electrogenic glutamate transporters in the CNS: molecular mechanism, pre-steady-state kinetics, and their impact on synaptic signaling. J Membr Biol 203:1–20

Gu Z, Liu W, Yan Z (2009) β-Amyloid impairs AMPA receptor trafficking and function by reducing Ca2+/calmodulin-dependent protein kinase II synaptic distribution. J Biol Chem 284(16):10639–10649

Guo C, Sun L, Chen X, Zhang D (2013) Oxidative stress, mitochondrial damage and neurodegenerative diseases. Neural Regen Res 8(21):2003–2014

Guo M-Y, Shang L, Hu Y-Y, Jiang L-P, Wan Y-Y, Zhou Q-Q, Zhang K, Liao H-F, Yi J-L, Han X-J (2018) The role of Cdk5-mediated Drp1 phosphorylation in Aβ1-42 induced mitochondrial fission and neuronal apoptosis. J Cell Biochem 119(6):4815–4825

Haleagrahara N, Siew CJ, Ponnusamy K (2013) Effect of quercetin and desferrioxamine on 6-hydroxydopamine (6-OHDA) induced neurotoxicity in striatum of rats. J Toxicol Sci 38(1):25–33

Han S, Jeong YY, Sheshadri P, Cai Q (2020a) Mitophagy coordination with retrograde transport ensures the integrity of synaptic mitochondria. Autophagy 16(10):1925–1927

Han S, Jeong YY, Sheshadri P, Su X, Cai Q (2020b) Mitophagy regulates integrity of mitochondria at synapses and is critical for synaptic maintenance. EMBO Rep 21(9):e49801

Hanslik KL, Ulland TK (2020) The role of microglia and the Nlrp3 inflammasome in Alzheimer's disease. Front Neurol 11:570711

Hartmann A, Troadec J-D, Hunot S, Kikly K, Faucheux BA, Mouatt-Prigent A, Ruberg M, Agid Y, Hirsch EC (2001) Caspase-8 is an effector in apoptotic death of dopaminergic neurons in Parkinson's disease, but pathway inhibition results in neuronal necrosis. J Neurosci 21(7):2247–2255

He H-J, Xiong X, Zhou S, Zhang X-R, Zhao X, Chen L, Xie C-L (2022) Neuroprotective effects of curcumin via autophagy induction in 6-hydroxydopamine Parkinson's models. Neurochem Int 155:105297

Heneka MT, Kummer MP, Stutz A, Delekate A, Schwartz S, Vieira-Saecker A, Griep A, Axt D, Remus A, Tzeng T-C (2013) NLRP3 is activated in Alzheimer's disease and contributes to pathology in APP/PS1 mice. Nature 493(7434):674–678

Heras-Sandoval D, Pérez-Rojas JM, Hernández-Damián J, Pedraza-Chaverri J (2014) The role of PI3K/AKT/mTOR pathway in the modulation of autophagy and the clearance of protein aggregates in neurodegeneration. Cell Signal 26(12):2694–2701

Herculano-Houzel S (2009) The human brain in numbers: a linearly scaled-up primate brain. Front Hum Neurosci 3:857

Hickey MA, Chesselet M-F (2003) Apoptosis in Huntington's disease. Prog Neuro-Psychopharmacol Biol Psychiatry 27(2):255–265

Höhn A, Tramutola A, Cascella R (2020) Proteostasis failure in neurodegenerative diseases: focus on oxidative stress. Oxid Med Cell Longevity 2020:5497046

Hong DK, Kho AR, Lee SH, Kang BS, Park MK, Choi BY, Suh SW (2023) Pathophysiological roles of transient receptor potential (Trp) channels and zinc toxicity in brain disease. Int J Mol Sci 24(7):6665

Hosokawa Y, Hosokawa I, Ozaki K, Matsuo T (2021) Nobiletin inhibits inflammatory reaction in interleukin-1β-stimulated human periodontal ligament cells. Pharmaceutics 13(5):667

Hsieh H, Boehm J, Sato C, Iwatsubo T, Tomita T, Sisodia S, Malinow R (2006) AMPAR removal underlies Aβ-induced synaptic depression and dendritic spine loss. Neuron 52(5):831–843

Huang S, Chen Z, Fan B, Chen Y, Zhou L, Jiang B, Long H, Zhong W, Li X, Li Y (2021) A selective NLRP3 inflammasome inhibitor attenuates behavioral deficits and neuroinflammation in a mouse model of Parkinson's disease. J Neuroimmunol 354:577543

Hulse J, Bhaskar K (2022) Crosstalk between the NLRP3 inflammasome/ASC speck and amyloid protein aggregates drives disease progression in Alzheimer's and Parkinson's disease. Front Mol Neurosci 15:805169

Hung W-L, Chiu T-H, Wei G-J, Pan M-H, Ho C-T, Hwang LS, Hsu B-Y (2023) Neuroprotective effects of nobiletin and tangeretin against amyloid β1-42-induced toxicity in cultured primary rat neurons. Nutrire 48(2):56

Italia M, Ferrari E, Luca D, Monica and Gardoni, Fabrizio. (2021) GluA3-containing AMPA receptors: from physiology to synaptic dysfunction in brain disorders. Neurobiol Dis 161:105539

Jaganjac M, Cindrić M, Jakovčević A, Žarković K, Žarković N (2021) Lipid peroxidation in brain tumors. Neurochem Int 149:105118

Jahan S, Ansari UA, Siddiqui AJ, Iqbal D, Khan J, Banawas S, Alshehri B, Alshahrani MM, Alsagaby SA, Redhu NS (2022) Nobiletin ameliorates cellular damage and stress response and restores neuronal identity altered by sodium arsenate exposure in human iPSCs-derived hNPCs. Pharmaceuticals 15(5):593

Jahan S, Ansari UA, Srivastava AK, Aldosari S, Alabdallat NG, Siddiqui AJ, Khan A, Albadrani HM, Sarkar S, Khan B (2024) A protein–miRNA biomic analysis approach to explore neuroprotective potential of nobiletin in human neural progenitor cells (hNPCs). Front Pharmacol 15:1343569

Jeong KH, Jeon M-T, Kim HD, Jung UJ, Jang MC, Chu JW, Yang SJ, Choi IY, Choi M-S, Kim SR (2015) Nobiletin protects dopaminergic neurons in the 1-methyl-4-phenylpyridinium-treated rat model of Parkinson's disease. J Med Food 18(4):409–414

Jha SK, Jha NK, Kar R, Ambasta RK, Kumar P (2015) p38 MAPK and PI3K/AKT signalling cascades inParkinson's disease. Int J Mol Cell Med 4(2):67

Jiang H, Wu Y-C, Nakamura M, Liang Y, Tanaka Y, Holmes S, Dawson VL, Dawson TM, Ross CA, Smith WW (2007) Parkinson's disease genetic mutations increase cell susceptibility to stress: mutant α-synuclein enhances H2O2-and Sin-1-induced cell death. Neurobiol Aging 28(11):1709–1717

Johansen T, Lamark T (2011) Selective autophagy mediated by autophagic adapter proteins. Autophagy 7(3):279–296

Kabir MT, Shah MA, Alothaim AAS, Al-Ghamdi MS, Rahman MH (2023) Neuroprotective effects of saponins on neurodegenerative diseases. In: Phytonutrients and neurological disorders. Elsevier, pp 259–282

Kadriu B, Musazzi L, Johnston JN, Kalynchuk LE, Caruncho HJ, Popoli M, Zarate Jr CA (2021) Positive AMPA receptor modulation in the treatment of neuropsychiatric disorders: a long and winding road. Drug Discov Today 26(12):2816–2838

Kalpage HA, Wan J, Morse PT, Zurek MP, Turner AA, Khobeir A, Yazdi N, Hakim L, Liu J, Vaishnav A (2020) Cytochrome c phosphorylation: control of mitochondrial electron transport chain flux and apoptosis. Int J Biochem Cell Biol 121:105704

Kaushik A, Chauhan K, Singh S (2023) Neuroprotective potential of quercetin as a nutraceutical targeting fused neuroinflammation in neurological disease. In: Treatments, nutraceuticals, supplements, and herbal medicine in neurological disorders. Elsevier, pp 623–637

Keifer J (2022) Regulation of AMPAR trafficking in synaptic plasticity by BDNF and the impact of neurodegenerative disease. J Neurosci Res 100(4):979–991

Khan H, Ullah H, Aschner M, Cheang WS, Akkol EK (2019) Neuroprotective effects of quercetin in Alzheimer's disease. Biomolecules 10(1):59

Kitagishi Y, Nakanishi A, Ogura Y, Matsuda S (2014) Dietary regulation of PI3K/AKT/GSK-3β pathway in Alzheimer's disease. Alzheimer's Res Ther 6:1–7

Kofoed RH, Betzer C, Ferreira N, Jensen PH (2020) Glycogen synthase kinase 3 β activity is essential for polo-like kinase 2-and leucine-rich repeat kinase 2-mediated regulation of α-synuclein. Neurobiol Dis 136:104720

Koirala S, Guo Q, Kalia R, Bui HT, Eckert DM, Frost A, Shaw JM (2013) Interchangeable adaptors regulate mitochondrial dynamin assembly for membrane scission. Proc Natl Acad Sci 110(15):E1342–E1351

Krell-Roesch J, Syrjanen JA, Rakusa M, Vemuri P, Machulda MM, Kremers WK, Mielke MM, Lowe VJ, Jack Jr CR, Knopman DS (2021) Association of cortical and subcortical β-amyloid with standardized measures of depressive and anxiety symptoms in adults without dementia. J Neuropsychiatry Clin Neurosci 33(1):64–71

L'episcopo F, Serapide MF, Tirolo C, Testa N, Caniglia S, Morale MC, Pluchino S, Marchetti B (2011) A Wnt1 regulated Frizzled-1/β-catenin signaling pathway as a candidate regulatory circuit controlling mesencephalic dopaminergic neuron-astrocyte crosstalk: Therapeutical relevance for neuron survival and neuroprotection. Mol Neurodegener 6:1–29

Lautrup S, Lou G, Aman Y, Nilsen H, Tao J, Fang EF (2019) Microglial mitophagy mitigates neuroinflammation in Alzheimer's disease. Neurochem Int 129:104469

Lee D-s, Kim J-E (2018) PDI-mediated S-nitrosylation of DRP1 facilitates DRP1-S616 phosphorylation and mitochondrial fission in CA1 neurons. Cell Death Dis 9(9):869

Lee SB, Kim JJ, Han S-A, Fan Y, Guo L-S, Aziz K, Nowsheen S, Kim SS, Park S-Y, Luo Q (2019) The AMPK–Parkin axis negatively regulates necroptosis and tumorigenesis by inhibiting the necrosome. Nat Cell Biol 21(8):940–951

Li S, Selkoe DJ (2020) A mechanistic hypothesis for the impairment of synaptic plasticity by soluble Aβ oligomers from Alzheimer's brain. J Neurochem 154(6):583–597

Li Q, Chen L, Liu X, Li X, Cao Y, Bai Y, Qi F (2018) Pterostilbene inhibits amyloid-β-induced neuroinflammation in a microglia cell line by inactivating the NLRP3/caspase-1 inflammasome pathway. J Cell Biochem 119(8):7053–7062

Li G, Zhang N, Geng F, Liu G, Liu B, Lei X, Li G, Chen X (2019) High-throughput metabolomics and ingenuity pathway approach reveals the pharmacological effect and targets of Ginsenoside Rg1 in Alzheimer's disease mice. Sci Rep 9(1):7040

Li Y, Xia Y, Yin S, Wan F, Hu J, Kou L, Sun Y, Wu J, Zhou Q, Huang J (2021) Targeting microglial α-synuclein/TLRs/NF-kappaB/NLRP3 inflammasome axis in Parkinson's disease. Front Immunol 12:719807

Li N, Zhang R, Tang M, Zhao M, Jiang X, Cai X, Ye N, Su K, Peng J, Zhang X (2023a) Recent progress and prospects of small molecules for NLRP3 inflammasome inhibition. J Med Chem 66(21):14447–14473

Li X, Zhang H, Yang L, Dong X, Han Y, Su Y, Li W, Li W (2023b) Inhibition of NLRP1 inflammasome improves autophagy dysfunction and Aβ disposition in APP/PS1 mice. Behav Brain Funct 19(1):7
Li Y, Ma R, Hao X (2024) Therapeutic role of PTEN in tissue regeneration for management of neurological disorders: stem cell behaviors to an in-depth review. Cell Death Dis 15(4):268
Liang H-Y, Zhang P-P, Zhang X-L, Zheng Y-Y, Huang Y-R, Zheng G-Q, Lin Y (2021) Preclinical systematic review of ginsenoside Rg1 for cognitive impairment in Alzheimer's disease. Aging 13(5):7549
Lin Y-S, Peng W-H, Shih M-F, Cherng J-Y (2021) Anxiolytic effect of an extract of Salvia miltiorrhiza Bunge (Danshen) in mice. J Ethnopharmacol 264:113285
Lonskaya I, Shekoyan AR, Hebron ML, Desforges N, Algarzae NK, Moussa CE-H (2013) Diminished parkin solubility and co-localization with intraneuronal amyloid-β are associated with autophagic defects in Alzheimer's disease. J Alzheimer's Dis 33(1):231–247
Lv R, Du L, Zhou F, Yuan X, Liu X, Zhang L (2020) Rosmarinic acid alleviates inflammation, apoptosis, and oxidative stress through regulating miR-155-5p in a mice model of Parkinson's disease. ACS Chem Neurosci 11(20):3259–3266
Ma T (2023) Roles of eukaryotic elongation factor 2 kinase (eEF2K) in neuronal plasticity, cognition, and Alzheimer disease. J Neurochem 166(1):47–57
Ma MW, Wang J, Zhang Q, Wang R, Dhandapani KM, Vadlamudi RK, Brann DW (2017) NADPH oxidase in brain injury and neurodegenerative disorders. Mol Neurodegener 12:1–28
Ma L, Yang C, Zhang X, Li Y, Wang S, Zheng L, Huang K (2018) C-terminal truncation exacerbates the aggregation and cytotoxicity of α-Synuclein: a vicious cycle in Parkinson's disease. Biochim Biophys Acta Mol basis Dis 1864(12):3714–3725
Madiha S, Batool Z, Tabassum S, Liaquat L, Sadir S, Shahzad S, Naqvi F, Saleem S, Yousuf S, Nawaz A (2021) Quercetin exhibits potent antioxidant activity, restores motor and non-motor deficits induced by rotenone toxicity. PLoS One 16(11):e0258928
Mahul-Mellier A-L, Burtscher J, Maharjan N, Weerens L, Croisier M, Kuttler F, Leleu M, Knott GW, Lashuel HA (2020) The process of Lewy body formation, rather than simply α-synuclein fibrillization, is one of the major drivers of neurodegeneration. Proc Natl Acad Sci 117(9):4971–4982
Maiti P, Manna J, Dunbar GL (2017) Current understanding of the molecular mechanisms in Parkinson's disease: targets for potential treatments. Transl Neurodegener 6:1–35
Mao Y, Chen X, Xu M, Fujita K, Motoki K, Sasabe T, Homma H, Murata M, Tagawa K, Tamura T (2016) Targeting TEAD/YAP-transcription-dependent necrosis, TRIAD, ameliorates Huntington's disease pathology. Hum Mol Genet 25(21):4749–4770
Marchetti B (2018) Wnt/β-catenin signaling pathway governs a full program for dopaminergic neuron survival, neurorescue and regeneration in the MPTP mouse model of Parkinson's disease. Int J Mol Sci 19(12):3743
Marsh AP (2019) Molecular mechanisms of proteinopathies across neurodegenerative disease: a review. Neurol Res Pract 1:1–7
Maruyama T, Hama Y, Noda NN (2024) Mechanisms of mitochondrial reorganization. J Biochem 175(2):167–178
Matsunaga D, Sreekumar PG, Ishikawa K, Terasaki H, Barron E, Cohen P, Kannan R, Hinton DR (2016) Humanin protects RPE cells from endoplasmic reticulum stress-induced apoptosis by upregulation of mitochondrial glutathione. PLoS One 11(10):e0165150
Matsuzaki K, Ohizumi Y (2021) Beneficial effects of citrus-derived polymethoxylated flavones for central nervous system disorders. Nutrients 13(1):145
Matus S, Lopez E, Valenzuela V, Nassif M, Hetz C (2013) Functional contribution of the transcription factor ATF4 to the pathogenesis of amyotrophic lateral sclerosis. PLoS One 8(7):e66672
Mehta A, Prabhakar M, Kumar P, Deshmukh R, Sharma PL (2013) Excitotoxicity: bridge to various triggers in neurodegenerative disorders. Eur J Pharmacol 698(1–3):6–18
Miller WP, Sunilkumar S, Dennis MD (2021) The stress response protein REDD1 as a causal factor for oxidative stress in diabetic retinopathy. Free Radic Biol Med 165:127–136

Milton RH, Abeti R, Averaimo S, DeBiasi S, Vitellaro L, Jiang L, Curmi PMG, Breit SN, Duchen MR, Mazzanti M (2008) CLIC1 function is required for β-amyloid-induced generation of reactive oxygen species by microglia. J Neurosci 28(45):11488–11499
Mironova GD, Pavlov EV (2021) Mitochondrial cyclosporine A-independent palmitate/Ca2+-induced permeability transition pore (PA-mPT Pore) and its role in mitochondrial function and protection against calcium overload and glutamate toxicity. Cells 10(1):125
Montagna C, Cirotti C, Rizza S, Filomeni G (2020) When S-nitrosylation gets to mitochondria: from signaling to age-related diseases. Antioxid Redox Signal 32(12):884–905
Morris JL, Gillet G, Prudent J, Popgeorgiev N (2021) Bcl-2 family of proteins in the control of mitochondrial calcium signalling: an old chap with new roles. Int J Mol Sci 22(7):3730
Nakajima A, Ohizumi Y (2019) Potential benefits of nobiletin, a citrus flavonoid, against Alzheimer's disease and Parkinson's disease. Int J Mol Sci 20(14):3380
Naveen K, Bhattacharjee A (2021) Medicinal herbs as neuroprotective agents. WJPPS 10(4):675–689
Ni H-M, Chao X, Kaseff J, Deng F, Wang S, Shi Y-H, Li T, Ding W-X, Jaeschke H (2019) Receptor-interacting serine/threonine-protein kinase 3 (RIPK3)–mixed lineage kinase domain-like protein (MLKL)–mediated necroptosis contributes to ischemia-reperfusion injury of steatotic livers. Am J Pathol 189(7):1363–1374
Ni Y-Q, Xu H, Liu Y-S (2022) Roles of long non-coding RNAs in the development of aging-related neurodegenerative diseases. Front Mol Neurosci 15:844193
Nita M, Grzybowski A (2016) The role of the reactive oxygen species and oxidative stress in the pathomechanism of the age-related ocular diseases and other pathologies of the anterior and posterior eye segments in adults. Oxid Med Cell Longevity 2016:3164734
Obata T, Kubota S, Yamanaka Y (2001) Allopurinol suppresses para-nonylphenol and 1-methyl-4-phenylpyridinium ion (MPP+)-induced hydroxyl radical generation in rat striatum. Neurosci Lett 306(1–2):9–12
Ochneva A, Zorkina Y, Abramova O, Pavlova O, Ushakova V, Morozova A, Zubkov E, Pavlov K, Gurina O, Chekhonin V (2022) Protein misfolding and aggregation in the brain: common pathogenetic pathways in neurodegenerative and mental disorders. Int J Mol Sci 23(22):14498
Ong G, Ragetli R, Mnich K, Doble BW, Kammouni W, Logue SE, Death C (2024) IRE1 signaling increases PERK expression during chronic ER stress. Disease 15(4):276
Orhan IE (2021) Cholinesterase inhibitory potential of quercetin towards Alzheimer's disease-a promising natural molecule or fashion of the day?-A narrowed review. Curr Neuropharmacol 19(12):2205
Orjuela A, Lakey-Beitia J, Mojica-Flores R, Hegde ML, Lans I, Alí-Torres J, Rao KS (2021) Computational evaluation of interaction between curcumin derivatives and amyloid-β monomers and fibrils: relevance to Alzheimer's disease. J Alzheimer's Dis 82(s1):S321–S333
Pádua MS, Guil-Guerrero JL, Prates JAM, Lopes PA (2024) Insights on the use of transgenic mice models in Alzheimer's disease research. Int J Mol Sci 25(5):2805
Parameshwaran K, Dhanasekaran M, Suppiramaniam V (2008) Amyloid beta peptides and glutamatergic synaptic dysregulation. Exp Neurol 210(1):7–13
Park H, Kang J-H, Lee S (2020) Autophagy in neurodegenerative diseases: a hunter for aggregates. Int J Mol Sci 21(9):3369
Passlick S, Rose CR, Petzold GC, Henneberger C (2021) Disruption of glutamate transport and homeostasis by acute metabolic stress. Front Cell Neurosci 15:637784
Payne A, Nahashon S, Taka E, Adinew GM, Soliman KFA (2022) Epigallocatechin-3-Gallate (EGCG): new therapeutic perspectives for neuroprotection, aging, and neuroinflammation for the modern age. Biomolecules 12(3):371
Perrett RM, Alexopoulou Z, Tofaris GK (2015) The endosomal pathway in Parkinson's disease. Mol Cell Neurosci 66:21–28
Pike AF, Szabò I, Veerhuis R, Bubacco L (2022) The potential convergence of NLRP3 inflammasome, potassium, and dopamine mechanisms in Parkinson's disease. NPJ Parkinson's Dis 8(1):32

Puspita L, Chung SY, Shim J-w (2017) Oxidative stress and cellular pathologies in Parkinson's disease. Mol Brain 10:1–12

Qu W, Yuan B, Liu J, Liu Q, Zhang X, Cui R, Yang W, Li B (2021) Emerging role of AMPA receptor subunit GluA1 in synaptic plasticity: implications for Alzheimer's disease. Cell Prolif 54(1):e12959

Rakovic A, Ziegler J, Mårtensson CU, Prasuhn J, Shurkewitsch K, König P, Paulson HL, Klein C (2019) PINK1-dependent mitophagy is driven by the UPS and can occur independently of LC3 conversion. Cell Death Differ 26(8):1428–1441

Rama R, García JC (2016) Excitotoxicity and oxidative stress in acute stroke. In: Ischemic stroke updates. IntechOpen, Rijeka, pp 17–33

Ramakrishna K, Nalla LV, Naresh D, Venkateswarlu K, Viswanadh MK, Nalluri BN, Chakravarthy G, Duguluri S, Singh P, Rai SN (2023) WNT-β catenin signaling as a potential therapeutic target for neurodegenerative diseases: current status and future perspective. Diseases 11(3):89

Ramezani M, Meymand AZ, Khodagholi F, Kamsorkh HM, Asadi E, Noori M, Rahimian K, Shahrbabaki AS, Talebi A, Parsaiyan H (2023) A role for flavonoids in the prevention and/or treatment of cognitive dysfunction, learning, and memory deficits: a review of preclinical and clinical studies. Nutr Neurosci 26(2):156–172

Rana T, Behl T, Sehgal A, Mehta V, Singh S, Bhatia S, Al-Harrasi A, Bungau S (2021) Exploring the role of autophagy dysfunction in neurodegenerative disorders. Mol Neurobiol 58(10):4886–4905

Read A, Schröder M (2021) The unfolded protein response: an overview. Biology 10(5):384

Regoni M, Valtorta F, Sassone J (2024) Dopaminergic neuronal death via necroptosis in Parkinson's disease: a review of the literature. Eur J Neurosci 59(6):1079–1098

Ren C, Gu X, Li H, Lei S, Wang Z, Wang J, Yin P, Zhang C, Wang F, Liu C (2019) The role of DKK1 in Alzheimer's disease: a potential intervention point of brain damage prevention? Pharmacol Res 144:331–335

Rios L, Pokhrel S, Li S-J, Heo G, Haileselassie B, Mochly-Rosen D (2023) Targeting an allosteric site in dynamin-related protein 1 to inhibit Fis1-mediated mitochondrial dysfunction. Nat Commun 14(1):4356

Ripoli C, Cocco S, Puma DD, Li P, Roberto M, Alessia S, Federico P, Daniela D'A, Marcello, Grassi C (2014) Intracellular accumulation of amyloid-β (Aβ) protein plays a major role in Aβ-induced alterations of glutamatergic synaptic transmission and plasticity. J Neurosci 34(38):12893–12903

Sahu MR, Mondal AC (2020) The emerging role of Hippo signaling in neurodegeneration. J Neurosci Res 98(5):796–814

Sarawi WS, Alhusaini AM, Fadda LM, Alomar HA, Albaker AB, Aljrboa AS, Alotaibi AM, Hasan IH, Mahmoud AM (2021) Curcumin and nano-curcumin mitigate copper neurotoxicity by modulating oxidative stress, inflammation, and akt/gsk-3β signaling. Molecules 26(18):5591

Sawant N, Morton H, Kshirsagar S, Reddy AP, Reddy PH (2021) Mitochondrial abnormalities and synaptic damage in Huntington's disease: a focus on defective mitophagy and mitochondria-targeted therapeutics. Mol Neurobiol 58:1–28

Seo J, Nam YW, Kim S, Oh D-B, Song J (2021) Necroptosis molecular mechanisms: recent findings regarding novel necroptosis regulators. Exp Mol Med 53(6):1007–1017

Sesaki H, Adachi Y, Kageyama Y, Itoh K, Iijima M (2014) In vivo functions of Drp1: lessons learned from yeast genetics and mouse knockouts. Biochim Biophys Acta Mol basis Dis 1842(8):1179–1185

Shacham T, Patel C, Lederkremer GZ (2021) PERK pathway and neurodegenerative disease: to inhibit or to activate? Biomolecules 11(3):354

Shevtsova EF, Maltsev AV, Vinogradova DV, Shevtsov PN, Bachurin SO (2021) Mitochondria as a promising target for developing novel agents for treating Alzheimer's disease. Med Res Rev 41(2):803–827

Shi M, Chai Y, Zhang J, Chen X (2022) Endoplasmic reticulum stress-associated neuronal death and innate immune response in neurological diseases. Front Immunol 12:794580

Shi H-Z, Sun J-F, Chen H (2023) The role of endoplasmic reticulum stress in melanoma. Int J Dermatol Venereol 6(3):150–156

Simpson DSA, Oliver PL (2020) ROS generation in microglia: understanding oxidative stress and inflammation in neurodegenerative disease. Antioxidants 9(8):743

Singh P, Barman B, Thakur MK (2022) Oxidative stress-mediated memory impairment during aging and its therapeutic intervention by natural bioactive compounds. Front Aging Neurosci 14:944697

Sirianni AC, Jiang J, Zeng J, Mao LL, Zhou S, Sugarbaker P, Zhang X, Li W, Friedlander RM, Wang X (2015) N-acetyl-l-tryptophan, but not N-acetyl-d-tryptophan, rescues neuronal cell death in models of amyotrophic lateral sclerosis. J Neurochem 134(5):956–968

Siwecka N, Rozpędek-Kamińska W, Wawrzynkiewicz A, Pytel D, Diehl JA, Majsterek I (2021) The structure, activation and signaling of IRE1 and its role in determining cell fate. Biomedicine 9(2):156

Siwicka-Gieroba D, Robba C, Gołacki J, Badenes R, Dabrowski W (2022) Cerebral oxygen delivery and consumption in brain-injured patients. J Personal Med 12(11):1763

Smith-Dijak AI, Nassrallah WB, Zhang LYJ, Geva M, Hayden MR, Raymond LA (2019) Impairment and restoration of homeostatic plasticity in cultured cortical neurons from a mouse model of Huntington disease. Front Cell Neurosci 13:209

Song B, Lai B, Zheng Z, Zhang Y, Luo J, Wang C, Chen Y, Woodgett JR, Li M (2010) Inhibitory phosphorylation of GSK-3 by CaMKII couples depolarization to neuronal survival. J Biol Chem 285(52):41122–41134

Song W, Chen J, Petrilli A, Liot G, Klinglmayr E, Zhou Y, Poquiz P, Tjong J, Pouladi MA, Hayden MR (2011) Mutant huntingtin binds the mitochondrial fission GTPase dynamin-related protein-1 and increases its enzymatic activity. Nat Med 17(3):377–382

Song W, Zhang L, Cui X, Wang R, Ma J, Xu Y, Jin Y, Wang D, Lu Z (2024) Nobiletin alleviates cisplatin-induced ototoxicity via activating autophagy and inhibiting NRF2/GPX4-mediated ferroptosis. Sci Rep 14(1):7889

Soraci L, Gambuzza ME, Biscetti L, Laganà P, Russo L, Carmela B, Annamaria B, Giada C, Andrea L, Fabrizia, Lorello G (2023) Toll-like receptors and NLRP3 inflammasome-dependent pathways in Parkinson's disease: mechanisms and therapeutic implications. J Neurol 270(3):1346–1360

Sorce S, Krause K-H (2009) NOX enzymes in the central nervous system: from signaling to disease. Antioxid Redox Signal 11(10):2481–2504

Srinivasan E, Chandrasekhar G, Chandrasekar P, Anbarasu K, Vickram AS, Karunakaran R, Rajasekaran R, Srikumar PS (2021) Alpha-synuclein aggregation in Parkinson's disease. Front Med 8:736978

Stancu IC, Lodder C, Lucena B, Pablo V, Sarah G, de Ravé M, Terwel D, Bottelbergs A, Dewachter I (2022) The NLRP3 inflammasome modulates tau pathology and neurodegeneration in a tauopathy model. Glia 70(6):1117–1132

Stoka V, Turk V, Turk B (2016) Lysosomal cathepsins and their regulation in aging and neurodegeneration. Ageing Res Rev 32:22–37

Sweeney P, Park H, Baumann M, Dunlop J, Frydman J, Kopito R, McCampbell A, Leblanc G, Venkateswaran A, Nurmi A (2017) Protein misfolding in neurodegenerative diseases: implications and strategies. Transl Neurodegener 6:1–13

Swerdlow RH (2023) The Alzheimer's disease mitochondrial cascade hypothesis: a current overview. J Alzheimer's Dis 92(3):751–768

Taghizadieh M, Modiri N, Salmaninejad A, Khayami R, Taheri F, Moghoofei M, Khatami A, Azhdari S, Bayat M, Samadi-khouzani A (2023) Neuroprotective effects of flavonoids. In: Phytonutrients and neurological disorders. Elsevier, pp 133–200

Tahir MS, Almezgagi M, Zhang Y, Bashir A, Abdullah HM, Gamah M, Wang X, Zhu Q, Shen X, Ma Q (2021) Mechanistic new insights of flavonols on neurodegenerative diseases. Biomed Pharmacother 137:111253

Taoufik E, Probert L (2008) Ischemic neuronal damage. Curr Pharm Design 14(33):3565–3573

Teleanu DM, Niculescu A-G, Lungu II, Radu CI, Vladâcenco O, Roza E, Costăchescu B, Grumezescu AM, Teleanu RI (2022) An overview of oxidative stress, neuroinflammation, and neurodegenerative diseases. Int J Mol Sci 23(11):5938

Telesford QK, Simpson SL, Burdette JH, Hayasaka S, Laurienti PJ (2011) The brain as a complex system: using network science as a tool for understanding the brain. Brain Connect 1(4):295–308

Thawkar BS, Kaur G (2019) Inhibitors of NF-κB and P2X7/NLRP3/Caspase 1 pathway in microglia: novel therapeutic opportunities in neuroinflammation induced early-stage Alzheimer's disease. J Neuroimmunol 326:62–74

Timmons S, Coakley MF, Moloney AM, O'Neill C (2009) Akt signal transduction dysfunction in Parkinson's disease. Neurosci Lett 467(1):30–35

van Vliet AR, Agostinis P (2018) Mitochondria-associated membranes and ER stress. In: Coordinating organismal physiology through the unfolded protein response. Springer, Cham, pp 73–102

Vandenabeele P, Declercq W, Van Herreweghe F, Vanden Berghe T (2010) The role of the kinases RIP1 and RIP3 in TNF-induced necrosis. Sci Signal 3(115):re4-re4

Venegas C, Kumar S, Franklin BS, Dierkes T, Brinkschulte R, Tejera D, Vieira-Saecker A, Schwartz S, Santarelli F, Kummer MP (2017) Microglia-derived ASC specks cross-seed amyloid-β in Alzheimer's disease. Nature 552(7685):355–361

Vidal RL, Figueroa A, Court FA, Thielen P, Molina C, Wirth C, Caballero B, Kiffin R, Segura-Aguilar J, Cuervo AM (2012) Targeting the UPR transcription factor XBP1 protects against Huntington's disease through the regulation of FoxO1 and autophagy. Hum Mol Genet 21(10):2245–2262

Volgyi K, Juhász G, Kovács Z, Penke B (2015) Dysfunction of endoplasmic reticulum (ER) and mitochondria (MT) in Alzheimer's disease: the role of the ER-MT cross-talk. Curr Alzheimer Res 12(7):655–672

Walter C, Clemens LE, Müller AJ, Fallier-Becker P, Proikas-Cezanne T, Riess O, Metzger S, Nguyen HP (2016) Activation of AMPK-induced autophagy ameliorates Huntington disease pathology in vitro. Neuropharmacology 108:24–38

Wang Y-T, Lu J-H (2022) Chaperone-mediated autophagy in neurodegenerative diseases: molecular mechanisms and pharmacological opportunities. Cells 11(14):2250

Wang Y, Dong X-X, Cao Y, Liang Z-Q, Han R, Wu J-C, Gu Z-L, Qin Z-H (2009) p53 induction contributes to excitotoxic neuronal death in rat striatum through apoptotic and autophagic mechanisms. Eur J Neurosci 30(12):2258–2270

Wang H, Sun L, Su L, Rizo J, Liu L, Wang L-F, Wang F-S, Wang X (2014) Mixed lineage kinase domain-like protein MLKL causes necrotic membrane disruption upon phosphorylation by RIP3. Mol Cell 54(1):133–146

Wang B, Yu Y, Wei L, Zhang Y (2020) Inhibition of ER stress improves progressive motor deficits in a REEP1-null mouse model of hereditary spastic paraplegia. Biol Open 9(9):bio054296

Wang Q, Dong X, Zhang R, Zhao C (2021) Flavonoids with potential anti-amyloidogenic effects as therapeutic drugs for treating Alzheimer's disease. J Alzheimer's Dis 84(2):505–533

Wang H, Guo M, Wei H, Chen Y (2023) Targeting p53 pathways: mechanisms, structures, and advances in therapy. Signal Transduct Target Ther 8(1):92

World Health Organization (2006) Neurological disorders: public health challenges. World Health Organization

Wróbel-Biedrawa D, Grabowska K, Galanty A, Sobolewska D, Podolak I (2022) A flavonoid on the brain: quercetin as a potential therapeutic agent in central nervous system disorders. Life 12(4):591

Wu Q-L, Gao Y, Li J-T, Ma W-Y, Chen N-H (2022) The role of AMPARs composition and trafficking in synaptic plasticity and diseases. Cell Mol Neurobiol 42(8):2489–2504

Xiang C, Wang Y, Zhang H, Han F (2017) The role of endoplasmic reticulum stress in neurodegenerative disease. Apoptosis 22:1–26

Xie C, Aman Y, Adriaanse BA, Cader MZ, Plun-Favreau H, Xiao J, Fang EF (2020) Culprit or bystander: defective mitophagy in Alzheimer's disease. Front Cell Dev Biol 7:391

Xu H, Wang E, Chen F, Xiao J, Wang M (2021) Neuroprotective phytochemicals in experimental ischemic stroke: mechanisms and potential clinical applications. Oxidative Med Cell Longev 2021:1–45

Yacoubian TA (2017) Neurodegenerative disorders: why do we need new therapies? In: Drug discovery approaches for the treatment of neurodegenerative disorders. Elsevier, pp 1–16

Yadav V, Guin S, Nayak S, Mishra A (2024) Herbal approaches for the Management of Neurological Disorders. In: Drug delivery strategies in neurological disorders: challenges and opportunities. Springer, pp 219–247

Yamanishi E, Hasegawa K, Fujita K, Ichinose S, Yagishita S, Murata M, Tagawa K, Akashi T, Eishi Y, Okazawa H (2017) A novel form of necrosis, TRIAD, occurs in human Huntington's disease. Acta Neuropathol Commun 5:1–10

Yamashima T (2004) Ca2+-dependent proteases in ischemic neuronal death: a conserved 'calpain–cathepsin cascade'from nematodes to primates. Cell Calcium 36(3–4):285–293

Yamashima T (2013) Reconsider Alzheimer's disease by the 'calpain–cathepsin hypothesis'—a perspective review. Prog Neurobiol 105:1–23

Yan Y, Jiang W, Liu L, Wang X, Ding C, Tian Z, Zhou R (2015) Dopamine controls systemic inflammation through inhibition of NLRP3 inflammasome. Cell 160(1):62–73

Yang Y, Gehrke S, Imai Y, Huang Z, Ouyang Y, Wang J-W, Yang L, Beal MF, Vogel H, Lu B (2006) Mitochondrial pathology and muscle and dopaminergic neuron degeneration caused by inactivation of Drosophila Pink1 is rescued by Parkin. Proc Natl Acad Sci 103(28):10793–10798

Yang L-Y, Greig NH, Tweedie D, Jung YJ, Chiang Y-H, Hoffer BJ, Miller JP, Chang K-H, Wang J-Y (2020) The p53 inactivators pifithrin-μ and pifithrin-α mitigate TBI-induced neuronal damage through regulation of oxidative stress, neuroinflammation, autophagy and mitophagy. Exp Neurol 324:113135

Yang X, Zhuang J, Song W, Shen W, Wu W, Shen H, Han S (2023) Mitochondria-associated endoplasmic reticulum membrane: overview and inextricable link with cancer. J Cell Mol Med 27(7):906–919

Yin K, Lee J, Liu Z, Kim H, Martin DR, Wu D, Liu M, Xue X (2021) Mitophagy protein PINK1 suppresses colon tumor growth by metabolic reprogramming via p53 activation and reducing acetyl-CoA production. Cell Death Differ 28(8):2421–2435

Yoo J-M, Lee BD, Sok D-E, Ma JY, Kim MR (2017) Neuroprotective action of N-acetyl serotonin in oxidative stress-induced apoptosis through the activation of both TrkB/CREB/BDNF pathway and Akt/Nrf2/antioxidant enzyme in neuronal cells. Redox Biol 11:592–599

Yu SP, Jiang MQ, Shim SS, Pourkhodadad S, Wei L (2023) Extrasynaptic NMDA receptors in acute and chronic excitotoxicity: implications for preventive treatments of ischemic stroke and late-onset Alzheimer's disease. Mol Neurodegener 18(1):43

Yuan S, She D, Jiang S, Deng N, Peng J, Ma L (2024) Endoplasmic reticulum stress and therapeutic strategies in metabolic, neurodegenerative diseases and cancer. Mol Med 30(1):40

Yue Z, Friedman L, Komatsu M, Tanaka K (2009) The cellular pathways of neuronal autophagy and their implication in neurodegenerative diseases. Biochim Biophys Acta, Mol Cell Res 1793(9):1496–1507

Zare N, Khalifeh S, Khodagholi F, Shahamati SZ, Motamedi F, Maghsoudi N (2015) Geldanamycin reduces Aβ-associated anxiety and depression, concurrent with autophagy provocation. J Mol Neurosci 57:317–324

Zeller V, Mariana D, Diogo S, Ana M, Valente CA (2021) NLRP3 inflammasome: a starring role in amyloid-β-and tau-driven pathological events in Alzheimer's disease. J Alzheimer's Dis 83(3):939–961

Zhang H, Bramham CR (2020) Bidirectional dysregulation of AMPA receptor-mediated synaptic transmission and plasticity in brain disorders. Front Synaptic Neurosci 12:26

Zhang H, Zhang C, Vincent J, Zala D, Benstaali C, Sainlos M, Grillo-Bosch D, Daburon S, Coussen F, Cho Y (2018a) Modulation of AMPA receptor surface diffusion restores hippocampal plasticity and memory in Huntington's disease models. Nat Commun 9(1):4272

Zhang Q, Han X, Chen J, Xie X, Xu J, Zhao Y, Shen J, Hu L, Xu P, Song H (2018b) Yes-associated protein (YAP) and transcriptional coactivator with PDZ-binding motif (TAZ) mediate cell density–dependent proinflammatory responses. J Biol Chem 293(47):18071–18085

Zhang S, Li W, Xu Y, Li T, Ek J, Zhang X, Wang Y, Song J, Zhu C, Wang X (2023) Alpha1-antitrypsin protects the immature mouse brain following hypoxic-ischemic injury. Front Cell Neurosci 17:1137497

Zhang J, Chen Y, Chen B, Sun D, Sun Z, Liang J, Liang J, Xiong X, Yan H (2024) The dual effect of endoplasmic reticulum stress in digestive system tumors and intervention of Chinese botanical drug extracts: a review. Front Pharmacol 15:1339146

Zheng H, Youdim MBH, Weiner LM, Fridkin M (2005) Novel potential neuroprotective agents with both iron chelating and amino acid-based derivatives targeting central nervous system neurons. Biochem Pharmacol 70(11):1642–1652

Zheng Z, Sabirzhanov B, Keifer J (2010) Oligomeric amyloid-β inhibits the proteolytic conversion of brain-derived neurotrophic factor (BDNF), AMPA receptor trafficking, and classical conditioning. J Biol Chem 285(45):34708–34717

Zheng T, Jiang T, Huang Z, Ma H, Wang M (2023) Role of traditional Chinese medicine monomers in cerebral ischemia/reperfusion injury: a review of the mechanism. Front Pharmacol 14:1220862

Zhu M, Cortese GP, Waites CL (2018) Parkinson's disease-linked Parkin mutations impair glutamatergic signaling in hippocampal neurons. BMC Biol 16:1–20

Zingale VD, Gugliandolo A, Mazzon E (2021) MiR-155: an important regulator of neuroinflammation. Int J Mol Sci 23(1):90

Zou J, Chen Z, Wei X, Chen Z, Fu Y, Yang X, Chen D, Wang R, Jenner P, Lu J-H (2017) Cystatin C as a potential therapeutic mediator against Parkinson's disease via VEGF-induced angiogenesis and enhanced neuronal autophagy in neurovascular units. Cell Death Dis 8(6):e2854–e2854

Zuccato CF, Asad AS, Candia N, Javier A, Gottardo MF, Ayala M, Alejandra M, Theas MS, Seilicovich A, Candolfi M (2019) Mitochondrial-derived peptide humanin as therapeutic target in cancer and degenerative diseases. Expert Opin Ther Targets 23(2):117–126

Natural Compounds and Neuroprotection

2

Vipin K Yadav and Chetan Singh Rajpurohit

Abstract

Neuroprotection involves preserving the integrity and function of neurons against damage resulting from injuries or neurodegenerative conditions. Diseases such as Alzheimer's, Parkinson's, Huntington's, and amyotrophic lateral sclerosis affect millions of individuals worldwide, with aging recognized as the leading risk factor. Each disorder impacts specific neuronal populations or brain regions and is marked by unique molecular and pathological hallmarks. To unravel the underlying mechanism and identify effective treatments, researchers have developed and employed a variety of disease specific models in both in vitro and in vivo settings. In recent decades, therapeutic development has largely concentrated on addressing factors such as oxidative stress, protein aggregation, neuroinflammation, and disruptions in neurotransmitter systems particularly involving dopamine and acetylcholine. This chapter highlights key examples of neuroprotective agents derived from plants, focusing on their structural categories and modes of action. Many of these natural compounds exhibit promise as therapeutic candidates due to their ability to influence multiple critical signaling pathways, including AMPK, ERK1/2, Akt, and JNK, which are essential to neuronal survival and function.

Keywords

Neurons · Neurodegeneration · Neurodegenerative diseases · Natural compounds · Neuroprotection · Herbal Extracts · Neuroinflammation · Personalized medicine

V. K. Yadav (✉)
Department of Tumor Microenvironment and Metastasis, Moffitt Cancer Center and Research Institute, Tampa, FL, USA
e-mail: Vipin.yadav@moffitt.org

C. S. Rajpurohit
Huffington Center on Aging, Baylor College of Medicine, Houston, TX, USA

S. Jahan et al. (eds.), *A Mechanistic Exploration of Natural Compounds for Neuronal Health*, https://doi.org/10.1007/978-981-95-0125-0_2

2.1 Introduction

Neuroprotection encompasses a range of strategies and biological mechanisms designed to preserve the central nervous system (CNS) by preventing or reducing neuronal damage resulting from both chronic neurodegenerative disorders such as Alzheimer's and Parkinson's diseases and acute neurological events like stroke. These conditions, regardless of their onset, involve the progressive deterioration and loss of CNS neurons, ultimately leading to impairments in cognitive and intellectual functions. The onset of symptoms associated with neurodegenerative diseases is typically gradual and progressive, manifesting as short-term memory loss, learning difficulties, impaired motor coordination, and other functional deficits (Selkoe 2001). Aging, a complex physiological process characterized by progressive morphological and biochemical changes, has been identified as a significant risk factor for neurodegenerative diseases (Wyss-Coray 2016). Other contributing factors include hypertension, genetic predispositions, environmental influences, and infections. With advancing age, pathological processes common to neurodegenerative diseases, such as protein aggregation, inflammation, oxidative stress, and neurotransmitter loss, become more prevalent (Mattson and Magnus 2006). Dysregulation of autophagy has been specifically linked in the pathogenesis of neurodegenerative diseases (Yadav et al. 2022). LncRNAs (long non-coding RNAs) such as BACE1-AS and 51A regulate amyloid-beta production and tau phosphorylation, contributing to neurotoxicity, while SNHG1 and MALAT1 are associated with alpha-synuclein aggregation, mitochondrial dysfunction, and oxidative stress, exacerbating dopaminergic neuron death (Yadav et al. 2021; Bhattacharyya et al. 2021).

Nature has long been a rich source of medicinal compounds for humanity. Many essential drugs in modern medicine, including vincristine, artemisinin, and gentamicin, are derived from natural sources or inspired by naturally occurring molecular structures (Newman and Cragg 2020). Traditional medicine also remains a critical component of healthcare worldwide, with estimates suggesting that it accounts for up to 80% of primary healthcare in some developing nations such as India, China, Ghana, and Nigeria. The rising challenges of drug resistance (particularly with antimalarials and antibiotics), high costs, side effects, and reduced efficacy of existing medications after prolonged use have renewed interest in developing new drugs from natural sources. For example, current drugs for managing neurodegenerative diseases such as donepezil, amantadine, selegiline, memantine, rivastigmine, and galantamine primarily provide symptomatic relief and may slow disease progression. However, synthetic options like donepezil are often associated with adverse side effects (Hampel, Toschi et al. 2018). Consequently, significant attention has shifted toward exploring herbs and natural products used in traditional medicine for age-related CNS disorders (Howes and Perry 2011). Various natural products, especially plant extracts, have been traditionally used for neuroprotection, memory enhancement, and anti-aging purposes. Examples include *Ginkgo biloba* (Mahadevan and Park 2008), *Panax ginseng* (Reay et al. 2005), *Curcuma longa* (Small et al. 2018), *Bacopa monnieri* (Stough et al. 2001), and *Salvia officinalis*

(Akhondzadeh et al. 2003). Scientific studies have sought to validate these traditional claims with particular emphasis on understanding the mechanisms underlying their neuroprotective effects. This book chapter provides a concise overview of neurodegenerative diseases, highlights key pathological processes associated with these conditions, and examines therapeutic options focusing on prominent plant-derived neuroprotective agents.

2.2 Neurodegenerative Diseases

Neurodegenerative diseases are characterized by a progressive and typically irreversible deterioration of cognitive and intellectual abilities, often associated with the aging process. Conditions like Parkinson's disease (PD), Alzheimer's disease (AD), and stroke are among the most prevalent neurodegenerative disorders (Fig. 2.1). These diseases not only lead to significant deterioration in the quality of life but also pose substantial healthcare challenges globally.

2.2.1 Aging and Oxidative Stress

As the global population continues to age, neurodegenerative diseases are emerging as a significant public health challenge. Age-dependent degenerative disorders encompass conditions where the structure and function of tissues and organs progressively deteriorate over time. These conditions include neurodegenerative disorders, cardiovascular diseases, a decline in immune function known as immunosenescence, and muscle wasting referred to as sarcopenia (Lopez-Otin et al. 2013). Aging itself is a multifaceted biological phenomenon marked by morphological and biochemical alterations at both the cellular and systemic levels. Among the many theories proposed to explain the underlying mechanism of aging, the oxidative stress or free radical theory has received considerable scientific interest (Emerit et al. 2004). Oxidative stress arises when there is a disruption in the balance of the cellular redox system, resulting in excessive generation of reactive oxygen species (ROS). These ROS, comprising both non radical and radical oxygen derivatives, are highly reactive and capable of inflicting damage on vital biomolecules, including protein, DNA, and lipids (Sies et al. 2017). Such oxidative harm contributes significantly to the onset and progression of age-associated diseases, notably neurodegenerative conditions (Emerit et al. 2004). The brain, although it represents just about 2% of the body's mass, utilizes nearly 20% of the body's oxygen supply, making it exceptionally susceptible to oxidative injury (Barnham et al. 2004). This vulnerability is further heightened by the presence of polyunsaturated fatty acids (PUFAs) in neuronal membranes, which are prone to lipid peroxidation. This degradation process leads to the formation of toxic aldehydes such as malondialdehyde (MDA) and 4-hydroxynonenal (HNE), compounds that intensify oxidative stress and cellular damage (Zarkovic 2003). To defend against ROS and sustain redox balance, organisms have evolved intricate antioxidant systems. These include

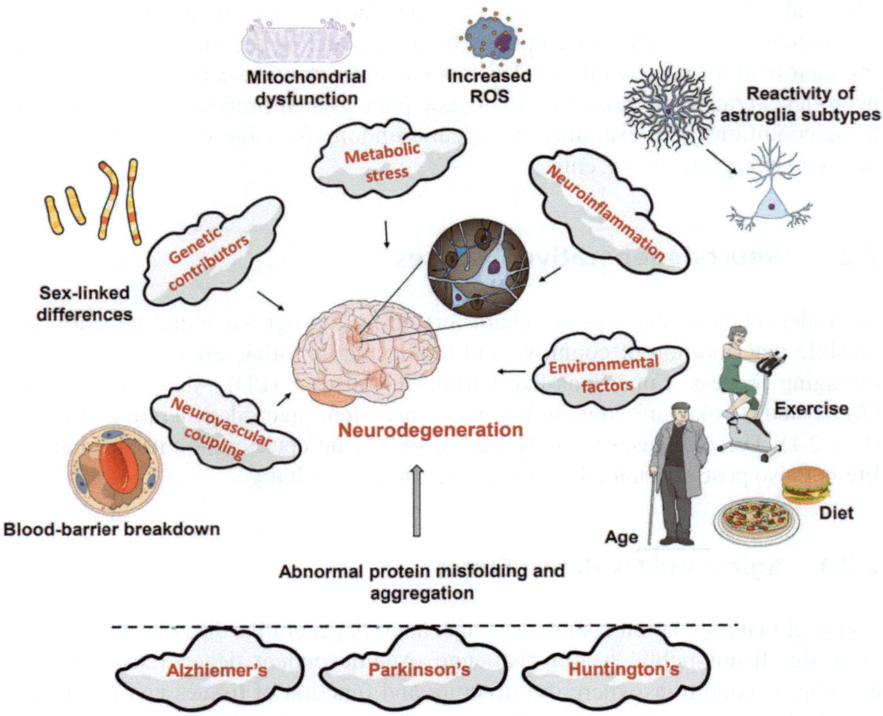

Fig. 2.1 Schematic illustration of the multifactorial mechanisms underlying neurodegeneration and potential therapeutic modifiers. Central processes include oxidative stress (increased ROS), mitochondrial dysfunction, neuroinflammation, and abnormal protein misfolding and aggregation, which collectively contribute to neuronal damage. Additional contributing factors include genetic predispositions, sex-linked differences, blood-brain barrier breakdown, neurovascular coupling impairments, metabolic stress, and environmental factors such as lifestyle and diet. These mechanisms ultimately lead to progressive neurodegeneration, exemplified in diseases such as Huntington's, Parkinson's, and Alzheimer's

enzymatic antioxidant like catalase, superoxide dismutase, and glutathione peroxidase, as well as non-enzymatic molecules such as vitamins C and E, along with plant-based phytochemicals (Birben et al. 2012). Dietary antioxidants are especially important as they can neutralize ROS effectively without generating harmful secondary products, thereby playing a vital role in reducing oxidative stress and its detrimental effects (Pham-Huy et al. 2008).

2.2.2 Alzheimer's and Parkinson's Diseases

Among the wide range of neurodegenerative disorders, Alzheimer's diseases (AD) and Parkinson's diseases (PD) stand out as the most common and debilitating, particularly among older adults. AD is primarily associated with the decline of memory and cognitive abilities, whereas PD is characterized mainly by motor dysfunction. Both diseases severely affect the quality of life of those diagnosed and contribute to significant healthcare costs, placing a heavy economic strain on families and healthcare systems alike.

2.2.2.1 Epidemiology of AD and PD

The risk of developing neurodegenerative disorder such as AD and PD increases markedly with advancing age. According to the World Health Organization (WHO), these conditions are expected to become the second leading cause of death globally by the mid-twenty-first century, overtaking cancer. In the year 2000, approximately 4.5 million individuals in the United States were living with AD, and the annual cost of managing the disease was estimated at around $100 billion. (Brookmeyer et al. 1998). Without the advent of effective treatments, projections indicate that the number of AD cases in the US could rise between 11 and 16 million by 20250. AD is the most common cause of dementia, contributing to 60–80% of cases among older adults (Prince et al. 2016). PD also shows a strong age-related prevalence, affecting about 1% of people between 65 and 69 years old, and rising to nearly 3% among those aged 80 and above (de Lau and Breteler 2006). Interestingly, men are about 1.5 times more likely to develop PD than women (Van Den Eeden et al. 2003). Variations in incidence have also been noted across different geographic and ethnic groups, with higher rates reported in northern latitudes compared to regions near the equator and in white populations compared to black populations (Dorsey et al. 2007).

2.2.2.2 Neuropathology of AD and PD

Despite presenting with distinct clinical manifestations, both AD and PD share common neuropathological features, including progressive neuronal degeneration and the accumulation of misfolded protein aggregates. The main pathological features of AD include amyloid-beta (Aβ) plaques outside cells and neurofibrillary tangles (NDTs) inside cells. The formation of Aβ plaques from APP through enzymatic cleavage results in fibrillar Aβ peptides (Hardy and Higgins 1992), but tau protein hyperphosphorylation leads to the formation of NFTs inside neurons (Iqbal et al. 2016). The brain regions vital for memory and cognitive functions such as hippocampus and neocortex along with the limbic system show the highest concentration of pathological structures (Braak and Braak 1991). The main characteristic of PD involves the step-by-step destruction of dopaminergic neurons that reside in the substantia nigra pars compacta which controls motor functions. Lewy bodies represent a major pathological feature of PD because they consist of abnormal alpha synuclein protein aggregates which form intracytoplasmic inclusions. The accumulation of proteins directly causes neuronal dysfunction leading to death in patients with PD (Spillantini et al. 1997).

2.2.2.3 Etiopathogenesis of AD and PD

The development of neurodegenerative disorders such as AD and PD is driven by a combination of genetic and environmental influences. While familial cases linked to inherited genetic mutations are relatively uncommon, the majority of AD and PD cases are sporadic and primarily associated with aging (Bekris et al. 2010). Oxidative stress is a key factor contributing to the progression of both conditions. In AD, it is closely connected to amyloid beta toxicity, impaired mitochondrial function, and imbalances in essential brain metals like zinc, iron, and copper, which can exacerbate neuronal damage (Barnham et al. 2004). In PD, oxidative stress largely results from dopamine metabolism, a process that naturally generates reactive oxygen species. The substantia nigra, a brain region rich in dopamine, is particularly susceptible to oxidative injury, making it a focal point of neurodegeneration in PD (Jenner 2003).

2.2.2.4 Current Pharmacological Interventions

Currently, no curative treatments exist for AD or PD; available therapies focus on symptom management. In AD, cholinesterase inhibitors such as donepezil are commonly prescribed to enhance acetylcholine levels in the brain, while NMDA receptor antagonists like memantine are used to prevent excitotoxicity caused by glutamate (Tariot et al. 2004). For PD, levodopa (L-DOPA) remains the gold standard treatment, replenishing dopamine levels in the brain (Olanow et al. 2009). Research into antioxidant and anti-inflammatory therapies is ongoing, though conclusive evidence supporting their efficacy is lacking (Weinreb et al. 2016).

2.3 Stroke

A stroke, also referred to as an acute cerebrovascular event, occurs when the blood supply to a part of the brain is interrupted, either due to ischemia where oxygen and glucose delivery is blocked or as a result of hemorrhage. Ischemic strokes are typically triggered by a blood clot obstructing a cerebral artery, whereas hemorrhagic strokes involve bleeding within the brain, often linked to high blood pressure (Donkor 2018). Stroke stands as the second most common cause of death in industrialized nations and is a leading contributor to long-term disability (Feigin et al. 2017). Alarmingly, approximately one quarter of individuals who experience a stroke develop dementia within 3 months. Poststroke examinations often reveal deposits of amyloid beta and amyloid precursor protein in both cortical and subcortical brain regions, suggesting a pathological overlap between stroke and neurodegenerative disorder (Kalaria et al. 2016; Venkat et al. 2016). This correlation points to a shared molecular basis and highlights the importance of continued research into the interconnected pathways between cerebrovascular events and neurodegeneration.

2.4 Development of Neuroprotective Agents

The exploration of natural products as sources for neuroprotective agents has emerged as a promising strategy in the search for novel treatments for neurodegenerative disorders such as AD, PD, and other conditions involving neuronal injury.

Many of these bioactive compounds, predominantly derived from plants, exhibit strong neuroprotective potential due to their antioxidant, anti-inflammatory, and neurorestorative properties (Lim et al. 2024), (Hussain et al. 2018). These natural substances are classified into various groups of secondary metabolites, including alkaloids, phenolics, terpenoids, and flavonoids. Such compounds are commonly found in a wide range of fruits, vegetables, and medicinal herbs (Fig. 2.2). Among them, flavonoids like quercetin, naringenin, and resveratrol have received considerable scientific attention for their robust antioxidant effects (Ayaz et al. 2019). By counteracting oxidative stress, a key contributor to the progression of neurodegenerative disease, these flavonoids help protect neuronal cells (Lim et al. 2024). Quercetin, for example, has shown beneficial effects in experimental models of

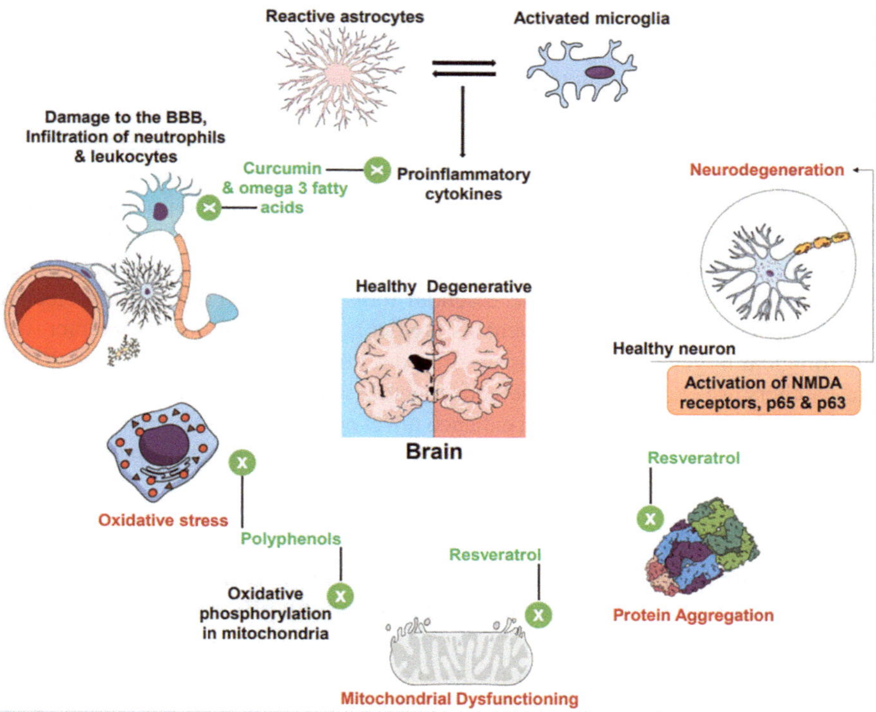

Fig. 2.2 Schematic representation of neurodegeneration mechanisms and the neuroprotective actions of natural compounds. Key pathological features include damage to the blood-brain barrier (BBB), infiltration of neutrophils and leukocytes, activation of reactive astrocytes and microglia, mitochondrial dysfunction, oxidative stress, protein aggregation, and activation of NMDA receptors. These processes collectively lead to neurodegeneration, as shown by the transition from a healthy to a degenerative brain. Natural products such as curcumin and omega-3 fatty acids mitigate inflammation by suppressing proinflammatory cytokine production. Polyphenols alleviate oxidative stress and enhance mitochondrial oxidative phosphorylation. Resveratrol inhibits protein aggregation and supports mitochondrial function. These findings highlight potential therapeutic targets and natural interventions for neurodegenerative diseases

Parkinson's disease by reducing oxidative damage and enhancing dopamine receptor function. Research has shown that naringenin exhibits anti-inflammatory activity alongside neuronal protection capabilities (Nouri et al. 2019). The well-known polyphenol resveratrol demonstrates protective effects in the central nervous system which helps to reduce ischemic damage and control oxidative stress-related signaling pathways thus providing potential benefits to brain and spinal cord health (Sun, Wang et al. 2010).

Glycyrrhetinic acid together with allicin and sulforaphane represent plant-derived compounds which demonstrate significant neuroprotective activity. The compound glycyrrhetinic acid shows activity against dopamine receptor D3 (Mirza, Mirza et al. 2015), and allicin from garlic and onion displays antioxidant properties with free-radical scavenging capabilities which protects neurons (Nadeem et al. 2021). Sulforaphane exhibits neuronal protection capabilities against both high oxidative stress conditions and mitochondrial toxins thereby demonstrating the wide range of natural compounds with neuroprotective potential (Klomparens and Ding 2019). The marine compounds dieckol and phlorofucofuroeckol derived from algae demonstrate memory-enhancing properties along with acetylcholinesterase inhibition which makes them suitable for Alzheimer's disease prevention (Hannan et al. 2020). The therapeutic potential of natural products for neurological disorder treatment becomes apparent through compounds such as hypericin and pseudo hypericin from *Hypericum perforatum* (St. John's Wort) which exhibit antidepressant activities (Karioti and Bilia 2010). Virtual screening and computational models through in silico techniques have become essential tools for discovering neuroprotective compounds because they help scientists predict how these substances interact with neurological targets which speeds up drug development (Vicidomini et al. 2024). Researchers use computational methods together with laboratory testing of samples in vitro and in vivo to identify and enhance novel neuroprotective compounds which hold promise for developing new neurodegenerative disease treatments (Gotina et al. 2021).

2.4.1 Therapeutic Potential of Natural Compounds: Mechanisms, Evidence, and Disease Applications

Natural compounds derived from plants and other biological sources possess significant therapeutic potential for various diseases, owing to their antioxidant, anti-inflammatory, and neuroprotective properties (Tables 2.1 and 2.2). These bioactive substances interact with multiple biochemical and signaling pathways, making them valuable in the prevention and treatment of chronic conditions, including neurodegenerative diseases, metabolic disorders, and inflammation-related ailments.

2.4.1.1 Mechanisms of Neuroprotection by Natural Compounds

(a) *Antioxidant Activity*

Table 2.1 Neuroprotective plant species: major compounds and modes of action

Serial number	Plant species	Family	Major neuroprotective compounds	Mode of action	References
1	*Ashwagandha (Withania somnifera)*	*Solanaceae*	Withanolides, somniferine	Antioxidant, anti-inflammatory, reduces oxidative stress, promotes neurogenesis, enhances cognitive function, and supports neuronal repair	Wongtrakul et al. (2021) Akhoon et al. (2016)
2	*Corydalis yanhusuo*	*Papaveraceae*	Tetrahydropalmatine, corydaline	Neuroprotective, reduces neuronal cell death, enhances neurotransmitter balance, modulates pain perception, supports neurogenesis	Sun et al. (2018)
3	*Ginseng* (*Panax ginseng*)	*Araliaceae*	Ginsenosides, panaxans	Antioxidant, improves circulation to the brain, modulates neurotrophic factors, enhances cognitive function, reduces inflammation and oxidative damage	Zheng et al. (2018)
4	*Rhodiola rosea*	*Crassulaceae*	Rosavins, salidroside	Adaptogenic, neuroprotective, enhances neuronal survival, increases resistance to stress, supports cognitive function, and regulates neurotransmitter levels	Zhong et al. (2018)
5	*Moringa oleifera*	*Moringaceae*	Moringine, quercetin	Antioxidant, anti-inflammatory, protects against neurodegeneration, supports memory and cognitive function, reduces oxidative stress	Costa et al. (2016) Azlan et al. (2022)
6	*Gotu kola* (*Centella asiatica*)	*Apiaceae*	Asiaticoside, madecassoside	Antioxidant, anti-inflammatory, enhances brain function, promotes neuronal regeneration, increases blood flow to the brain, reduces oxidative damage	Bandopadhyay et al. (2023)
7	*Bacopa monnieri*	*Plantaginaceae*	Bacopasides, bacosides	Enhances synaptic communication, improves memory and learning, antioxidant, anti-inflammatory, protects neurons from oxidative damage	Sekhar et al. (2019)
8	*Saffron* (*Crocus sativus*)	*Iridaceae*	Crocin, safranal	Antioxidant, anti-inflammatory, protects against neurodegeneration, enhances memory, improves mood, and alleviates symptoms of depression	Soeda et al. (2016)

(continued)

Table 2.1 (continued)

Serial number	Plant species	Family	Major neuroprotective compounds	Mode of action	References
9	*Lemon balm* (*Melissa officinalis*)	*Lamiaceae*	Rosmarinic acid, ursolic acid	Antioxidant, anxiolytic, reduces neuroinflammation, protects brain cells from oxidative stress, improves cognitive function, and enhances memory	Mirza and Zahid (2022)
10	*Turmeric* (*Curcuma longa*)	*Zingiberaceae*	Curcumin	Anti-inflammatory, antioxidant, modulates amyloid-beta plaques, inhibits neuroinflammation, protects against neurodegeneration and promotes brain function	Cole et al. (2007)
11	*Peppermint* (*Mentha piperita*)	*Lamiaceae*	Menthol, menthone	Antioxidant, anti-inflammatory, improves blood circulation to the brain, supports cognitive function, and relieves stress	Du et al. (2020)
12	*St. John's Wort* (*Hypericum perforatum*)	*Hypericaceae*	Hypericin, pseudohypericin	Antidepressant, neuroprotective, inhibits reuptake of serotonin, dopamine, and norepinephrine, promotes neuronal regeneration	Klemow et al. (2011), Wang et al. (2010)
13	*Schisandra chinensis*	*Schisandraceae*	Schisandrin, schisandrol	Antioxidant, anti-inflammatory, promotes neuronal survival, enhances memory and cognitive function	Wei et al. (2018)
14	*Green tea* (*Camellia sinensis*)	*Theaceae*	Catechins (EGCG, ECG), theaflavins	Antioxidant, neuroprotective, promotes neurogenesis, reduces amyloid plaque formation, protects dopaminergic neurons, modulates oxidative stress	Musial et al. (2020)
15	*Holy Basil* (*Ocimum sanctum*)	*Lamiaceae*	Eugenol, ursolic acid	Antioxidant, anti-inflammatory, modulates oxidative stress, and protects against neurodegeneration	Liu et al. (2022)
16	*Ginkgo biloba* (*Ginkgo biloba*)	*Ginkgoaceae*	Ginkgolides, bilobalide	Increases blood flow to the brain, neuroprotective, antioxidant, inhibits platelet aggregation, enhances cognitive function and memory	Singh et al. (2019)
17	*Fennel* (*Foeniculum vulgare*)	*Apiaceae*	Anethole, fenchone	Antioxidant, anti-inflammatory, protects neurons from oxidative damage, supports cognitive health	Habiba et al. (2024)
18	*Lemon verbena* (*Aloysia citrodora*)	*Verbenaceae*	Verbascoside, rosmarinic acid	Antioxidant, anti-inflammatory, modulates brain neurotransmitters, protects against neuroinflammation	Fachel et al. (2019)

19	*Passionflower* (*Passiflora incarnata*)	*Passifloraceae*	Harman, passiflorine	Anxiolytic, antioxidant, reduces neuroinflammation, supports GABAergic neurotransmission, provides neuroprotective effects against stress-induced neuronal damage	Janda et al. (2020)
20	*Coriander* (*Coriandrum sativum*)	*Apiaceae*	Linalool, coriandrol	Antioxidant, anti-inflammatory, alleviates oxidative stress in the brain, protects against neurodegeneration	Hosseini et al. (2021)
21	*Jasmine* (*Jasminum officinale*)	*Oleaceae*	Jasmin, linalool	Anxiolytic, neuroprotective, antioxidant, supports neurotransmitter balance and modulates stress responses in the brain	Ota et al. (2017) Lu et al. (2023)
22	*Marjoram* (*Origanum majorana*)	*Lamiaceae*	Terpinen-4-ol, linalool	Antioxidant, neuroprotective, modulates GABAergic activity, reduces neuroinflammation, promotes cognitive health	Nobrega et al. (2014) Wang and Heinbockel (2018)
23	*Ylang-ylang* (*Cananga odorata*)	*Annonaceae*	Linalool, germacrene D	Antioxidant, neuroprotective, enhances blood circulation to the brain, modulates serotonin, dopamine, and acetylcholine levels	Park et al. (2016)

Table 2.2 Neuroprotective compounds: sources, functions, and chemical profiles

Serial number	Compound	Source	Functions	Parent compound	Chemical structure	References
1	Curcumin	Turmeric (Curcuma longa)	Antioxidant, anti-inflammatory, reduces amyloid-beta plaques, modulates oxidative stress	Curcuminoids	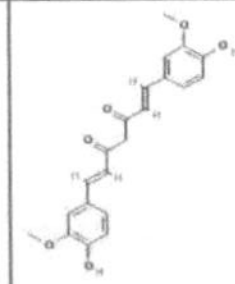	Priyadarsini (2014)
2	Epigallocatechin Gallate (EGCG)	Green tea (Camellia sinensis)	Antioxidant, prevents mitochondrial dysfunction, reduces neuroinflammation	Catechins (flavonoids)	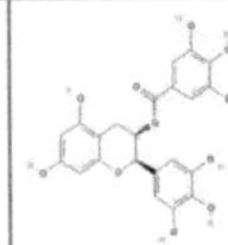	Nagle et al. (2006)
3	Resveratrol	Grapes, red wine, peanuts (Vitis vinifera)	Activates SIRT1, anti-inflammatory, mitigates oxidative stress	Stilbenoids	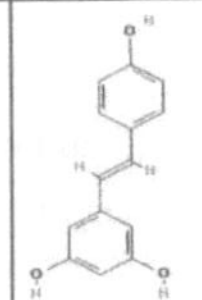	Akinwumi et al. (2018)
4	Quercetin	Onions, apples, and berries	Reduces neuroinflammation, scavenges free radicals, inhibits neurotoxic enzyme activity	Flavonoids	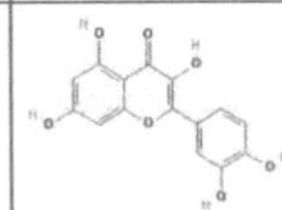	Dias et al. (2021)
5	Huperzine A	Chinese club moss (Huperzia serrata)	Acetylcholinesterase inhibitor, enhances cholinergic signaling, protects against neurodegeneration	Alkaloid (huperzine)	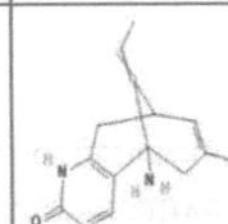	Friedli and Inestrosa (2021)

6	Berberine	Barberry, goldenseal (Berberis spp.)	Anti-inflammatory, antioxidant, improves mitochondrial function	Alkaloids (isoquinoline-type)	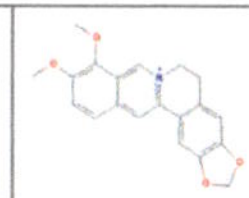	Zhang et al. (2019)
7	Nicotine	Tobacco (Nicotiana tabacum)	Neuroprotective in low doses by modulating acetylcholine receptors, though toxic in excess	Alkaloid	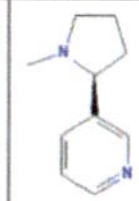	Brooks and Henderson (2021) Ross and Petrovitch (2001)
8	Ginkgolides and Bilobalides	Ginkgo biloba	Improve cerebral blood flow, reduce neuroinflammation, prevent oxidative stress	Terpenoids (ginkgolides, bilobalides)	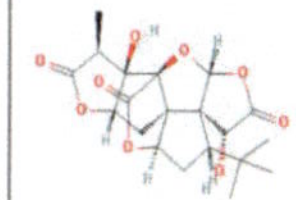	Feng et al. (2019)
9	Asiaticoside	Gotu kola (Centella asiatica)	Enhances cognitive function, promotes neuronal repair	Triterpenoids (saponins)	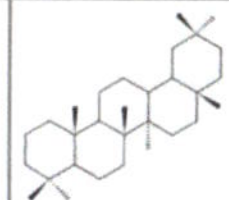	Zhang et al. (2023)
10	Cannabidiol (CBD)	Cannabis (Cannabis sativa)	Antioxidant, anti-inflammatory, neuroprotective against excitotoxicity	Cannabinoids	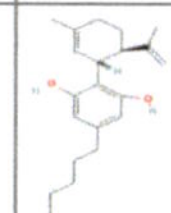	Bow and Rimoldi (2016)
11	Ginsenosides	Ginseng (Panax ginseng)	Antioxidant, improves mitochondrial function, reduces inflammation	Saponins	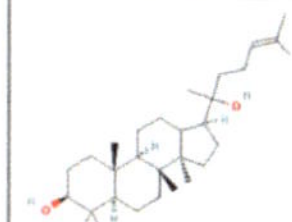	Peng et al. (2012)

(continued)

Table 2.2 (continued)

Serial number	Compound	Source	Functions	Parent compound	Chemical structure	References
12	Withanolides	Ashwagandha (Withania somnifera)	Enhances neuronal regeneration, reduces oxidative stress, modulates stress responses	Withanolides (steroidal lactones)	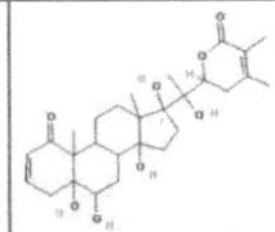	Williams et al. (2011)
13	Apigenin	Chamomile, parsley	Reduces neuroinflammation, modulates GABA receptors, enhances neurogenesis	Flavonoids	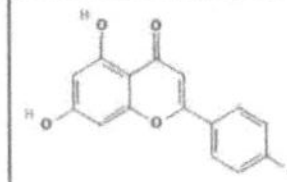	Dias et al. (2021)
14	Diosmin	Citrus fruits	Reduces oxidative stress and improves cerebral circulation	Flavonoids (hesperidin derivative)	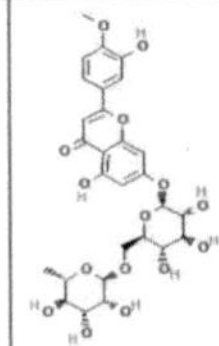	Habib et al. (2022)
15	Luteolin	Celery, broccoli	Anti-inflammatory, reduces microglial activation, enhances neuronal survival	Flavonoids	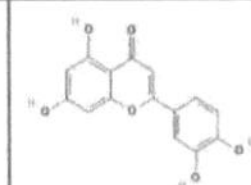	Nabavi et al. (2015)

- *Mechanism*: Reduces oxidative stress by scavenging free radicals, enhancing endogenous antioxidant systems, and reducing lipid peroxidation
- *Example*: Curcumin, a natural compound found in turmeric, helps neutralize reactive oxygen species and boosts glutathione levels in neuronal cells, thereby supporting cellular defense mechanisms against oxidative stress

(b) *Anti-inflammatory Effects*

- *Mechanism*: Regulates inflammation by blocking the release of pro-inflammatory cytokines, decreasing microglial activation, and inhibiting the NF-κB signaling
- *Example*: Resveratrol from grapes reduces neuroinflammation in models of Alzheimer's disease

(c) *Anti-apoptotic Action*

- *Mechanism*: Prevents neuronal death by stabilizing mitochondrial membranes, reducing caspase activation, and upregulating anti-apoptotic proteins
- *Example*: Epigallocatechin gallate (EGCG) from green tea protects neurons from programmed cell death

(d) *Neurogenesis and Synaptic Plasticity*

- *Mechanism*: Promotes the growth and repair of neurons, supports synaptic function, and enhances brain-derived neurotrophic factor (BDNF)
- *Example*: Quercetin, a flavonoid, stimulates BDNF expression and synaptic plasticity

(e) *Metal Chelation*

- *Mechanism*: Interacts with and sequesters excess metal ions such as iron and copper, which are known to facilitate the generation of harmful free radicals
- *Example*: Catechins found in green tea act as metal chelators, particularly binding to iron, thereby helping to minimize oxidative stress associated with Parkinson's disease

(f) *Amyloid Aggregation Inhibition*

- *Mechanism*: Inhibits the accumulation or enhances the removal of harmful protein aggregates, including beta-amyloid and tau, which are implicated in Alzheimer's disease
- *Example*: Berberine has been shown to prevent the aggregation of amyloid-beta peptides, thereby reducing their neurotoxic effects

2.4.2 Natural Compounds with Neuroprotective Properties

2.4.2.1 Polyphenols

Polyphenols are widely recognized for their potent antioxidant and signaling-modulating effects. *Curcumin*, extracted from turmeric (*Curcuma longa*), has demonstrated the ability to reduce amyloid-beta plaques and modulate oxidative stress, which are critical factors in Alzheimer's disease. Preclinical studies reveal that curcumin crosses the blood-brain barrier, decreases neurotoxic aggregate accumulation, and improves memory and cognitive function in Alzheimer's models by modulating the PI3K/Akt and MAPK signaling pathways. *Epigallocatechin gallate (EGCG)*, derived from green tea (*Camellia sinensis*), prevents mitochondrial dysfunction and reduces neuroinflammation, primarily by regulating the NF-κB and Nrf2 pathways. In Parkinson's disease models, EGCG prevents dopaminergic neuron degeneration and enhances motor function. *Resveratrol*, a polyphenol found in grapes and red wine, activates SIRT1, a longevity-associated gene, reducing inflammation and oxidative stress through the AMPK and mTOR signaling pathways. Studies in diabetic models highlight its efficacy in extending lifespan and improving glucose homeostasis. *Quercetin*, present in onions and berries, scavenges free radicals and inhibits neurotoxic enzyme activity, reducing neuroinflammation by modulating the MAPK and JAK/STAT pathways, as observed in models of neurodegenerative diseases.

2.4.2.2 Alkaloids

Alkaloids exhibit therapeutic efficacy by targeting specific signaling mechanisms. *Huperzine A*, derived from Chinese club moss (*Huperzia serrata*), is a potent acetylcholinesterase inhibitor that enhances cholinergic signaling in the brain. Clinical trials demonstrate its efficacy in improving cognitive function and memory in Alzheimer's patients. *Berberine*, sourced from barberry (*Berberis* spp.), exhibits anti-inflammatory and antioxidant properties, improving mitochondrial function through the AMPK and Akt signaling pathways. Preclinical studies show that berberine ameliorates insulin resistance and hyperglycemia in type 2 diabetic models. *Nicotine*, although toxic at high doses, modulates acetylcholine receptors and enhances neurotransmitter signaling. In Parkinson's disease models, it improves dopaminergic transmission and alleviates motor symptoms.

2.4.2.3 Terpenoids

Terpenoids offer neuroprotective effects by modulating signaling pathways associated with vascular and neuronal health. *Ginkgolides* and *bilobalides* from *Ginkgo biloba* improve cerebral blood flow, reduce neuroinflammation, and protect against oxidative damage by modulating the PI3K/Akt and VEGF signaling pathways. Clinical studies suggest cognitive improvements in patients with mild cognitive impairment and vascular dementia. *Asiaticoside*, derived from *Gotu kola* (*Centella asiatica*), enhances neuronal repair and cognitive function through the activation of the ERK1/2 and JNK pathways. Preclinical models demonstrate recovery from cognitive deficits following traumatic brain injury. *Cannabidiol (CBD)* from *Cannabis sativa* exhibits antioxidant, anti-inflammatory, and neuroprotective effects by regulating the CB1/CB2 and TRPV1 signaling pathways. Clinical trials highlight its efficacy in reducing seizure frequency in epilepsy patients and its potential in managing chronic pain.

2.4.2.4 Saponins

Saponins, such as ginsenosides from ginseng (*Panax ginseng*), improve mitochondrial function and reduce inflammation. Studies in chronic fatigue models demonstrate enhanced energy metabolism and reduced oxidative stress mediated through the Nrf2 and AMPK pathways. *Withanolides*, derived from Ashwagandha (*Withania somnifera*), promote neuronal regeneration and counteract oxidative damage by modulating the HSP70 and Akt/mTOR signaling pathways. Preclinical evidence suggests that withanolides reduce stress and enhance cognitive function in aging and neurodegenerative models.

2.4.2.5 Flavonoids

Flavonoids are recognized for their ability to modulate neuroprotective and anti-inflammatory pathways. Apigenin, found in chamomile and parsley, reduces neuroinflammation, modulates GABA receptor signaling, and enhances neurogenesis. Studies indicate its anxiolytic effects and its ability to promote hippocampal neurogenesis in animal models. Luteolin, present in celery and broccoli, reduces microglial activation and enhances neuronal survival by obstructing the MAPK and NF-κB pathways. Experimental models have demonstrated its protective effects against neurotoxicity in AD and PD. Diosmin, a flavonoid derived from citrus fruits, improves cerebral circulation, and reduces oxidative stress by influencing the VEGF and PI3K/Akt pathways. Clinical trials suggest its potential role in managing vascular cognitive impairments. Cumulative findings from in vivo and in vitro studies affirm that these natural compounds modulate critical signaling pathways, including AMPK, PI3K/Akt, Nrf2, SIRT1, NF-κB, and mTOR. These signaling pathways contribute to reducing oxidative stress, protecting mitochondria, suppressing inflammation, and promoting neuronal repair. Such mechanisms position flavonoids as promising candidates for therapeutic interventions against chronic and degenerative diseases (Jahan et al. 2023). Continued research into their bioavailability, pharmacokinetics, and synergistic effects is essential to pave the way for their integration into modern medicine.

2.4.2.6 Fatty Acids

Omega-3 fatty acids, predominantly sourced from fish oil, flaxseed, and chia seeds, are polyunsaturated fatty acids with well-documented neuroprotective properties. These bioactive lipids exert anti-inflammatory effects by modulating the manufacture of eicosanoids and pro-inflammatory cytokines, enhancing membrane fluidity by integrating into phospholipid bilayers, and promoting synaptic function through improved signal transduction and neurotransmitter release. The therapeutic potential of these mechanisms exists for managing neuropsychiatric and neurodegenerative disorders including Alzheimer's disease and bipolar disorder and depression. Omega-3 fatty acids protect neuronal membrane structure and decrease neuroinflammation which enables them to effectively treat cognitive decline and mood disorders. The neurotherapeutic applications of *Withania somnifera*-derived steroidal lactones known as withanolides have received substantial scientific interest. The compounds demonstrate strong antioxidant behavior through ROS scavenging which helps reduce oxidative stress that causes neuronal damage and aging. The anti-apoptotic effect of these compounds helps protect neurons from programmed cell death which helps maintain cellular survival in neurodegenerative diseases. The compounds enhance neuronal connectivity through their ability to stimulate dendritic development and synaptogenesis which leads to better neuronal communication and plasticity. The diverse range of mechanisms makes these compounds suitable for treating neurodegenerative diseases such as AD through their antioxidant effects and synaptic protection as well as anxiety disorders by controlling the HPA axis and lowering stress-induced neurotoxicity.

2.5 Conclusions

Plants, including spices, herbs, and other foods, are an inexhaustible source of bioactive molecules with great potential to improve human health. Each plant harbors hundreds or even thousands of secondary metabolites, a chemical diversity that has not only supported the evolutionary success of plants but has also enabled their adaptation to dynamic and challenging environments (Cheynier et al. 2013). These bioactive compounds, often termed phytochemicals, play diverse roles in maintaining health and preventing diseases. Nonetheless, it is an oversimplification to credit the health benefits of a medicinal herb or plant-based food to just one molecule or a specific group of compounds. In reality, the interplay of multiple phytochemicals often results in additive or synergistic effects, which can amplify, diminish, or modulate their individual activities, thus underscoring the importance of studying these compounds in their natural combinations (Liu 2004). Historically, much of our understanding of CNS-active plants has been derived from ethnobotanical and ethnopharmaceutical studies. These approaches have provided critical insights into the traditional uses of plants and their active constituents, many of which have become the foundation of modern medicine (Fabricant and Farnsworth 2001). Notable examples include *Atropa belladonna* and *Hyoscyamus niger*, which produce alkaloids such as atropine and hyoscyamine; *Erythroxylum coca*, the source of cocaine; *Papaver somniferum*, known for yielding morphine and codeine; and Ephedra sinica, which contains the stimulant ephedrine. These discoveries highlight the power of ethnobiological studies in identifying neuroactive natural products with therapeutic potential. Such studies continue to serve as a guiding framework for the discovery of novel compounds from plants used in diverse cultural and traditional contexts as medicinal herbs, spices, or foods. Despite the well-documented bioactivity of many dietary phytochemicals, understanding their bioavailability remains a critical area of research. For these compounds to exert their therapeutic effects, they must be absorbed, distributed, metabolized, and excreted within the human body (Manach et al. 2004). Although human studies have provided evidence of the absorption and urinary excretion of certain phytochemicals following dietary intake, detailed biokinetic data remain scarce. The ability of these compounds to cross physiological barriers, such as the blood-brain barrier (BBB), is particularly significant for neuroactive agents (Youdim et al. 2002). Only a limited number of studies, primarily conducted in vitro or in animal models, have examined the mechanisms by which phytochemicals, such as flavonoids, traverse the endothelial cell layer of the BBB. Factors such as lipophilicity and interactions with specific membrane transporters or receptors are thought to influence their permeability. Polyphenols bind to transporters or receptors located on the plasma membrane which leads to the activation of signaling pathways inside the cell. The only known receptor for epigallocatechin gallate (EGCG) in vascular cells remains unidentified to date (Duda-Chodak and Tarko 2023). The bioavailability of phytochemicals particularly polyphenols depends on food matrix complexity and compound-specific chemical structures in fruits and vegetables (Scalbert and Williamson 2000). Experimental results become harder to interpret and treatment strategies become more difficult to develop because of differences in how substances are absorbed and metabolized by the body. Polyphenols demonstrate multiple target capabilities although researchers have not established definitive evidence for brain region-specific polyphenol receptors or transporters (Williams et al. 2004). Their capacity to influence multiple cellular pathways simultaneously makes them attractive candidates for complex disease treatment including neurodegenerative disorders. The study of bioactive compounds derived from plants holds vast potential to enhance human health particularly through neuroprotective effects. Researchers can discover new treatment approaches by studying the bioavailability and biokinetics of these substances as well as their interactions with physiological barriers such as the blood brain barrier. New therapeutic approaches for diseases that do not respond to single-target treatments could emerge through the combination of phytochemicals and their ability to affect multiple pathways simultaneously. The identification of therapeutic applications for plant-based substances will become possible through additional research in this field which will also enhance our knowledge about their health benefits.

2.6 Challenges and Future Directions

2.6.1 Challenges

The therapeutic potential of natural compounds remains limited because bioavailability issues along with standardization problems create obstacles for clinical efficacy. The poor bioavailability of curcumin and resveratrol occurs because these compounds have low solubility and fast metabolism and limited systemic circulation. Advanced drug delivery systems including nanoparticles and liposomes and micelles together with structural modifications and bioenhancers like piperine serve to enhance absorption and metabolic stability and in vivo half-life. The unpredictable nature of natural compound content creates challenges for reproducibility and regulatory approval because it depends on plant sources and cultivation practices and processing methods. The implementation of HPLC and mass spectrometry and NMR spectroscopy quality control methods combined with standardized cultivation and extraction procedures leads to batch consistency. The failure of traditional in vitro and animal models to duplicate human physiology explains why there exists a wide gap between preclinical and clinical success. The combination of predictive models including organ-on-chip systems with humanized animal models and pharmacokinetic and pharmacodynamic studies during early clinical trials will boost translational research reliability. The field can unlock complete therapeutic potential of natural compounds for clinical use by implementing these strategies to address essential barriers.

2.6.2 Future Directions

The therapeutic potential of natural bioactive compounds receives advancement through the combined power of nanotechnology, synthetic biology, combination therapies, and omics approaches. Nanotechnology enables the creation of nanoparticles and liposomes which improve compound stability and bioavailability while enabling targeted delivery to specific tissues including blood-brain barrier crossing for neuroprotection and sustained release mechanisms that enhance treatment effectiveness. Through synthetic biology, researchers can optimize microbial metabolic pathways to produce complex natural products and novel analogs with improved efficacy and reduced toxicity at lower costs. Natural compounds used in combination therapies with conventional drugs produce enhanced therapeutic effects and reduced side effects when used together to fight conditions that cause oxidative stress and neurodegeneration and inflammation. Omics technologies which include genomics proteomics and metabolomics provide high-throughput tools to discover new bioactive compounds while revealing their molecular targets and action mechanisms. Advanced computational modelling and bioinformatics work together with these approaches to speed up the development of innovative therapeutic strategies which solve essential problems in natural compound research and enable precision medicine.

References

Akhondzadeh S, Noroozian M, Mohammadi M, Ohadinia S, Jamshidi AH, Khani M (2003) Salvia officinalis extract in the treatment of patients with mild to moderate Alzheimer's disease: a double blind, randomized and placebo-controlled trial. J Clin Pharm Ther 28(1):53–59

Akhoon BA, Pandey S, Tiwari S, Pandey R (2016) Withanolide A offers neuroprotection, ameliorates stress resistance and prolongs the life expectancy of Caenorhabditis elegans. Exp Gerontol 78:47–56

Akinwumi BC, Bordun KM, Anderson HD (2018) Biological activities of stilbenoids. Int J Mol Sci 19(3):792

Alzheimer's disease facts and figures (2021) Alzheimers Dement 17(3):327–406

Ayaz M, Sadiq A, Junaid M, Ullah F, Ovais M, Ullah I, Ahmed J, Shahid M (2019) Flavonoids as prospective neuroprotectants and their therapeutic propensity in aging associated neurological disorders. Front Aging Neurosci 11:155

Azlan UK, Khairul Annuar NA, Mediani A, Aizat WM, Damanhuri HA, Tong X, Yanagisawa D, Tooyama I, Wan Ngah WZ, Jantan I, Hamezah HS (2022) An insight into the neuroprotec-

tive and anti-neuroinflammatory effects and mechanisms of Moringa oleifera. Front Pharmacol 13:1035220
Bandopadhyay S, Mandal S, Ghorai M, Jha NK, Kumar M, Radha, Ghosh A, Prockow J, de la Lastra JMP, Dey A (2023) Therapeutic properties and pharmacological activities of asiaticoside and madecassoside: a review. J Cell Mol Med 27(5):593–608
Barnham KJ, Masters CL, Bush AI (2004) Neurodegenerative diseases and oxidative stress. Nat Rev Drug Discov 3(3):205–214
Bekris LM, Yu CE, Bird TD, Tsuang DW (2010) Genetics of Alzheimer disease. J Geriatr Psychiatry Neurol 23(4):213–227
Bhattacharyya N, Pandey V, Bhattacharyya M, Dey A (2021) Regulatory role of long non coding RNAs (lncRNAs) in neurological disorders: from novel biomarkers to promising therapeutic strategies. Asian J Pharm Sci 16(5):533–550
Birben E, Sahiner UM, Sackesen C, Erzurum S, Kalayci O (2012) Oxidative stress and antioxidant defense. World Allergy Organ J 5(1):9–19
Bow EW, Rimoldi JM (2016) The structure-function relationships of classical cannabinoids: CB1/CB2 modulation. Perspect Medicin Chem 8:17–39
Braak H, Braak E (1991) Neuropathological stageing of Alzheimer-related changes. Acta Neuropathol 82(4):239–259
Brookmeyer R, Gray S, Kawas C (1998) Projections of Alzheimer's disease in the United States and the public health impact of delaying disease onset. Am J Public Health 88(9):1337–1342
Brooks AC, Henderson BJ (2021) Systematic review of nicotine exposure's effects on neural stem and progenitor cells. Brain Sci 11(2):172
Cheynier V, Comte G, Davies KM, Lattanzio V, Martens S (2013) Plant phenolics: recent advances on their biosynthesis, genetics, and ecophysiology. Plant Physiol Biochem 72:1–20
Cole GM, Teter B, Frautschy SA (2007) Neuroprotective effects of curcumin. Adv Exp Med Biol 595:197–212
Costa LG, Garrick JM, Roque PJ, Pellacani C (2016) Mechanisms of neuroprotection by quercetin: counteracting oxidative stress and more. Oxidative Med Cell Longev 2016:2986796
de Lau LM, Breteler MM (2006) Epidemiology of Parkinson's disease. Lancet Neurol 5(6):525–535
Dias MC, Pinto D, Silva AMS (2021) Plant flavonoids: chemical characteristics and biological activity. Molecules 26(17):5377
Donkor ES (2018) Stroke in the 21(st) century: a snapshot of the burden, epidemiology, and quality of life. Stroke Res Treat 2018:3238165
Dorsey ER, Constantinescu R, Thompson JP, Biglan KM, Holloway RG, Kieburtz K, Marshall FJ, Ravina BM, Schifitto G, Siderowf A, Tanner CM (2007) Projected number of people with Parkinson disease in the most populous nations, 2005 through 2030. Neurology 68(5):384–386
Du J, Liu D, Zhang X, Zhou A, Su Y, He D, Fu S, Gao F (2020) Menthol protects dopaminergic neurons against inflammation-mediated damage in lipopolysaccharide (LPS)-evoked model of Parkinson's disease. Int Immunopharmacol 85:106679
Duda-Chodak A, Tarko T (2023) Possible side effects of polyphenols and their interactions with medicines. Molecules 28(6):2536
Emerit J, Edeas M, Bricaire F (2004) Neurodegenerative diseases and oxidative stress. Biomed Pharmacother 58(1):39–46
Fabricant DS, Farnsworth NR (2001) The value of plants used in traditional medicine for drug discovery. Environ Health Perspect 109 Suppl 1(Suppl 1):69–75
Fachel FNS, Schuh RS, Veras KS, Bassani VL, Koester LS, Henriques AT, Braganhol E, Teixeira HF (2019) An overview of the neuroprotective potential of rosmarinic acid and its association with nanotechnology-based delivery systems: a novel approach to treating neurodegenerative disorders. Neurochem Int 122:47–58
Feigin VL, Norrving B, Mensah GA (2017) Global burden of stroke. Circ Res 120(3):439–448
Feng Z, Sun Q, Chen W, Bai Y, Hu D, Xie X (2019) The neuroprotective mechanisms of ginkgolides and bilobalide in cerebral ischemic injury: a literature review. Mol Med 25(1):57
Friedli MJ, Inestrosa NC (2021) Huperzine A and its neuroprotective molecular signaling in Alzheimer's disease. Molecules 26(21):6531

Gao HM, Hong JS (2008) Why neurodegenerative diseases are progressive: uncontrolled inflammation drives disease progression. Trends Immunol 29(8):357–365

Gotina L, Seo SH, Kim CW, Lim SM, Pae AN (2021) Pharmacophore-based virtual screening of novel competitive inhibitors of the neurodegenerative disease target Kynurenine-3-monooxygenase. Molecules 26(11):3314

Habib CN, Mohamed MR, Tadros MG, Tolba MF, Menze ET, Masoud SI (2022) The potential neuroprotective effect of diosmin in rotenone-induced model of Parkinson's disease in rats. Eur J Pharmacol 914:174573

Habiba SU, Choi HJ, Munni YA, Yang IJ, Haque MN, Moon IS (2024) Neurotrophic effects of Foeniculum vulgare ethanol extracts on hippocampal neurons: role of anethole in neurite outgrowth and synaptic development. Int J Mol Sci 25(23):12701

Hampel H, Toschi N, Babiloni C, Baldacci F, Black KL, Bokde ALW, Bun RS, Cacciola F, Cavedo E, Chiesa PA, Colliot O, Coman CM, Dubois B, Duggento A, Durrleman S, Ferretti MT, George N, Genthon R, Habert MO, Herholz K, Koronyo Y, Koronyo-Hamaoui M, Lamari F, Langevin T, Lehericy S, Lorenceau J, Neri C, Nistico R, Nyasse-Messene F, Ritchie C, Rossi S, Santarnecchi E, Sporns O, Verdooner SR, Vergallo A, Villain N, Younesi E, Garaci F, Lista S, I. Alzheimer Precision Medicine (2018) Revolution of Alzheimer precision neurology. Passageway of systems biology and neurophysiology. J Alzheimers Dis 64(s1):S47–S105

Hannan MA, Dash R, Haque MN, Mohibbullah M, Sohag AAM, Rahman MA, Uddin MJ, Alam M, Moon IS (2020) Neuroprotective potentials of marine algae and their bioactive metabolites: pharmacological insights and therapeutic advances. Mar Drugs 18(7):347

Hardy JA, Higgins GA (1992) Alzheimer's disease: the amyloid cascade hypothesis. Science 256(5054):184–185

Hosseini M, Boskabady MH, Khazdair MR (2021) Neuroprotective effects of Coriandrum sativum and its constituent, linalool: a review. Avicenna J Phytomed 11(5):436–450

Howes MJ, Perry E (2011) The role of phytochemicals in the treatment and prevention of dementia. Drugs Aging 28(6):439–468

Hussain G, Rasul A, Anwar H, Aziz N, Razzaq A, Wei W, Ali M, Li J, Li X (2018) Role of plant derived alkaloids and their mechanism in neurodegenerative disorders. Int J Biol Sci 14(3):341–357

Iqbal K, Liu F, Gong CX (2016) Tau and neurodegenerative disease: the story so far. Nat Rev Neurol 12(1):15–27

Jahan S, Kumar D, Mukherjee S, Roy K, Siddiqui AJ, Khan A, Rajpurohit CS, Aldosari S (2023) Stem cells, derived exosomes, and associated signaling molecules in neuroprotection. In: Applications of stem cells and derived exosomes in neurodegenerative disorders. Springer, Singapore, pp 259–290

Janda K, Wojtkowska K, Jakubczyk K, Antoniewicz J, Skonieczna-Zydecka K (2020) Passiflora incarnata in neuropsychiatric disorders-a systematic review. Nutrients 12(12):3894

Jenner P (2003) Oxidative stress in Parkinson's disease. Ann Neurol 53 Suppl 3:S26–S36; discussion S36–28

Kalaria RN, Akinyemi R, Ihara M (2016) Stroke injury, cognitive impairment and vascular dementia. Biochim Biophys Acta 1862(5):915–925

Karioti A, Bilia AR (2010) Hypericins as potential leads for new therapeutics. Int J Mol Sci 11(2):562–594

Klemow KM, Bartlow A, Crawford J, Kocher N, Shah J, Ritsick M (2011) Medical attributes of St. John's Wort (Hypericum perforatum). In: Benzie IFF, Wachtel-Galor S (eds) Herbal medicine: biomolecular and clinical aspects. CRC Press/Taylor & Francis, Boca Raton

Klomparens EA, Ding Y (2019) The neuroprotective mechanisms and effects of sulforaphane. Brain Circ 5(2):74–83

Lim DW, Lee JE, Lee C, Kim YT (2024) Natural products and their neuroprotective effects in degenerative brain diseases: a comprehensive review. Int J Mol Sci 25(20):11223

Liu RH (2004) Potential synergy of phytochemicals in cancer prevention: mechanism of action. J Nutr 134(12 Suppl):3479S–3485S

Liu KM, Huang Y, Wan PP, Lu YH, Zhou N, Li JJ, Yu CY, Chou JJ, Zhang L, Zhang C, Qiang YY, Zhang R, Guo L (2022) Ursolic acid protects neurons in temporal lobe epilepsy and cognitive impairment by repressing inflammation and oxidation. Front Pharmacol 13:877898

Lopez-Otin C, Blasco MA, Partridge L, Serrano M, Kroemer G (2013) The hallmarks of aging. Cell 153(6):1194–1217

Lu J, Zeng X, Feng Y, Li S, Wang Y, Liu Y, Chen F, Guan Z, Chen T, Wei F (2023) Inhibitory effects of Jasminum grandiflorum L. essential oil on lipopolysaccharide-induced microglia activation-integrated characteristic analysis of volatile compounds, network pharmacology, and BV-2 cell. Front Pharmacol 14:1180618

Mahadevan S, Park Y (2008) Multifaceted therapeutic benefits of Ginkgo biloba L.: chemistry, efficacy, safety, and uses. J Food Sci 73(1):R14–R19

Manach C, Scalbert A, Morand C, Remesy C, Jimenez L (2004) Polyphenols: food sources and bioavailability. Am J Clin Nutr 79(5):727–747

Mattson MP, Magnus T (2006) Ageing and neuronal vulnerability. Nat Rev Neurosci 7(4):278–294

Mirza FJ, Zahid S (2022) Ursolic acid and rosmarinic acid ameliorate alterations in hippocampal neurogenesis and social memory induced by amyloid beta in mouse model of Alzheimer's disease. Front Pharmacol 13:1058358

Mirza MU, Mirza AH, Ghori NU, Ferdous S (2015) Glycyrrhetinic acid and E.resveratroloside act as potential plant derived compounds against dopamine receptor D3 for Parkinson's disease: a pharmacoinformatics study. Drug Des Devel Ther 9:187–198

Musial C, Kuban-Jankowska A, Gorska-Ponikowska M (2020) Beneficial properties of green tea catechins. Int J Mol Sci 21(5):1744

Nabavi SF, Braidy N, Gortzi O, Sobarzo-Sanchez E, Daglia M, Skalicka-Wozniak K, Nabavi SM (2015) Luteolin as an anti-inflammatory and neuroprotective agent: a brief review. Brain Res Bull 119(Pt A):1–11

Nadeem MS, Kazmi I, Ullah I, Muhammad K, Anwar F (2021) Allicin, an antioxidant and neuroprotective agent. Ameliorates cognitive impairment. Antioxidants (Basel) 11(1):87

Nagle DG, Ferreira D, Zhou YD (2006) Epigallocatechin-3-gallate (EGCG): chemical and biomedical perspectives. Phytochemistry 67(17):1849–1855

Newman DJ, Cragg GM (2020) Natural products as sources of new drugs over the nearly four decades from 01/1981 to 09/2019. J Nat Prod 83(3):770–803

Nobrega FF, Salvadori MG, Masson CJ, Mello CF, Nascimento TS, Leal-Cardoso JH, de Sousa DP, Almeida RN (2014) Monoterpenoid terpinen-4-ol exhibits anticonvulsant activity in behavioural and electrophysiological studies. Oxidative Med Cell Longev 2014:703848

Nouri Z, Fakhri S, El-Senduny FF, Sanadgol N, Abd-ElGhani GE, Farzaei MH, Chen JT (2019) On the neuroprotective effects of naringenin: pharmacological targets, signaling pathways, molecular mechanisms, and clinical perspective. Biomolecules 9(11):690

Olanow CW, Stern MB, Sethi K (2009) The scientific and clinical basis for the treatment of Parkinson disease (2009). Neurology 72(21 Suppl 4):S1–S136

Ota M, Sato N, Sone D, Ogura J, Kunugi H (2017) (-)-Linalool influence on the cerebral blood flow in healthy male volunteers revealed by three-dimensional pseudo-continuous arterial spin labeling. Indian J Psychiatry 59(2):225–227

Park H, Seol GH, Ryu S, Choi IY (2016) Neuroprotective effects of (-)-linalool against oxygen-glucose deprivation-induced neuronal injury. Arch Pharm Res 39(4):555–564

Peng D, Wang H, Qu C, Xie L, Wicks SM, Xie J (2012) Ginsenoside Re: its chemistry, metabolism and pharmacokinetics. Chin Med 7:2

Pham-Huy LA, He H, Pham-Huy C (2008) Free radicals, antioxidants in disease and health. Int J Biomed Sci 4(2):89–96

Prince M, Ali GC, Guerchet M, Prina AM, Albanese E, Wu YT (2016) Recent global trends in the prevalence and incidence of dementia, and survival with dementia. Alzheimers Res Ther 8(1):23

Priyadarsini KI (2014) The chemistry of curcumin: from extraction to therapeutic agent. Molecules 19(12):20091–20112

Reay JL, Kennedy DO, Scholey AB (2005) Single doses of Panax ginseng (G115) reduce blood glucose levels and improve cognitive performance during sustained mental activity. J Psychopharmacol 19(4):357–365
Ross GW, Petrovitch H (2001) Current evidence for neuroprotective effects of nicotine and caffeine against Parkinson's disease. Drugs Aging 18(11):797–806
Scalbert A, Williamson G (2000) Dietary intake and bioavailability of polyphenols. J Nutr 130(8S Suppl):2073S–2085S
Sekhar VC, Viswanathan G, Baby S (2019) Insights into the molecular aspects of neuroprotective Bacoside A and Bacopaside I. Curr Neuropharmacol 17(5):438–446
Selkoe DJ (2001) Alzheimer's disease: genes, proteins, and therapy. Physiol Rev 81(2):741–766
Sies H, Berndt C, Jones DP (2017) Oxidative stress. Annu Rev Biochem 86:715–748
Singh SK, Srivastav S, Castellani RJ, Plascencia-Villa G, Perry G (2019) Neuroprotective and antioxidant effect of Ginkgo biloba extract against AD and other neurological disorders. Neurotherapeutics 16(3):666–674
Small GW, Siddarth P, Li Z, Miller KJ, Ercoli L, Emerson ND, Martinez J, Wong KP, Liu J, Merrill DA, Chen ST, Henning SM, Satyamurthy N, Huang SC, Heber D, Barrio JR (2018) Memory and brain amyloid and tau effects of a bioavailable form of curcumin in non-demented adults: a double-blind, placebo-controlled 18-month trial. Am J Geriatr Psychiatry 26(3):266–277
Soeda S, Aritake K, Urade Y, Sato H, Shoyama Y (2016) Neuroprotective activities of saffron and crocin. Adv Neurobiol 12:275–292
Spillantini MG, Schmidt ML, Lee VM, Trojanowski JQ, Jakes R, Goedert M (1997) Alpha-synuclein in Lewy bodies. Nature 388(6645):839–840
Stough C, Lloyd J, Clarke J, Downey LA, Hutchison CW, Rodgers T, Nathan PJ (2001) The chronic effects of an extract of Bacopa monniera (Brahmi) on cognitive function in healthy human subjects. Psychopharmacology 156(4):481–484
Sun AY, Wang Q, Simonyi A, Sun GY (2010) Resveratrol as a therapeutic agent for neurodegenerative diseases. Mol Neurobiol 41(2-3):375–383
Sun R, Song Y, Li S, Ma Z, Deng X, Fu Q, Qu R, Ma S (2018) Levo-tetrahydropalmatine attenuates neuron apoptosis induced by cerebral ischemia-reperfusion injury: involvement of c-Abl activation. J Mol Neurosci 65(3):391–399
Tariot PN, Farlow MR, Grossberg GT, Graham SM, McDonald S, Gergel I, G. Memantine Study (2004) Memantine treatment in patients with moderate to severe Alzheimer disease already receiving donepezil: a randomized controlled trial. JAMA 291(3):317–324
Van Den Eeden SK, Tanner CM, Bernstein AL, Fross RD, Leimpeter A, Bloch DA, Nelson LM (2003) Incidence of Parkinson's disease: variation by age, gender, and race/ethnicity. Am J Epidemiol 157(11):1015–1022
Venkat P, Chopp M, Chen J (2016) New insights into coupling and uncoupling of cerebral blood flow and metabolism in the brain. Croat Med J 57(3):223–228
Vicidomini C, Fontanella F, D'Alessandro T, Roviello GN (2024) A survey on computational methods in drug discovery for neurodegenerative diseases. Biomolecules 14(10):1330
Wang ZJ, Heinbockel T (2018) Essential oils and their constituents targeting the GABAergic system and sodium channels as treatment of neurological diseases. Molecules 23(5):1061
Wang Y, Shi X, Qi Z (2010) Hypericin prolongs action potential duration in hippocampal neurons by acting on K+ channels. Br J Pharmacol 159(7):1402–1407
Wei BB, Liu MY, Chen ZX, Wei MJ (2018) Schisandrin ameliorates cognitive impairment and attenuates Abeta deposition in APP/PS1 transgenic mice: involvement of adjusting neurotransmitters and their metabolite changes in the brain. Acta Pharmacol Sin 39(4):616–625
Weinreb O, Amit T, Bar-Am O, Youdim MB (2016) Neuroprotective effects of multifaceted hybrid agents targeting MAO, cholinesterase, iron and beta-amyloid in ageing and Alzheimer's disease. Br J Pharmacol 173(13):2080–2094
Williams RJ, Spencer JP, Rice-Evans C (2004) Flavonoids: antioxidants or signalling molecules? Free Radic Biol Med 36(7):838–849
Williams P, Sorribas A, Howes MJ (2011) Natural products as a source of Alzheimer's drug leads. Nat Prod Rep 28(1):48–77

Wongtrakul J, Thongtan T, Kumrapich B, Saisawang C, Ketterman AJ (2021) Neuroprotective effects of Withania somnifera in the SH-SY5Y Parkinson cell model. Heliyon 7(10):e08172
Wyss-Coray T (2016) Ageing, neurodegeneration and brain rejuvenation. Nature 539(7628):180–186
Yadav VK, Kumar A, Tripathi PP, Gupta J (2021) Long noncoding RNAs in intestinal homeostasis, regeneration, and cancer. J Cell Physiol 236(11):7801–7813
Yadav VK, Awasthi P, Behl R, Kumar A (2022) HSc70 interactome reveal major role of macroautophagy and minor role of chaperone mediated autophagy in K-Ras G12V cell proliferation and survival. J Proteome 264:104614
Youdim KA, Spencer JP, Schroeter H, Rice-Evans C (2002) Dietary flavonoids as potential neuroprotectants. Biol Chem 383(3-4):503–519
Zarkovic N (2003) 4-hydroxynonenal as a bioactive marker of pathophysiological processes. Mol Asp Med 24(4-5):281–291
Zhang YT, Yu YQ, Yan XX, Wang WJ, Tian XT, Wang L, Zhu WL, Gong LK, Pan GY (2019) Different structures of berberine and five other protoberberine alkaloids that affect P-glycoprotein-mediated efflux capacity. Acta Pharmacol Sin 40(1):133–142
Zhang R, Zeng M, Zhang X, Zheng Y, Lv N, Wang L, Gan J, Li Y, Jiang X, Yang L (2023) Therapeutic candidates for Alzheimer's disease: saponins. Int J Mol Sci 24(13):10505
Zheng M, Xin Y, Li Y, Xu F, Xi X, Guo H, Cui X, Cao H, Zhang X, Han C (2018) Ginsenosides: a potential neuroprotective agent. Biomed Res Int 2018:8174345
Zhong Z, Han J, Zhang J, Xiao Q, Hu J, Chen L (2018) Pharmacological activities, mechanisms of action, and safety of salidroside in the central nervous system. Drug Des Devel Ther 12:1479–1489

Natural Compounds and Their Mechanism of Action: Therapeutic Targets of Neuroprotection and Neurorestoration in Neurodegenerative Diseases

3

Ehraz M. Siddiqui, Mashoque Ahmad Rather, Sadaf Jahan, Arif Jamal Siddiqui, and Andleeb Khan

Abstract

Endogenous antioxidant defenses are insufficient; therefore, oxidative stress is generated due to excessive production of reactive species, especially reactive oxygen (ROS) and reactive nitrogen (RNS) species, and degradation of cell structures, lipids, proteins, and genetic material. Oxidative stress is a key mechanism underlying the death of neurons, which is an essential component of the pathophysiology of neurodegeneration. One of the medically prescribed antidotes to this skewed oxidation-antioxidation state is the profession of antioxidants. Recently, natural substances with the properties to both scavenge free radicals and protect cells against oxidative threat as well as possess neuroprotective potential have attracted great interest.

This chapter summarized the state-of-research efforts attempting to identify natural chemical derived from medicinal plants that possess protecting action against neurodegenerative processes at molecular levels by controlling the death

E. M. Siddiqui
Department of Pharmacology, Buddha Institute of Pharmacy, Gorakhpur, UP, India

M. A. Rather
Department of Molecular Pharmacology and Physiology, University of South Florida, Tampa, FL, USA

S. Jahan
Department of Medical Laboratory Sciences, College of Applied Medical Sciences, Majmaah University, Al Majmaah, Saudi Arabia

A. J. Siddiqui
Department of Biology, College of Science, University of Ha'il, Ha'il, Saudi Arabia

A. Khan (✉)
Department of Biosciences, Faculty of Science, Integral University, Lucknow, India

S. Jahan et al. (eds.), *A Mechanistic Exploration of Natural Compounds for Neuronal Health*, https://doi.org/10.1007/978-981-95-0125-0_3

of neurons through inhibiting a variety of neurotoxins. Natural regulators of reactive species and mitochondrial activity are lignans, quinones, 1MeTIQ, resveratrol, curcumin, *Gingko biloba*, alkaloids, glycosides, vitamin C, flavonoids, polyphenols, coumarins, etc. They could decrease the neurotoxic agent-induced oxidative damage and apoptosis, by decreasing ROS/RNS generation, lipid peroxidation, caspase-3 and caspase-9 activities, LDH release, the ratio of Bax/Bcl-2, Ca^{2+} influx, cytochrome c release, increasing MMP, and restoring reverse endogenous enzymatic activities.

They also prevented mitochondria-dependent apoptosis pathways and modulated the oxidative plunges of several signaling pathways to produce neuroprotective effects against apoptosis. The current study examines the part that reactive species play in neurodegeneration, emphasizing the possible anti-oxidation properties of natural substances as a viable means of creating novel neuroprotective tactics.

Keywords

Neuroprotection · Neurodegeneration · Cell death · Oxidative stress · Reactive species

3.1 Introduction

One of the most dangerous medical conditions, neurodegenerative diseases affect a large number of people globally (Zhou et al. 2018). The frequency of these diseases is dramatically increasing as individuals live longer. These diseases are a varied group of chronic, progressive disorders characterized by a gradual loss of neurons in the central nervous system, leading to deficits in specific brain functions. The most common neurodegenerative disorders are AD, PD, ALS, MS, and HD (Campbell et al. 2019).

As of right now, there are no therapeutic options that can reverse or even slow the progression of these conditions; instead, the majority of treatments are palliative. In this regard, natural products are seen to be promising substitutes for the treatment of neurodegeneration due to their wide range of pharmacological and biological actions, since they may aid in the process of developing new drugs (Cui and Wan 2019). Numerous research has demonstrated the potential health benefits of using natural ingredients as neurodegenerative treatments. Many biological actions, such as anti-inflammatory, antioxidant, and anti-apoptotic properties, have been documented for natural substances (Zhu et al. 2021). Furthermore, it has recently been demonstrated that natural substances can modify autophagy and proteasome activity in addition to preventing protein misfolding (Chen et al. 2018) (Fig. 3.1).

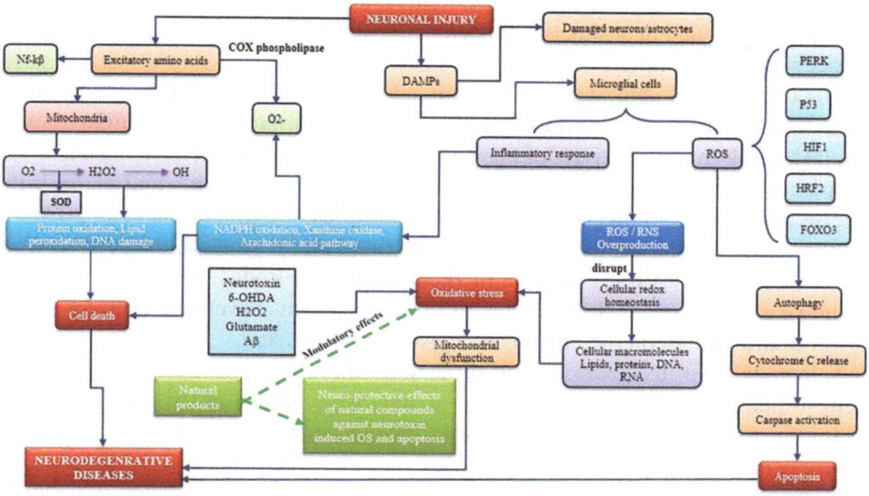

Fig. 3.1 Cascade reaction process for neurodegeneration. Injury of neurons results in decreased ATP production in brain, followed by failure of Na/K pump and membrane depolarization. This leads to action potential firing, resulting in a pathophysiological state of continuous elevation of intracellular calcium and glutamate release leading to activation of signaling pathways ultimately culminating in cell death pathways such as apoptosis, autophagy, and inflammation. Increased calcium in cells leads to mitochondrial dysfunction and the production of ROS, which in turn leads to lipid peroxidation, protein degeneration, enzyme inactivation, and mitochondrial injury, and thus facilitates cytochrome-C release, cell signaling pathway disorders, and the establishment of subsequent apoptotic cascade reactions

Neurodegenerative disorders in humans are primarily caused by disruptions to brain redox balance. Oxidative stress (OS) is characterized as an imbalance between the body's natural antioxidant system and harmful ROS. Furthermore, it is thought that OS is a crucial component of the aging process (Yu et al. 2017). One universal mechanism that leads to cell death is oxidative stress. Different processes, such as altered iron buildup inside the SN, altered α-syn accumulation and proteolysis, altered calcium carrier function, and protein anomalies, generate oxidative stress in PD (Lin et al. 2020). Furthermore, it has been noted that the substantia nigra of PD patients contains higher levels of hydroperoxides, malondialdehyde, and protein oxidation. Furthermore, the CSF of Parkinson's disease patients showed a buildup of lipid peroxidation products (Liu et al. 2019).

OS and nitrosative stress are caused by the overproduction of RNS and ROS, which disturbs normal cellular homeostasis. Many proteins, lipids, and DNA are damaged by high concentrations of these reactive species, but modest concentrations of these molecules serve as signaling molecules during normal cellular metabolism (Barthels and Das 2020). Because dopaminergic neurons contain neuromelanin, they are the most vulnerable to oxidative stress. Black pigment called neuromelanin has ferrous ions (Fe^{2+}) in it. The Fenton reaction is aided by the presence of Fe^{2+} (Fluri et al. 2015). Apart from the organic elements that lead to oxidative stress, like heavy metals, pesticides, and herbicides in the environment, ultraviolet light, and heat shock, DP itself also contributes significantly to this harmful occurrence (Sommer 2017).

The two types of natural products included in the published articles are extracted compounds and isolated compounds (Xu et al. 2021). However, it might also be crucial to look into the activity of particular natural products that have been isolated in order to comprehend their cellular and molecular mechanisms and identify the precise bioactive ingredients that are present in foods or extracts (Su et al. 2020). A growing number of scientists have been utilizing natural compounds derived from plants in an effort to find effective medications for neurodegenerative illnesses in recent years (Shao et al. 2020). Many of these substances have been used for centuries in traditional medicine and have received great praise.

Scientists have just rediscovered these substances and are thoroughly studying them to determine their mechanism of action. We would want to highlight a few intriguing naturally occurring substances having neuroprotective potential in this review (Chen et al. 2017).

Numerous elements that promote inflammation are expressed when oxygen FR increase. Apoptosis is linked to the inflammatory retort brought on by pro-inflammatory stimuli. One significant source of reactive oxygen species is mitochondria (Uzdensky 2019). The primary issue is neuronal damage-induced neurological dysfunction. Preserving and regaining the functionality of injured neurons is therefore the most important intervention job (Tian et al. 2018). Recent studies have shown that the primary pathophysiological mechanisms generating neuronal injury are neuronal auto-damage and the plunging response process between neurons and the damage to their surrounds. Excitatory AA release, calcium overload, excessive FR generation, inflammation, and procedural cell death are the outcomes of these processes (Qin et al. 2019).

Drug mediation techniques for neutralizing the metabolic and biochemical changes can be classified into two categories: neuro-restorative therapy and neuroprotective therapy, based on the combination of the pathophysiological features and the concept treatment time window. Many studies have shown that certain natural medications, which has been used, have beneficial effects with minimal side effects. Some of the processes via which the major active substances in natural products work include the encouragement of angiogenesis and neurogenesis, as well as the inhibition of oxidative stress, apoptosis, and inflammatory reactions (Zhu 2019).

The active component or effective site of Chinese herbal compounds has been the primary subject of investigation. Several studies have demonstrated the preventative and therapeutic properties of the active components, which include *Ginkgo biloba*, astragaloside IV, baicalin, resveratrol, 1MeTIQ, curcumin, vitamin C, glycosides, phenols, flavonoids, and terpenes (Alishahi et al. 2019). As a result, the creation of successful multi-targeted medications can significantly hasten the creation of novel medications to treat neurological illnesses. This chapter gives a summary of the efficacy and mechanisms of conventional natural drugs after reviewing cases of brain injury, neuroprotection, and neurorestoration. Additionally, we aimed to provide therapeutic modalities and drug choices for clinical research (Wang et al. 2018a, b, c).

3.2 Reactive Species

Numerous endogenous and external activities result in the production of RNS and ROS. Molecular entities or fragments that are paramagnetic and include a couple of unpaired electrons are called free radicals. The majority of FR are very reactive, unstable, and have a brief lifespan (Liu et al. 2015a, b). The most symbolic class of FR produced in biological systems are ROS, which are produced as a result of cellular metabolism, aerobic respiration, or pathogen defense (Xie et al. 2020). The oxygen molecule (O_2) is a FR in and of itself and has a distinct electrical configuration. O_2 can readily take on one electron to become O_2•− (Huang et al. 2019). The primary ROS is thought to be O_2•−. Through direct interactions with other molecules or through processes aided by metals or enzymes, O_2•− can also form more radical species (Roberts et al. 2016). One of the main energy sources is the mitochondrial ETC and RC. In mammalian cells, the ETC or RC is one of the main origins of ATP and the end product of aerobic respiration is ROS, which takes place in the mitochondria's inner membrane (Wen et al. 2020). O_2 is the last acceptor of protons and electrons during energy transduction. If electrons from a couple of the redox centers in the ETC leak out to O_2, the molecule may be partially reduced to O_2, followed by H_2O_2 and HO•. RNS refers to NO and its derivative molecules, ONOO− and NO•. RNS play a significant role in the production of oxidative stress (Zhou et al. 2017a, b).

Three isoforms of NOS—neuronal NOS (nNOS), endothelial NOS (eNOS), and inducible NOS (iNOS)—catalyze the catalytic process that produces NO from L-arginine in mitochondria. Numerous critical nervous system functions, such as the control of sleep-wake cycle, synaptic plasticity, cerebral blood flow, and neurosecretion, are impacted by NO (Seok et al. 2018). Higher doses of NO are clearly neurotoxic, although physiological levels of NO are often neuroprotective. Excessive NO release damages proteins, lipids, and nucleic acids in surrounding cells, impairing mitochondrial function and influencing cell survival and metabolism. Reactive species have the ability to react with one another to generate more active forms (Huo et al. 2017). For example, when the free form of NO combines with O_2, it produces the peroxynitrite radical, or ONOO•. Nitrogen dioxide radical (NO_2 •), HO•, and nitryl cation (NO^{2+}) are the products of the quick breakdown of ONOO−. They're all capable of harming nerve cells. These extremely reactive species attack several cell processes (lipid peroxidation, nucleic acid oxidation), as well as biomolecules (genetic material, enzymes, lipids, proteins, and hormones) (Wang et al. 2015). Cell membranes undergo structural and functional alterations as a result of reactive species. Lipid peroxidation's transition products lead to the breakdown of polyunsaturated fatty acids and the creation of receptive aldehydes, including 4-hydroxynonenal (HNE) and malondialdehyde (MDA) (Zhang et al., 2017a, b, c).

Following their interaction with DNA or protein molecules, these hazardous byproducts cause somatic mutation and cellular harm (Seto et al. 2016).

3.3 Neuroprotective Strategy

By obstructing the cell death cascade of penumbra and preventing the lyse of penumbric cells, neuroprotective drugs can intervene in the neuronal death cascade. While meaningful clinical outcomes for this neuron-preserving treatment technique have not yet been obtained, neuroprotective therapy is generally thought to be able to minimize neuronal damage and considerably extend the window of thrombolytic therapy. The neuroprotective mechanisms implicated in neurological illnesses have been evaluated in recent investigations.

3.3.1 Targeting Excitotoxicity

The quick and enormous discharge of the excitatory AA glutamate and reuptake blockage brought on by energy loss are referred to as excitatory toxicity (Hu et al. 2020). In addition, hypoxia and a shortage of glucose brought on by the disruption of blood flow to the brain will result in abnormalities in the metabolism of neuronal energy, depolarization of the presynaptic membrane, and an increase in glutamate release. Excitatory amino acid receptors, primarily NMDARs, are further activated by glutamate (Hu et al. 2020). This results in the opening of Ca channels and the influx of Ca++, raising the intracellular calcium concentration and setting off a chain of serious reactions in the nucleus and cytoplasm that eventually break down the cytoskeleton and cause necrosis in neurons (Wang and Jin 2015).

Furthermore, a calcium overload in the mitochondria causes mitochondrial degradation, triggers downstream apoptotic pathways, and causes apoptosis. As a result, controlling the activity of excitatory AA is a crucial component of therapy (Edelbrock et al. 2018). Currently, glutamate receptor modulators, GABA receptor agonists, 5-HT1 receptor agonists, and glutamate release inhibitors have all demonstrated anti-brain effects in basic trials by inhibiting the release of glutamate and regulating it (Shang et al. 2019). Controlling the downstream glutamate signaling pathways, such as Ca++ influx, is another strategy to lessen the damaging consequences of glutamate excitotoxicity. Certain natural substances have been shown to have a neuroprotective effect by inhibiting excitatory toxicity, including astragaloside IV and baicalin (Du et al. 2020).

3.3.2 Targeting Oxidative Stress

Because it has a larger membranous content than any other organ in the body, fewer antioxidant enzymes, and a higher oxidative metabolism, the brain is more vulnerable to OS than any other organ. The primary process causing brain injury is a state of OS brought on by the breakdown of the equilibrium between the oxi and anti-oxi systems of the body (Uyeda and Muramatsu 2020). The body's primary producers ROS are NADPH oxidase and SDH. Under typical physiological circumstances, ROS can be produced by the organism during aerobic metabolism, and the creation and removal of ROS are dynamically balanced (Venkat et al. 2018).

During the brain injuries, the activity and expression of certain enzymes, like xanthine (dehydrogenase, reductase, oxidase), NOS, COX, coenzyme II oxidase, and MAO, alters. ROS builds up in the body as a result of these alterations that upset the dynamic balance of ROS. Some other enzymes like SOD, CAT, and GSH-PX are also altered during the ROS production (Carmichael et al. 2017). Notably, brain ischemia causes neurons and the inflammatory response to consistently express high levels of iNOS, which generates a lot of NO. Overabundance of NO has the potential to combine with superoxide anions to generate potent oxidizing peroxynitrite, which can disrupt SOD's function and cause neuronal injury (Zhu et al. 2020).

3.3.3 Targeting Apoptosis

Apoptosis is a type of programmed cell death (PCD) in which cells undergo an autonomous death that is regulated by their genes. There are three primary pathways via which neuronal injury triggers neuronal apoptosis: the ER stress pathway, the DR pathway, and the MT pathway (Wang et al. 2021). This process involves a large number of genes and signaling pathways relevant to apoptosis, including as the JAK2/STAT3 signaling system, the caspase family, the p53 gene, the Bcl-2 protein family, and the Fas gene (Sun et al. 2020a, b). An endogenous neuroprotective agent called Bcl-2 can prevent neuronal death brought on by brain injury, which is especially common in cases of neurodegenerative injuries. Bcl-2 is highly expressed in the brain. Brain injury can cause p53 to pile up in the cellular membrane, causing activation of BAX followed by cell death (Zhang et al., 2019a, b, c, d).

Furthermore, p53 can sprightly cause a spike in the outer membrane of MT by creating a complex structure with the protein BCL-2, which releases cytochrome C. The family of CPS is essential to cell death. Cell apoptosis is initiated and carried out by CPS, and one of the main factors in this process is the renewing of its downstream protease CPS-3. After activating CPS-3 and triggering cell apoptosis, numerous intracellular components can be directly lysed by mitochondria and death receptor-mediated apoptosis pathways (Sun et al. 2020a, b). Fas is a cell surface death receptor that is present in many different types of tissue cells. In addition to inducing apoptosis by activating the CPS family, Fas receptors primarily arbitrate apoptosis via the Fas/FasL pathway (Liang et al. 2020).

3.3.4 Targeting Inflammation

The inflammatory response through immune system is another highly desirable curative target after the brain injury. The duration and intensity of the inflammatory response mostly determine its function in brain injury (Xue et al. 2019). In the first phase, for example, brain traumas cause neuro-inflammation and neuronal damage by inducing the formation of inflammatory cells and the overexpression of inflammatory fixers with small molecular structures (Li et al. 2021a, b). Neuronal death, glial cell growth, and brain tissue atrophy are the results of brain inflammation in the sub-acute and chronic phases. Leukocyte exudation and the initiation of astrocytes and microglia are important phases in the inflammatory response (Du et al. 2021).

Brain endogenous microglia are the foremost immune cells to respond after a stroke, and then neutrophils that worsen OS and BBB damage. Glial cells emit mediators during activation, which draw monocytes, neutrophils, and lymphocytes in addition (Li et al. 2019). Notably, brain injury results in the microglia's fast activation upon stimulation, which profoundly alters their shape and function from a resting to an energized state (Shi et al. 2021). A variety of proteins, inflammatory cytokines, and other physiologically active materials are secreted or released by activated microglia, which mediates neurotoxicity and results in secondary brain injury. Another subtype of immune cell is an astrocyte. Astrocytes are crucial for the CNS's normal metabolism, nutritional support, and preservation of neuronal function under normal conditions (Yang et al. 2021).

Astrocytes play two important roles in brain damage. They worsen neuronal damage by releasing inflammatory substances, on the one hand. However, during brain injury, astrocytes become more active and release neurotrophic substances, which aid in the healing of injured neurons. Inflammatory mediators, which are produced and released by activated inflammatory cells, can in turn trigger other activation of inflammatory cells, creating a vicious cycle that exacerbates brain damage. Cerebral inflammatory injury is primarily caused by excessively elevated levels of inflammatory cytokines, including TNF-α, IL-1β, and IL-6. These cytokines are also important curative targets (Qiu et al. 2019).

These elements have the potential to affect the vascular endothelium wall, induce cerebral edema, and damage the BBB's tight connections' structure and functionality. By accelerating the metabolism of arachidonic acid and exacerbating the damage caused by OS, they can also enhance the formation of FRs. Nevertheless, IL-13 and IL-10, anti-inflammatory cytokines, can also be produced during the inflammatory feedback process (Zhang et al., 2019a, b, c, d). It is suppressed by the creation of these anti-inflammatory substances, protecting brain tissue by clearing away and rebuilding damaged cells. Understanding the connection between inflammasomes, brain injury, and NLRP has advanced recently. The most representative inflammasome among them is NLRP3 (Wang et al. 2018a, b, c).

Through domain contact, adapter protein (apoptosis-associated speck-like protein containing a CARD, ASC) can be recruited by NLRP3 during brain injury. Through CARD-CARD synergy, the called-up ASC binds to caspase-1 to generate NLRP3 (Sui et al. 2019). The NLRP3/CPS-1/ASC plunging response pathway is then created as a result of CPS-1 activation, and it generates IL-18, IL-1β, and other inflammatory chemicals to promote tissue damage and inflammation (Luo et al. 2018).

3.3.5 Targeting Autophagy

Following stimulation, autophagy occurs when cells devour their cytoplasm or organelles and ultimately break down the phagosome in the lysosome. In controlling homeostasis and controlling cell survival, autophagy is crucial. Indeed, there is proof that during a brain injury, autophagy can be triggered by neurons, glial cells, and brain microvascular endothelial cells (Zhang et al. 2021). Moderate autophagy can cause cell death and apoptosis, eliminate aberrant proteins and bruised organelles from the body, and preserve homeostasis (Chu et al. 2019). Autophagy is recognized as a regulatory mechanism of cell homeostasis in vivo, and it has been linked to the onset and progression of brain injury via PINK1/Parkin, PKC/JNK, HiF-1α/BNIP3, AMPK/mTORC1, AKT-mTOR/PI3K, and other pathways (Chen et al. 2019).

One of the pathways involved in the pathological process of mitochondrial autophagy is PINK1, which is activated by oxygen glucose deprivation and reoxygenation. Furthermore, a number of factors may influence PINK1/Parkin pathway-mediated mitophagy (Li et al. 2017a, b, c). For instance, ROS can trigger the PINK1/Parkin pathway, which in turn triggers mitochondrial autophagy (Zheng et al. 2019). Apart from MT autophagy route, Parkin and PINK1 can also impact the movement of mitochondria by phosphorylating the mitochondrial adapter protein (Miro) through PINK1-mediated phosphorylation and degrading it with a proteasome through Parkin-mediated degradation (Zhong et al. 2019). This process promotes the removal of damaged mitochondria through mitochondrial autophagy (Li et al. 2020).

Moreover, autophagy via the BNIP3/HIF-1α/ pathway can eliminate bruised mitochondria, prevent Cyt C release and CPS-9/CPS-3 activation, stabilize MMP, and obstruct the mitochondria-induced apoptosis pathway, all of which protect against brain injury (Wang et al. 2020). Additionally, ROS can control autophagy in a variety of cell lines by activating it. Under casual conditions, the body can use the right quantity of ROS as signal molecules to development, support immunity, and cell growth (Zhang et al. 2018).

On the other hand, when brain injury happens, the abrupt drop in cerebral blood flow results in a shortage of oxygen and glucose, which causes anomalies in the structure of the mitochondria and illnesses related to oxidative phosphorylation (Zhang et al., 2019a, b, c, d). In OS, FOXO3A is the point where several signaling pathways converge (Wei et al. 2017a, b). As of right now, it is understood that FOXO3A stimulates autophagy through upregulating the gene expressions linked to autophagy, including LC3, BNIP3, Beclin1, Atg12, etc. (Liu et al. 2018a, b). On the other hand, elevated autophagy can lead to autophagic cell death, which is characterized by cell lysis. This apoptotic and necrotic cell death that frequently follows autophagic cell death exacerbates neuronal brain injury.

3.4 Neuro-Restorative Strategy

Angiogenesis, neurogenesis, and axonal sprouting are a few of the highly interactive processes involved in neurorestoration following a stroke (Dong et al. 2021). These mechanisms cooperate with one another and aid in the restoration of brain function. But compared to acute neuroprotective therapy, this neuro-restorative therapy might have a far longer window of time (Feng et al. 2021).

3.4.1 Targeting Angiogenesis

The act of creating new blood vessels from preexisting ones is known as angiogenesis. The maturation of the capillary network is achieved by the migration and multiplication of endothelial cells, which emerge from the main blood vessels (Newman and Cragg 2016). It is a critical physiological activity that is generated in response to hypoxia and is required for both embryonic development and wound healing (Miura et al. 2010). Importantly, angiogenesis depends on the intricate synchronization of angiogenesis inhibitors and angiogenic growth factors and to promote and sustain the proliferation and migration of endothelial cells needed for tissue repair within a short time frame (Liu et al. 2015a, b). In both human and animal stroke models, angiogenesis takes place 2–4 days following brain injury.

The penumbra region of the brain has more cerebral blood artery density than the usual area on the opposite side, according to an examination of the tissues of brain of animals and stroke patients (Olatunji et al. 2016). Additionally, people with a higher cerebral blood vessel density had higher survival and recovery rates than people with a relatively lower blood vessel density (Magalingam et al. 2014). After a brain injury, it has been discovered that certain these factors can cause angiogenesis; prostaglandins, nitric oxide (NO), MMP, platelet-derived growth factor (PDGF), basic fibroblast growth factor (bFGF), vascular endothelial growth factor (VEGF), and transforming growth factor (TGF) are some of these factors (Ren et al. 2019).

Of these angiogenic factors, VEGF, a cytokine, is widely produced and is thought to be the most significant in controlling the angiogenesis process (Kim et al. 2015). In order to function, VEGF must first adhere to its receptor, which then encourages endothelial cell migration and proliferation as well as angiogenesis (Zhang et al. 2015a, b). HIF-1α promotes angiogenesis in the ischemic brain model by increasing the production of VEGF (Jing et al. 2016). This suggests that the pathway including VEGF, VEGFR, and HIF-1α may be crucial for controlling angiogenesis (Zhang et al. 2017a, b, c). Another significant component that promotes angiogenesis is NO. NO has the ability to increase VEGF expression as well as endothelial cell proliferation, motility, and adherence to the extracellular matrix (Watanabe et al. 2018). Furthermore, the VEGF-induced angiogenesis can be attained via the Akt/PI3K route or the NAMPT/NAD/SIRT1 cascade. Cerebral edema is promoted by VEGF, increases permeability, and damages the BBB in addition to improving angiogenesis (Zhang et al., 2019a, b, c, d).

3.4.2 Targeting Neurogenesis

Neurogenesis is the cascade by which neural stem/progenitor cells differentiate into new functioning neurons. It involves the migration, proliferation, and differentiation of NSCs into fully developed neurons (Jha et al. 2015). Nowadays, it is widely acknowledged that the two primary locations for neurogenesis are the subgranular zone (SGZ) of the hippocampus dentate gyrus and the subventricular zone (SVZ) of the adult brain's middle and lateral ventricles (Park et al. 2015). Newly formed neurons do not, however, migrate just to these particular areas. In response to the injury, endogenous forerunner neurons slowly go to the cerebral cortex's ischemic core area, the striatum, and the hippocampal CA1 region. To finish the neuro-restorative process, they can combine with already-existing brain ambits there (Zhang et al. 2015a, b).

On the other hand, in pathological circumstances like brain injuries, neural stem cells proliferate much more in the SVZ region, move to the ischemic penumbra, and develop into adult glial cells and neurons, which aids in the restoration of neurological function. Important physiological roles for newborn neurons include emotion modulation, neuroplasticity, and learning and memory (Kim et al. 2017). Thus, controlling neurogenesis and restoring impaired neurological function is one of the primary goals of current brain injury research (Wang et al. 2014a, b). Recent research has demonstrated that several signal transduction pathways, such as the Wnt/β-Catenin pathway, brain-derived neurotrophic factor (BDNF) pathway, Notch pathway, Sonic Hedgehog (Shh) Access, and bone morphogenetic protein (BMP) pathway, are involved in the proliferation and differentiation of neural stem cells (Waseem and Parvez 2016).

Mammalian central nervous system development is significantly influenced by the BDNF signaling pathway (De Oliveira et al. 2015). Throughout adulthood, it continues to control a number of functions, such as adult neurogenesis, progenitor cell proliferation, and the preservation and self-renewal of stem cell pools (Cai et al. 2015). The neurotrophic factor BDNF interacts specifically to the receptors for tyrosine kinase B (TrkB). It can enhance synaptic plasticity and support neuronal differentiation and maturation during central nervous system development following brain injury (Wang et al. 2014a, b). Currently, it is thought that at various phases of neurogenesis and synaptic development, the TrkB/BDNF signaling pathway is involved in protein synthesis (Zhang et al. 2014).

Its activation can improve synaptic plasticity and encourage neural stem cell differentiation. The three signaling pathway reactions PLCγ pathway, phosphatidylinositol 3-kinase (PI3K), and mitogen-activated protein kinase (MAPK) that are reinforced by the combination of TrkB and BDNF that work together to accomplish these tasks (De Oliveira et al. 2016). The PI3K/Akt signaling system is essential for controlling cell growth, proliferation, and survival in both healthy and pathological settings (Zhou et al. 2014). Furthermore, the PI3K/Akt signaling pathway plays a role in axonal sprouting in cultured hippocampus neurons, which is a critical process for neurogenesis and the restoration of neurological function following a stroke (Li et al. 2013).

CREB, or cAMP-response element-binding protein, is also essential for neurogenesis and the formation of synapses (Ayissi et al. 2014). In addition to controlling neuronal development, CREB controls early neurogenesis, including the survival of early neural precursor cells (Cao et al. 2014). Reduction of CREB transcription is another effect of disrupting the BDNF-TrkB signaling cascade. Thus, it is possible to employ PI3K/Akt, CREB, and BDNF-TrkB as therapeutic targets and signaling pathways to encourage neurogenesis and restore neurological function following brain injury (Valenti et al. 2016).

3.4.3 Targeting Axonal Budding

The creation of synapses in between neurons is known as synaptogenesis. Strong plasticity exists in the nervous system and brain. Following a brain infarction, the axons of the injured neurons regenerate in a specific direction, form new synapses, and form functional connections, which leads to a number of intricate offsetting mechanisms (Fu et al. 2012). The three most crucial components of this process are alterations in synaptic threshold, dendritic collateral sprouting, and nerve cell axon renewal. The remodeling of nerve function can be completed by activating neural stem cells, rebuilding the neural network, and reactivating latent nerve cell function (Haleagrahara et al. 2013). Together with other neurotrophic factors, synaptophysin is a key regulator of the growth, maturation, and development of synapses (Bagli et al. 2016).

All nerve terminals include the presynaptic vesicle protein synaptophysin, which is frequently utilized as a synapse occurrence marker (Moridi Farimani et al. 2011). The surrounding infarct area's unharmed neurons are stimulated by hypoxia to proliferate and form new networks of synapse. Dendritic spines are rich in actin and are projections from dendrites that increase axon contact points and aid in synaptic transmission through synapses to facilitate neuronal communication and plasticity (Ansari and Khodagholi 2013). Consequently, the enhancement of rodent nerve function recovery is directly associated with therapy targeted at enhancing synapse occurrence following brain injury (Shi et al. 2011).

3.5 Natural Substances' Modulatory Effects on OS and Cell Death Caused by Neurotoxins

3.5.1 Natural Compounds

Natural compounds are pure and organic molecules originating from plants, animals, and microorganisms that have a limited molecular weight (typically less than 3000 Daltons). Here, natural substances derived from plants are our main concern (Ahmed et al. 2012). Plants have played a fundamental role in human existence for thousands of years. In addition to providing us with food, drink, cosmetics, and dyes, plants are also used by humans as medical medications. Historically, medicinal plants have proven to be valuable and offer therapeutic promise as suppliers of certain molecules (Lin et al. 2014). These substances are referred to as natural compounds, phytochemicals, or secondary metabolites of plants. Natural compounds are a diverse group of structures that can be used to create new, potent medications in the future (Singh and Singh 2020).

These chemicals' libraries usually provide a higher degree of "drug-likeness," according to recently published physicochemical features of natural compounds. Natural substances are more readily transported and disseminated across biological boundaries when they are created by living systems. These tiny chemical compounds have the ability to alter cellular processes that may alter in disease states by interfering with protein–protein interactions (Choi et al. 2011). Natural compounds are becoming significant sources of lead compounds for drug discovery efforts because of these unique qualities. Numerous biological actions, including antioxidative, anti-inflammatory, and anti-apoptotic properties, define natural substances (Chinta et al. 2013). Among these, a few natural substances have been shown to improve brain and central nervous system activity and to have neurotrophic and neuroprotective effects (Albarracin et al. 2012). Thus, there is great encouragement for the discovery of novel natural compounds having neuroprotective properties.

3.6 Natural Compounds Neuroprotective Goals

3.6.1 Mitochondrial Demise

Because neurons require a lot of energy to survive, they rely significantly on mitochondria. Numerous vital cellular processes, such as metabolism, apoptosis, and homeostasis of calcium, depend on mitochondria (Hwang 2013). They are a primary target for cellular damage caused by ROS and a significant source of ROS generation in relation to OS (Yuste et al. 2015). Changes in the biology of mitochondria are likely to have an impact on a wide range of disorders, including neurodegenerative ones, because of their important role in regulating both cellular death and life (Bastianetto et al. 2015).

It has also been shown that anticancer medication-induced neurotoxic symptoms are directly related to mitochondrial dysfunction (Walle 2011). A member of the sirtuin protein family, SIRT1, is a histone and protein deacetylase that is dependent on nicotinamide adenine dinucleotide (NAD)(+). SIRT1 over-expression and activation have been shown in numerous studies to have neuroprotective effects in both acute CNS injuries and chronic neurodegenerative diseases. Increased NAD+ levels in the cells cause AMP-activated protein kinase (AMPK) to deacetylate and alter the activity of SIRT1 targets downstream, including PPAR gamma co-activator 1 alpha (PGC-1α), a key regulator of the biogenesis of mitochondria (Kasiotis et al. 2013).

Thus, in mammalian cells, the regulation of mitochondrial activity and energy/redox status is closely linked to the AMPK/SIRT1/PGC-1α axis. One of the main sources of compounds that can enhance mitochondrial function is nature (Jing et al. 2013). Among them, polyphenols with positive effects on mitochondrial function have been well researched in vivo and in vitro. These include quercetin, epigallocatechin gallate, luteolin, and myricitrin (Caruana et al. 2016). Furthermore, proanthocyanidins, a class of polyphenolic bioflavonoids, moderate the production of ROS and mitochondrial superoxide while stimulating the MMP and RC complex IV to mitigate the effects of hydrogen peroxide-induced mitochondrial dysfunction. Hesperidin has also been shown to improve the enzymatic capability of the mitochondrial complex I–IV (Mohamed et al. 2014).

Conversely, resveratrol has the potential to positively impact the mitochondria either directly or indirectly. It helps in the regulation of genes involved in antioxidant defense, mitochondrial biogenesis, and oxidative phosphorylation, as well as cell signaling pathways, to modify mitochondrial function. It specifically controls the levels of uncoupling protein 2 (UCP2) in mitochondrial membranes and the expression of antioxidant enzymes like manganese dependent SOD. These processes are critical for the redox maintenance of the organelle and may be mediated by SIRT1 activation. Additionally, it has been demonstrated in various experimental settings (in vitro and in vivo) that resveratrol-induced neuroprotection is mostly dependent on the AMPK–SIRT1 pathway (Soni and Salh 2012).

Additionally, by activating the AMPK/Sirt1/PGC-1α axis, the co-administration of above natural compounds corrected the significant detriment of mitochondrial biogenesis and bioenergetics in hippocampus progenitor cells from an animal model of Down syndrome. In an HD model, viniferin, a stilbene resveratrol dimer, was having a neuroprotective potential that was mediated by the AMPK/SIRT3 cascade (Ataie et al. 2016). Natural substances have also been investigated with regard to neurotoxicity generated by antineoplastic agents. Co-administration of quercetin and curcumin reduced the peripheral neurotoxicity caused by oxaliplatin by restoring glutathione S-transferase, glutathione peroxidase, and SOD activity (Camarena and Wang 2016). Diabetic complications such as neurodegeneration are linked to methyl-glyoxal, a significant precursor of advanced glycation end products that primarily targets MT proteins. Numerous investigations, both in vitro and in vivo, have demonstrated that flavonoids may suppress the production of advanced glycation end and avert diabetes neurodegeneration. Furthermore, it was demonstrated that myricitrin mitigated MGO-induced mitochondrial dysfunction, potentially via nuclear factor kappa B/AGEs/receptor for AGEs (RAGE)/pathway regulation (Vissers et al. 2011).

3.6.2 Apoptosis

Patients with a variety of neurodegenerative disorders frequently experience apoptotic neuronal death, which can be intrinsic (mediated by mitochondria) or extrinsic (mediated by death receptors) and is regulated by a number of proteins. The Bcl-2 protein family is essential for controlling the intrinsic pathway because it senses the permeability of the mitochondrial membrane and the release of cytochrome c, a pro-apoptotic chemical that triggers CPS-9 activation. Members of the Bcl-2 protein family either induce or inhibit programmed cell death (BclxL, Bcl-w, Mcl-1, and Bcl-2) or both (Bax, Bad, Bcl-rambo, and Bak) (Goswami et al. 2014).

The distribution of these components dictates whether a cell will go through apoptosis or not. However, stress conditions that impact the folding of proteins in the ER lumen can also begin the caspase cascade at the endoplasmic reticulum (ER). Finally, a number of stress-sensing transcription factors are activated and linked to the pathophysiology of AD, in addition to mitochondria and ER-resident proteins. Many naturally occurring substances have been demonstrated to have neuroprotective effects by directly altering pathways involved in programmed cell death (Ahmad et al. 2016). The primary terpenoid found in *Ginkgo biloba* leaves, bilobalide, had strong neuroprotective effects as well as early prevention of ROS-induced apoptosis and attenuation of CPS-3, Bax, and p53, and also an increase in PC12 cells (Shah et al. 2015). Additionally, bilobalide altered mitochondrial activity by upregulating the component of cytochrome c oxidase. It has been demonstrated that the sesquiterpene alkaloid huperzine A inhibits apoptotic factors such p53, CPS-3, and Bax. Through cleaving poly ADP-ribose polymerase, decreasing ROS production, and blocking CPS-12 and -3 activation, the bioflavonoid apigenin eventually stopped ER stress-induced apoptosis (Harrison 2012).

3.6.3 Excitotoxicity

In the CNS, which includes the retina, glutamate is a major excitatory neurotransmitter. It is momentarily released from the presynaptic terminals and binds to a couple of receptor-linked ports in the post-synaptic membrane to cause an inflow of calcium ions and the start of an action potential. Kainite, N-methyl-D aspartate receptors, and α-amino-3-hydroxy-5-methyl-4-isoxazolepropionic acid are the names of three kinds of glutamate-gated ion channels (Du et al. 2021). Because of the high Ca level in the cytosol, glutamate can actually drive cells to apoptotic cell death when present in excess or for prolonged periods of time. Moreover, ineffective glutamate clearance or even incorrect glutamate release during acute and chronic neurodegenerative diseases might result in altered energy metabolism, raising cytosolic Ca concentrations and glutamate levels. In conclusion, excite-toxicity may occur even at normal glutamate levels if the NMDA receptor is triggered (Li et al. 2019). Numerous brain disorders have been linked to glutamate receptor-mediated excitotoxicity. However, several researches have demonstrated that inhibiting both NMDA and non-NMDA receptors at the same time provides the best protection against ischemic neurodegeneration. Natural substance kaitocephalin, which was taken out from the fungus *Eupenicillium shearii*, has been demonstrated to shield central nervous system neurons from excite-toxicity and to be a strong antagonist of specific glutamate receptor subtypes (AMPA and NMDA, but not kainate) (Shi et al. 2021).

Moreover, myricetin prevented glutamate-induced excitotoxicity in neurons by influencing the phosphorylation of the NMDAR receptor (NMDAR), which in turn reduced intracellular Ca overload. By interacting with it, myricetin also reduced the formation of ROS generated by glutamate and the activation of caspase-3. Acacetin has been shown to effectively inhibit kainic acid-induced in vivo excitotoxicity by inhibiting glutamate release from rat hippocampal synaptosomes and decreasing voltage-dependent Ca entry (Yang et al. 2021).

3.6.4 Inflammation

Neurodegenerative disorders have been linked to neuronal damage and death, as evidenced by chronic inflammation in the CNS. Activated microglia are believed to play a major role in neuro-inflammation and to have a role in neuronal death by generating reactive oxygen species (ROS) and reactive nitrogen species (RNS). The deposits of fibrillar β-amyloid peptide (Aβ) are also important in tissue neuro-inflammation. The dimeric, sequence-specific transcription factors known as NF-kB/Rel proteins are responsible for activating a remarkable amount of genes with respect to stressful stimuli like lipopolysaccharide, TNF-α, and other situations that need quick changes in gene expression (Qiu et al. 2019). In unstimulated cells, NF-B is linked to inhibitory I-kB proteins and sequestered, inactive, in the cytoplasm. Following induction, I-kB is phosphorylated, ubiquitinylated, and eventually degraded by proteases. Following its release and translocation to the nucleus, NF-kB stimulates the transcription of specific genes of concern. NF-kB regulates the expression of multiple genes connected to OS and inflammation, such as COX-2, and boosts the manufacture of numerous facilitators, including NO (Zhang et al., 2019a, b, c, d).

The initiation and progression of neurodegenerative diseases are closely linked to the brisk of NF-kB, especially in the CNS, which has been demonstrated to initiate multicellular responses. These responses include the production of FRs in glial cells and the transactivation of inflammatory molecules (Wang et al. 2018a, b, c).

However, a number of factors, including cell type, stage of development, signal type, and type of activated NF-kB dimer, may ultimately determine how NF-kB activation affects neuronal death and survival. Anti-inflammatory drugs have been suggested as a way to lessen the summative effects of neuro-inflammation in the brain since, although it may not always be the main cause, inflammation does play a role in the ongoing loss of CNS neurons. It has been shown that flavonoids reduce inflammation by hindering with a wide range of inflammatory molecular pathways, including those that trigger microglia activation and the production of pro-inflammatory cytokines (Sui et al. 2019).

Daidin along with other isoflavones are believed to be neuroprotective because of their ability to block the oxidative debut of the p38 mitogen-activated protein kinase (MAPK)-NF-kB pathway, thereby reducing the onset of microglial production and resulting release of pro-inflammatory factors. Furthermore, it has been observed that flavonoids, such as EGCG, partially regulate T cell response by inhibiting NF-kappa B activation. Furthermore, EGCG-treated transgenic mice with amyotrophic lateral sclerosis showed reduced microglial activation and enhanced neuron survival (Zhang et al. 2021).

Numerous investigations have demonstrated that flavonoids downregulate NF-kappa B signal transmission by either activating extracellular signal-related ERK 1/2 or inhibiting the activity of Ikappa B kinase (Chu et al. 2019). Given that COX-2 is produced via the NF kappa B-treated pathway, COX-2 production may be another significant target of flavonoids that contributes to their anti-inflammatory properties. It has been discovered that a large number of flavonoids are effective COX inhibitors. However, it has been shown that several polyphenols and catechins influence the expression of nuclear NF-kappa B and chromatin remodeling via modifying the activities of DNA methyl transferase and histone deacetylase (Chen et al. 2019).

Interestingly, research on curcumin has demonstrated that it can also function as an agonist of the PPAR-γ to suppress NF-kB signaling. Flavonoids have also been demonstrated to prevent human monocytes and astrocytes from producing inflammatory cytokines in response to lipopolysaccharide. Furthermore, they have been shown to impede the function of kinases that are crucial collaborators in inflammatory signal transduction pathways, including p38 MAPK, stress-activated protein kinase, and protein kinase C (Li et al. 2017a, b, c).

3.6.5 Oxidative Stress

The detrimental effects of neuro-toxicants are mediated by oxidative stress, which is also acknowledged as a critical component in a number of neurodegenerative illnesses and as a mechanism for age-related degenerative processes. When ROS build up in cells as a result of either excessive production or inadequate neutralization as a result of a shortcoming between anti-oxidant defense systems and ROS production, OS results in damage to proteins, lipids, and DNA. Production of RNS and ROS can cause neuronal cell death and irreversible harmful alteration of macromolecules like proteins. Because the brain cannot effectively negate the effects of these reactive species, it is susceptible to their accumulation (Zheng et al. 2019).

To be more precise, it maintains high levels of ascorbic acid, high aerobic metabolism, and concentrations of transition metals, all of which contribute to the production of RNS/ROS and oxidized products. On the other hand, it lacks a functional antioxidant defense system, which is indicated by low glutathione levels and moderate activity of antioxidant enzymes like catalase and SOD. Ultimately, a variety of neurotransmitters undergo autoxidation, producing ROS and RNS such as NO. In inclusion to these, inflammation brought on by activated microglia also plays a role in OS in neurodegenerative illnesses (Zhong et al. 2019).

Through the ARE, Nrf2 is a critical transcription factor involved in the transcriptional brisk of phase II detoxifying enzymes. Additionally, it has been documented that Nrf2 initiation in microglia and macrophages, suppresses inflammatory responses linked to NF-kappa B. Strong antioxidants found in natural chemicals have drawn a lot of attention; polyphenols have been the subject of the greatest research (Li et al. 2020). By using a variety of scavenging assays, phenolic substances including flavonoids, ferulic acid, rosmarinic acid, chlorogenic acid, caffeic acid, vanillic acid, p-hydroxybenzoic acid, p-coumaric acid, and protocatechuic acid were found to contribute to the antioxidant potential (Wang et al. 2020).

Antioxidants made from nature function via a variety of pathways, including electron donation, hydrogen atom disposal, metal the chelation process, direct radical scavenging, antioxidant enzyme activation, singlet oxygen quenching, and Nrf2 pathway activation. They also boost the efficacy of antioxidant regulation of genes (Zhang et al. 2018).

3.6.6 Protein Misfolding

A characteristic of most neurodegenerative illnesses is similar. In the afflicted tissue, an accumulation of an insoluble protein occurs. For every specific illness, the proteins' identities and the locations of the deposits are distinctive (Zhang et al., 2019a, b, c, d). Nonetheless, under healthy settings, most of the impacted proteins or its fragments are unfolded partially, and they all have a common intra-molecular cross-beta sheet outline that results in the production of structures that are insoluble and fibrillated. The procedure of seeded polymerization, which is akin to 1-D crystal development, is the mechanism by which amyloid is formed in protein misfolding disorders. Clusters of a specific size or shape catalyze and stabilize the addition of monomers to produce large fibrillar clusters (Wei et al. 2017a, b).

Extracellular protein deposits in AD are the result of processing of the amyloid precursor protein, while intraneuronal deposits in the form of neurofibrillary tangles are created by the tau protein associated with microtubules that are highly phosphorylated (Liu et al. 2018a, b). While tau, a highly phosphorylated protein linked with microtubules, forms intraneuronal deposits in the form of neurofibrillary tangles, AD is defined by deposits, the other common protein misfolding illness, PD, is caused by intra-neurial deposits. Alpha-synuclein protein predominates in deposits found in dopaminergic neurons of the substantia nigra in PD, another common protein misfolding disorder. Tau and Aβ are secondary components (Dong et al. 2021).

Through its antioxidant action and facilitation of APP breakdown into non-amyloidogenic peptides, polyphenolic extract demonstrated to reduce the generation of the deposits of amyloid and to greatly mitigate cognitive impairment in a mouse model of ADHD. Additionally, it has been observed that by binding directly to a number of proteins linked to illnesses involving misfolding of proteins, EGCG inhibits their fibrillization (Gui et al. 2013). Additionally, it encourages the development of spherical, stable, and non-cytotoxic aggregates that do not catalyze the creation of fibrils and have a lower β-sheet concentration than fibrils. It has been observed that orcein derivatives deplete the oligomeric and protofibrillar forms of the AD-related Aβ peptide in addition to speeding up the peptide's fibril formation (Feng et al. 2021).

The leaves and drupes of Olea europaea contain the phenylethanoid oleuropein aglycone moiety, which is produced when oleuropein is deglycosylated. By encouraging fibril and plaque disaggregation and reducing the production of harmful oligomers during the amyloid aggregation of the Aβ peptide, it has been proven to ameliorate AD-like symptoms in a number of model species (Wang et al. 2016a, b).

In vitro, Olea europaea exhibits anti-tau and anti-human islet amyloid polypeptide (hIAPP) aggregation properties. Ultimately, Aβ aggregation was considerably suppressed by RSV derivatives, which also had an antioxidant activity and disintegrated fibrils produced by Cu (II)- and self-induced Aβ aggregation (Zhou et al. 2017a, b).

3.6.7 Autophagy

The multi-step process of autophagy involves the creation of autophagosomes, which are double-membrane structures that merge with lysosomes. Hydrolytic enzymes then break down the contents of these structures, which include misfolded proteins and cellular metabolic waste (Li et al. 2017a, b, c). Under physiological settings, autophagy plays a critical role in mitophagy, the process that involves mitochondrial turnover (Li et al. 2021a, b). The CNS's integrity is fundamentally dependent on normal basal autophagy. Consequently, increasing autophagy in the central nervous system would result in neuroprotection, which may account for the beneficial properties of many substances. It has been demonstrated that quercetin-induced autophagy reduces the harm that high hyperglycemia causes to Schwann cells (Mo et al. 2020). It was also discovered that in a mouse model of Aβ deposition, Olea europaea reduced inflammation and induced autophagy. Resveratrol was also demonstrated lately to have an impact on mitochondrial fission and mitophagy as well (Dai et al. 2021).

3.6.8 Epigenetics

The normal development and maintenance of adult life depend on epigenetic process such genomic imprinting, DNA methylation, microRNA regulation, and histone modification. Dysregulation of these mechanisms may also increase the complexity and vulnerability of neurodegenerative disorders. In fact, AD-related molecular pathways require DNA methylation (Liu et al. 2018a, b). Numerous natural substances, such as alkaloids, polyphenols, organosulfur, and terpenoids compounds, have been shown to have the ability to modify histone modifications and DNA methylation, which can result in the activation or silencing of certain genes. These substances may be crucial in regulating these epigenetic modifications. Numerous polyphenols exhibit their neuroprotective qualities via activating HDACs, while anti-inflammatory polyphenols have been shown to alter NF-kB and remodeling of chromatin network through modulation of both DNMTs and HDACs (Xiong et al. 2016).

3.7 Other Targets

It has been demonstrated that neurodegenerative disorders are associated with a decrease in the activity of cholinergic neurons. Galantamine, an alkaloid that is used to treat many symptoms, activates muscarinic acetylcholine receptors, which has a neuroprotective effect. It has been reported that flavonoids and other naturally occurring substances inhibit the mitochondrial enzymes known as monoamine oxidases which has been linked to a number of processes that lead to neurodegeneration, including OS, neuro-inflammation, apoptosis triggers, and glial activation. Some of the case and clinical studies have been summarized in Table 3.1 indicating therapeutic potential of the given below natural compounds.

3.7.1 1-Methyl-1, 2, 3, 4-Tetrahydroisoquinoline

1MeTIQ exhibits a broad, brain-modulating effect. It demonstrated antidepressant-like, neuroprotective, and anti-addictive qualities. Reports state that 1MeTIQ promotes the production of neuro-trophic factors (brain-derived neuro-trophic factor and nerve-growth factor) (Hongyun et al. 2017). It is important to emphasize that its primary and unique impact on the metabolism of monoamines is closely linked to the CNS neuroprotection induced by 1MeTIQ (Wang et al. 2018a, b, c). Both of these mechanisms shield dopaminergic cells in particular from FR damage. In addition, 1MeTIQ's structure allows it to comb FRs (Tao et al. 2017). 1MeTIQ is an effective inhibitor of the MAO-dependent oxidative pathway that is reversible and both MAOA and MAOB. Furthermore, 1MeTIQ stimulates the O-methylation pathway that is dependent on COMT (Pengyue et al. 2017).

Consequently, this substance has the potential to be an antioxidant. 1MeTIQ inhibits the production of FRs and decreases the Fenton reaction's production of H_2O_2 from DP. As previously noted, after chronic treatment, there was no evidence of resistance to its neuroprotective qualities (the degree of impact is equivalent to that of a single dosage of 1MeTIQ) (Cheng et al. 2012). Certain results suggested that distinct MAO inhibitors, such rasagiline and selegiline, mitigated MPTP toxicity by obstructing MAOB activity, yet both medications demonstrated neuro-trophic, antioxidant, and anti-glutamatergic effects (Li et al. 2012). Clinical research showed that the beneficial effects of early MAO inhibitor treatment were greater than those of later therapy introduction. Our earlier findings showed that 1MeTIQ displaces the DA receptor agonist [3H] apomorphine from DA D2 receptors (Jiang 2017).

Furthermore, 1MeTIQ inhibited calcium influx and stopped glutamate-induced cell death. The aforementioned results indicates that 1MeTIQ exhibits antagonistic tendencies towards the glutamatergic system and can reduce excitotoxicity. However, its affinity for the NMDA receptor was approximately 1000 times weaker at μmolar doses compared to MK-80. In summary, 1MeTIQ has the capacity to halt numerous neurotoxic processes, including inhibiting MAO, scavenging FRs, and inhibiting glutamate-induced excitotoxicity (Hou et al. 2018).

Table 3.1 Summarized effects and mechanisms of several natural products on different targets related to neurodegenerative diseases

S/N#	Drug	Model	Dose/ Conc.	Effects	Mechanism	References
1.	1MeTIQ	Neurons OGD/R model	10, 20, 40, 60, and 80 mmol/L	Dec. caspase-3, LDH Inc. SOD	ROS scavenger, antioxidant, MAO inhibitor	Pengyue et al. (2017)
		Rats MCAO/R model	25 mg/kg, i.p.	Inc. p-AMPK, ATP, SIRT1, and cAMP	Regulating the AMPK/cAMP/SIRT1 pathway	Cheng et al. (2012)
		Cortical neuron OGD/R model	1, 5, and 25 μmol/L	Inc. GAP-43, shh, Ptc-1, Gli-1, Sirt1	Inc. neurite outgrowth and synaptogenesis	Li et al. (2012)
2.	Resveratrol	Rats MCAO/R model	35 mg/kg, i.p.	Inc. neurological function Dec. neuronal damage	ROS scavenger, antioxidant, GSH and SOD enhancer	Hou et al. (2018)
		Mice MCAO model	35 mg/kg, i.p.	Dec. CD32 and IL-1β Inc. CD206 and arginase-1	Inc. M2 polarization of microglia Dec. neuro-inflammation	Ma et al. (2020)
		Rats MCAO/R model	35 mg/kg, i.p. 45 μmol/L	Dec. infarct volume Inc. neurological function	Inc. shh signaling pathway	Yu et al. (2017)
3.	Curcumin	Mice dMCAO model LPS and IFN-γ-induced BV2 microglia	155 mg/kg, i.p. 12.5 and 30 μmol/L	Dec. cerebral ischemic damage Inc. M2 microglia/macrophage markers	Inc. M2 microglial polarization Dec. microglia-mediated pro-inflammatory responses	Liu et al. (2017)
		Rats MCAO model	200 mg/kg, i.p.	Dec. brain infarction Inc. neurological function	Mediating Akt/PI3K/mTOR pathway regulating TLR4/MAPK/p38 pathway	Huang et al. (2018)
		Rats MCAO model	300 mg/kg, i.p.	Inc. neurobehavioral deficits Inc. BrdU/DCX-positive cells	Inc. notch signaling pathway	Xie et al.

(continued)

Table 3.1 (continued)

S/N#	Drug	Model	Dose/ Conc.	Effects	Mechanism	References
4.	Vitamin C	Microglia and neurons OGD/R model	0.20, 0.5, and 2 mg/ L	Dec. IL-1β, IL-6, and TNF-α Dec. TUNEL-positive cells	Dec. IRE1α-XBP1 pathway	Lu et al. (2018)
		Rats MCAO model	65 mg/kg, i.g.	Dec. infarction volume Inc. VEGF and BDNF	Inc. angiogenesis and neurogenesis	Miao et al. (2016)
		Rats MCAO/R model	65 mg/kg, i.g.	Dec. infarction volume Inc. VEGF and BDNF	Inc. angiogenesis and neurogenesis	Li et al. (2016a, b)
5.	*Gingko biloba*	Mice tMCAO model	1, 3, and 6 mg/kg, i.p.	Dec. neurological deficits, water content and infarct volume Dec. caspase-3	Inc. BDNF level	Xu et al. (2017)
		Mice tMCAO model	1.7, 3.5, and 7.5 mg/kg, i.p. 0.3, 1 and 10 uM	Dec. cerebral ischemic Dec. CCL3	Modulating macrophage and microglia polarization	Wei et al. (2017a, b)
		Rats MCAO model	1, 3, and 6 mg/kg, i.p. 30 mg/L	Dec. infarct volume Dec. ROS Inc. SOD, HO-1	Inc. Nrf2/Akt pathway	Shu et al. (2016)

6.	Glycosides	Mice photothrombosis model	3 mg/kg, i.v. 100 nM	Dec. cognitive deficits Inc. hippocampal	Dec. IL-17 expression	Wang et al. (2019)
		Rats MCAO model	25 mg/kg, i.p. 60, 100, 150, and 200 μmol/L	Dec. neurological score Dec. infarct volume Inc. cell viability balance	Dec. apoptosis	Zhu et al. (2019a, b)
		Mice photochemical brain ischemia model NSCs	250 mg/kg, i.v. 25 μM	Inc. numbers of DCX/BrdU Dec. numbers of NeuN/TUNEL	Inc. p-Akt and p-PI3K Inc. GSK-3β/Akt pathway	Dong et al. (2016)
7.	Flavonoids	Microglia and neurons OGD/R model	0.3, 0.5, and 1 mg/ L	Dec. IL-1β, IL-6, and TNF-α Dec. TUNEL-positive cells	Dec. IRE1α-XBP1 pathway	Dong et al. (2016)
		Rats MCAO model	65 mg/kg, i.g.	Dec. infarction volume Inc. VEGF and BDNF	Inc. neurogenesis and angiogenesis	Zhu et al. (2019a, b)
		Rats MCAO/R model	65 mg/kg, i.g.	Dec. infarct volume Dec. TNF-α, IL-6, cleaved-caspase 3	Dec. NF-κB activation	Zhu et al. (2019a, b)

(continued)

Table 3.1 (continued)

S/N#	Drug	Model	Dose/ Conc.	Effects	Mechanism	References
8.	Phenols	Rats MCAO/R model	100 and 300 mg/kg, i.p.	Dec. cerebral infarction Inc. cognitive dysfunction	Regulating RELA p65/miR-7 axis	Xu et al. (2019)
		Rats MCAO/R model	30 mg/kg, i.p.	Dec. brain edema and water content dec. TNF-α and IL-6	Dec. inflammation Inc. mitochondrial dysfunction	Zhang et al. (2017a, b, c)
		Cortical neurons OGD/R model	0.5–10 μM		Regulating ERK and flotillin-1 pathway	Lu et al. (2018)
9.	Terpene	Rats electrocoagulation model	3, 5, and 7 mg/kg, i.p. 0.5, 1, 10, and 100 μg/mL	Dec. neurological score Inc. neuronal cell activity	Inc. axonal regeneration	Wang et al. (2019)
		Rats pMCAO model	5.5 and 10 mg/kg	Dec. neurological deficits Inc. BrdU+-nestin+ and BrdU+-Tuj-1+	Inc. BDNF signaling	Dong et al. (2016)
		Rats MCAO model	3, 5, and 10 mg/kg, i.v. 30, 60, and 100 μM	Dec. neurological deficit scores and infarct volume Inc. CD31+/EDU+	Inc. VEGF/HIF-1α pathway	Zhu et al. (2019a, b)

Abbreviation: *Inc.* increase, *Dec.* decrease

3.7.2 Resveratrol

Resveratrol's capacity to readjust antioxidant enzymes, such as GSH-Px and SOD, is linked to its antioxidant qualities. Resveratrol affected the expression of genes related to redox pathways in the ischemia model by upregulating the transcription factor erythroid 2 (Nrf2) (Ma et al. 2020). According to certain research, resveratrol protected neuroblastoma cells from amyloid beta and beta-metal complex exposure by acting as a scavenger of FRs (Yang et al. 2018). Resveratrol showed protective qualities for the mitochondrial respiration process, according to an in vitro investigation. Through the activation of sirtuin 1, it had a significant impact on mitochondrial biogenesis and ROS generation (SIRT1) (Wan et al. 2016). AMP-activated protein kinase (5′ AMPK) is activated in response to SIRT 1 stimulation (Tang et al. 2017). Moreover, resveratrol alters respiratory chain complexes; that is, at low concentrations, it stimulates ATP synthase enzyme activity, whereas at high concentrations, it inhibits it. Mice treated with resveratrol showed a significant reduction in MPTP-induced toxicity. It also avoided DP decrease in the mouse brain and shielded striatal neurons from overproduction of hydroxyl radicals (Shen et al. 2016). Resveratrol corrected the cellular redox status in the dopaminergic neurons and increased the activity of antioxidant enzymes in the 6-OHDA model of PD (Lei and Chen 2018).

Resveratrol's neuroprotective mechanism of action is incredibly intricate. Numerous cellular factors are triggered by it, including nuclear factor erythroid-derived 2 (Nrf2), AMP-activated kinase (AMPK), and SIRT (Li et al. 2016a, b). Resveratrol has the capacity to directly scour FRs due to its molecular structure. Moreover, resveratrol therapy reduced lipid peroxidation in several rodent models by raising GSH levels and lowering NO concentrations. Numerous research showed that using resveratrol administration reduced inflammation by lowering microglial initiation and release of pro-inflammatory cytokines inhibition (Liu et al. 2017).

3.7.3 Curcumin

Highly lipophilic, curcumin readily penetrates the blood-brain barrier. Nevertheless, this substance's bioavailability is not very good. It's interesting to note that this molecule inhibits the aggregation of amyloid-β peptides by binding to plaques. According to several scientists, curcumin's potent antioxidant qualities reduced the levels of intracellular calcium, reduced the increase in tau hyperphosphorylation, and reduced the amount of amyloid-β. Curcumin therapy concurrently increased the amounts of antioxidant enzymes in PC12 cell culture (Huang et al. 2018). Furthermore, curcumin was shown to improve cell survival in another in vitro investigation by lowering ROS generation and obstructing pro-apoptotic signals. According to an in vivo investigation, curcumin prevented cerebral ischemia in the rat brain in a model of cerebral artery closure. Additionally, curcumin dramatically inhibited peroxidation of lipid and decreased the amounts of superoxide anion

(SOA) and malondialdehyde (MDA) in the hippocampus in the homocysteine model of neurotoxicity (Liu et al. 2016).

According to several scientists, curcumin blocked several inflammatory cascade processes, including JNK, protein-1 transcription, and iNOS activation. Moreover, curcumin demonstrated antioxidant activity through the generation of NO-related radicals: It prevented the NOS gene from being expressed. Curcumin's antioxidant qualities have been demonstrated multiple times in various animal models of cerebral ischemia, AD, and brain damage. Furthermore, in both the acute inflammatory and chronic AD models, this drug reduced IL-1β (Li et al. 2017a, b, c). Curcumin significantly inhibited the increase in iNOS protein and mRNA generated by LPS, as demonstrated by Weber and colleagues, by limiting the activity of transcription factors. Another crucial benefit of curcumin is its inhibition of fibril and oligomer production, which reduces plaque pathogenesis. Furthermore, curcumin inhibited ROS and β-secretase activity (Xie et al. 2018).

3.7.4 Vitamin C

Through collagen crosslinking and the creation of components of the basement membrane, vit. C is crucial for myelination. Numerous studies have documented how vit. C exhibits antioxidant qualities in several animal models by lowering lipid peroxidation and minimizing oxidative stress. Since ROS play a major role in neuroinflammation and neurodegeneration, vit. C therapy may prevent these conditions by inhibiting the production of ROS (Xu et al. 2019). Furthermore, β-carotene, GSH, and α-tocopherol are among the additional antioxidants that can be renewed by vit. C. According to some research, vit. C may be able to stop alcoholic or glutamate-induced neurotoxicity in rats. Additionally, when administered before paraquat, this molecule exhibited protective action against paraquat toxicity (Zhang et al. 2017a, b, c). The process behind this action has to do with vit. C's capacity to promote the redox cycling of ferrous and ferric metal ions. In colchicine-induced neurotoxicity and memory deficits, Sil and colleagues suggested that vit. C plays a neuroprotective effect through the regulation of several neuro-inflammatory variables (Lu et al. 2018). In those studies, the rat hippocampal neuro-inflammatory indicators (cytokines, ROS, and RNS) were decreased by vit. C therapy at 200 and 400 mg/kg. In contrast to rats given colchicine to produce AD, the greater dose of vit. C (600 mg/kg) resulted in an increase in the levels of neuro-inflammatory markers. The redox status of cells and concentration gradients appear to be important factors in these opposing actions of vit. C (Miao et al. 2016).

According to Schaus, the amount of vit. C in the cerebral cortex is substantially lower in those 80 years of age and older than in younger persons. Vit. C content is also relevant to neurodegenerative illnesses like AD and PD. Ide and colleagues found that lymphocyte vit. C levels were considerably lower in PD patients than in the control group (Li et al. 2016a, b). When given to senior PD patients, this chemical enhanced levodopa's bioavailability. By preventing the OS brought on by MPP+, vit. C has also demonstrated neuroprotective action in animal models of PD. Furthermore, several writers have established that preserving adequate vit. C levels has the ability to prevent AD.

3.7.5 Gingko biloba

A trademarked extract of *Gingko biloba* leaves (EGb-761) has demonstrated preventive qualities against vascular and neural damage, according to clinical and preclinical research (Fan et al. 2018). Ginkgo extracts have been shown in numerous clinical trials to exhibit therapeutic activity in treating a range of illnesses, including AD, age-related dementias, inadequate cerebral and ocular blood flow, and failing memory. Additionally, because gingko increases cerebral and peripheral circulation, it is used to treat peripheral vascular diseases and cerebrovascular dysfunctions. Evidence suggests that EGb-761 extract reduces apoptosis brought on by various stimuli, including hydroxyl radicals, serum deprivation, staurosporine, and olfactory nerve sectioning (Wang et al. 2016a, b).

Additionally, EGb 761 causes a rise in the release of catecholamines from its intracellular reserves. In the rat brain, long-term administration of EGb 761 causes a decrease in the density of cerebral β-adrenoceptors and an increase in noradrenaline release. *Gingko biloba* extract has demonstrated neuroprotective qualities against MPTP-induced damage. It prevents the striatal degeneration of dopaminergic neurons in mice. The neuroprotective effects of EGb 761 are attributed to its capacity to inhibit MAO-A and MAO-B activity as well as to prevent MPTP from being converted to MPP+ ions (Xu et al. 2017). Furthermore, EGb 761 therapy stopped the MPTP-induced decrease in tyrosine hydroxylase content. EGb 761's antioxidant potential was mentioned by certain writers. This extract lowers the production of FRs in developed neurons, avoiding the OS caused by $FeSO_4$ and H_2O_2.

3.7.6 Glycosides

One of the primary active components of *Astragalus membranaceus* is astragaloside IV (AST). Research indicates that its actions are well known for lowering cerebral infarction and encouraging the return of neurological function after brain ischemia or reperfusion. Additionally, a growing body of research suggests that AST has both in vitro and in vivo anti-brain ischemia properties. To assess AST's neuroprotective effect on ischemic stroke, for instance, rats were given intra-peritoneal injections of the protein (20 and 40 mg/kg) (Wei et al. 2017a, b).

AST enhanced microglia/macrophages M2 polarization, blocked neuronal cell death and parthanatos, lowered ROS production and Ca flow, and decreased the infarct volume and neurological impairments. According to recent in vitro research, AST therapy can shield cerebral cortex neurons exposed to OGD by modifying the CREB/PKA signaling pathway and maintaining MT activity. Additionally, it has been discovered that AST shields astrocytes from cell damage by blocking the CXCR4 receptor and reducing the p-JNK/JNK pathway's activation. Furthermore, a number of pharmacological investigations elucidating the mechanism of action of AST in the treatment of brain ischemia have demonstrated that the primary mechanism by which AST ameliorates brain ischemia is neurorestoration, a process that can stimulate the development of new neurons and blood vessels.

Elevation of the TrkB/BNDF signaling system, the downregulation of interleukin-17 production via the Wnt pathway, and the initiation of the Notch/VEGF/HIF pathway through miRNA-210 have been proposed as potential mechanisms behind this (Shu et al. 2016). The primary bioactive ingredient in the traditional Chinese herbal remedy Gastrodia is called gastrondin, or gas. It has demonstrated possible therapeutic and preventative benefits in ischemic stroke, reduces brain infarct volume and neurological behavioral impairments after microvascular collateral occlusion (MCAO) surgery, inhibits ROS production and inflammasome activation, and increases the expression of anti-inflammatory and anti-apoptotic proteins. This shows that the neuroprotective effects of gas therapy may be due to its anti-apoptotic, anti-inflammatory, and antioxidant bioactivities (Wang et al. 2019).

Furthermore, the HIF-1α, Akt/PI3K/mTOR, and VEGF signaling cascades are the mechanisms most directly linked to Rg1's pro-angiogenesis actions. The primary active component of Rhodiolae Crenulatae Radix et Rhizoma is salidroside, or sal. Its numerous pharmacological effects include memory enhancement and improved cerebral and cardiovascular system performance. Sal lowers cerebral ischemia/reperfusion injury and possesses anti-inflammatory, antiapoptotic, and pro-synaptogenesis characteristics (Zhu et al. 2019a, b). It also suppresses reactive astrogliosis and glial scar formation. Sal has a variety of biological effects, such as controlling the dopaminergic system, triggering the above signaling cascades, and blocking cyclin-dependent kinase 4 and Cyclin D1 in brain injury. According to recent research, Sal medication reduced the amount of brain glial scar tissue and lessened neuronal death. Sal also improved microglial phagocytosis and prevented the release of inflammatory cytokines produced by microglia (Dong et al. 2016).

3.7.7 Flavonoid

Baicalin (BA), a naturally occurring flavonoid with anti-inflammatory, anti-apoptotic, and antioxidant qualities, is derived from the roots of *Scutellaria baicalensis*. Numerous impacts of BA have been studied due to its outstanding activity. By suppressing ROS production, BA can boost mitochondrial membrane potential ($\Delta\psi$m), improve mitophagy, and drastically reduce infarct volume and neurological impairments. Notably, a study discovered that BA's anti-apoptosis action can reduce hypoxic-ischemic injury caused in the brains of newborn rats. Through the PI3K/PKB signaling pathway, it upregulates glutamate transporter 1, which mediates its neuroprotective effects (Zhu et al. 2019a, b).

The treatment of ICA (60 mg/kg) in a rat I/R model was seen to suppress the production of pro-inflammatory and proapoptotic proteins as well as I/R-induced NF-κB activation, indicating the anti-inflammatory and anti-apoptotic properties of ICA. Research has also demonstrated that ICA inhibits the IRE1α-XBP1 signaling pathway, thereby shielding neurons from ER stress-induced death following OGD/R damage. Apart from shielding against I/R damage, stem cell therapy plays a crucial role in uncovering the therapeutic mechanism of ICA (Gao et al. 2020).

Breviscapine is a flavonoid component that is isolated from *Erigeron breviscapus*. It improves microcirculation, dilates blood vessels, lowers blood viscosity, and increases heart's coronary, and cerebral blood flow. For clinical use, a 40-mg intravenous drip is used to administer the injections. According to clinical findings, platelet activating factor, plasma alpha granule membrane protein, and platelet aggregation rate can all be significantly decreased by its injection (Shen et al. 2016). It improves patient prognosis and quality of life by lowering the frequency of ischemic episodes and thus reducing cerebral infarction (Li et al. 2016a, b). To assess its preventive efficacy against ischemic stroke, intra-peritoneal injections at a dose of 20 mg/kg were commonly administered to rats.

It also contains a variety of bio-activities, comprising anti-inflammatory, anti-oxidative stress, and anti-apoptotic properties. It has been discovered that its therapy lowers the cerebral water content, infarct volume, and neurological score. Subsequent investigations shown that the administration of Bre therapy markedly attenuated the activation of NF-κB signaling and microglia, as well as the creation of ROS, pro-apoptosis protein expression, and inflammatory cytokine expression. It has also been demonstrated in a brain ischemia model to suppress astrocytic and neuronal autophagy. The LC3-I/LC3-II ratio is inhibited in vivo in part due to the anti-autophagy activity (Huang et al. 2018).

3.7.8 Phenols

The component of Salviae Miltiorrhizae Bunge that is the most soluble in water is called salvianolic acid B (SAB). According to current research, SAB suppresses inflammation to have strong neuroprotective benefits against brain injuries. Neu-N release from primary cortical neurons, pro-inflammatory cytokine responses, and NF-kB transcriptional activity were all observed to be improved by SAB treatment. In vivo, SAB also suppresses the expression of several interleukins, MCP-1, and ICAM-1. Additionally, it has been demonstrated that SAB therapy inhibits the activation of astrocytes and microglia and also reduces ROS generation. These findings imply that SAB protects brains by having a strong anti-inflammatory effect (Xie et al. 2018).

3.7.9 Terpene

Ginkgolide B is a naturally occurring antagonist of platelet activation factor receptors and a terpene lactone found in *Ginkgo biloba* extract (Xu et al. 2019). By altering the polarization of macrophages and microglia, it has beneficial anti-inflammatory effects on cerebral damage. In rat models of MCAO-induced SY5Y cell death and OGD/R-induced SY5Y cell death, it has been shown to significantly reduce ROS generation and infarct volume (Lu et al. 2018). Furthermore, it exhibits anti-apoptotic properties as evidenced by its ability to raise BDNF and decrease CPS-3 levels (Fan et al. 2018). *Rehmannia glutinosa* contains an iridoid chemical called catalpol.

Research on the pharmacological effects of *Rehmannia glutinosa* components, including catalpol, has demonstrated a range of effects, such as anti-cardiovascular, anti-diabetic, and anti-cerebrovascular illnesses (Xu et al. 2017).

Research on catalpol's pharmacological effects and mechanism of action has mostly concentrated on how it promotes neurorestoration to reduce ischemia in stroke patients (Wang et al. 2019). For instance, administering catalpol for 7 days considerably improved neurological deficiencies and induced neurogenesis, with the underlying processes supporting neonatal neuron survival through BDNF activation (Dong et al. 2016). It has also been claimed that catalpol enhances angiogenesis and safeguards the vascular structure. An in vitro investigation showed that in OGD-induced BMEC cells, catalpol-enhanced migration, proliferation tube formation, and VEGF expression. Catalpol has been consistently shown to improve stroke-induced STAT3 activation and cerebral blood flow, which in turn leads to the recovery of STAT3 binding to VEGF and the restoration of STAT3 activity. Furthermore, it was shown that the catalpol treatment increased the expression of p-S6 and GAP-43. Its neurorestorative effects are also facilitated by pro-axonal regeneration (Gao et al. 2020).

3.8 Conclusion

Neurodegenerative diseases are a complicated group of severe conditions defined by neuronal loss and increasing dysfunction. These cause a gradual but irreversible decline in brain function that affects over 30 million people globally. For a considerable amount of time, oxidative stress-induced cellular damage has been regarded as the primary cause of several neurodegenerative illnesses, including PD, AD, ALS, and HD. Unregulated reactive species damage proteins, lipids, DNA, and cellular structures during neurodegenerative processes, which ultimately result in brain cell apoptosis and degeneration. An increasing number of scientists have been looking for effective natural substances derived from medicinal plants to treat neurodegenerative illnesses in recent years. Most of these therapeutic herbs are well-known and have been utilized for millennia in conventional medicine. Because of their capacity to regulate the oxidative stress that contributes to neurodegeneration, natural substances have the potential to be neuroprotective. The natural compounds derived from medicinal plants that are effective in modifying the production of reactive species and mitochondrial function were outlined in this chapter. These naturally occurring compounds provided protection against neurotoxin-induced oxidative stress and cell apoptosis by decreasing the generation of ROS, lipid peroxidation, caspase-3 and caspase-9 protein activities, LDH release, the ratio of Bax/Bcl-2, Ca^{2+} influx, and cytochrome c release; increasing MMP; and restoring endogenous antioxidant enzymatic activities.

By regulating the oxidative cascades of several signaling pathways, they prevented mitochondria-dependent apoptotic pathways and displayed neuroprotective effects against cell damages and apoptosis. These days, a plethora of fresh discoveries have strengthened the case for the oxidative stress mechanism linking neurodegenerative diseases. For the purpose of neuro-regenerative medicine and the prevention and treatment of neurological disorders, the discovery of natural substances with antioxidant qualities is extremely valuable. The neuroprotective properties of natural compounds derived from medicinal plants are gaining prominence as disease-treating natural products receive more and more attention. More remarkable advancements in the field of neuroprotective natural chemicals are anticipated, opening up new avenues for the diagnosis, treatment, and prevention of neurobiological illnesses.

References

Ahmad A, Shah SA, Badshah H, Kim MJ, Ali TH, Yoon G et al (2016) Neuroprotection by vitamin C against ethanol-induced neuroinflammation associated neurodegeneration in developing rat brain. CNS Neurol Disord Drug Targets 15(3):360–370

Ahmed AH, Hamada M, Shinada T, Ohfune Y, Weerasinghe L, Garner PP, Oswald RE (2012) The structure of (−)-Kaitocephalin bound to the ligand binding domain of the (S)-α-amino-3-hydroxy-5-methyl-4-isoxazolepropionic acid (AMPA)/glutamate receptor, GluA2. J Biol Chem 287(49):41007–41013

Albarracin SL, Stab B, Casas Z, Sutachan JJ, Samudio I, Gonzalez J et al (2012) Effects of natural antioxidants in neurodegenerative disease. Nutr Neurosci 15(1):1–9

Alishahi M, Farzaneh M, Ghaedrahmati F, Nejabatdoust A, Sarkaki A, Khoshnam SE (2019) NLRP3 inflammasome in ischemic stroke: as possible therapeutic target. Int J Stroke 14(6):574–591

Ansari N, Khodagholi F (2013) Natural products as promising drug candidates for the treatment of Alzheimer's disease: molecular mechanism aspect. Curr Neuropharmacol 11(4):414–429

Ataie A, Shadifar M, Ataee R (2016) Polyphenolic antioxidants and neuronal regeneration. Basic Clin Neurosci 7(2):81

Ayissi VBO, Ebrahimi A, Schluesenner H (2014) Epigenetic effects of natural polyphenols: a focus on SIRT1-mediated mechanisms. Mol Nutr Food Res 58(1):22–32

Bagli E, Goussia A, Moschos MM, Agnantis N, Kitsos G (2016) Natural compounds and neuroprotection: mechanisms of action and novel delivery systems. In Vivo 30(5):535–547

Barthels D, Das H (2020) Current advances in ischemic stroke research and therapies. Biochim Biophys Acta Mol Basis Dis 1866(4):165260

Bastianetto S, Ménard C, Quirion R (2015) Neuroprotective action of resveratrol. Biochim Biophys Acta 1852(6):1195–1201

Cai Z, Zeng W, Tao K, Lu F, Gao G, Yang Q (2015) Myricitrin alleviates MPP+-induced mitochondrial dysfunction in a DJ-1-dependent manner in SN4741 cells. Biochem Biophys Res Commun 458(2):227–233

Camarena V, Wang G (2016) The epigenetic role of vitamin C in health and disease. Cell Mol Life Sci 73:1645–1658

Campbell BC, De Silva DA, Macleod MR, Coutts SB, Schwamm LH, Davis SM, Donnan GA (2019) Ischaemic stroke. Nat Rev Dis Primers 5(1):70

Cao K, Zheng A, Xu J, Li H, Liu J, Peng Y et al (2014) AMPK activation prevents prenatal stress-induced cognitive impairment: modulation of mitochondrial content and oxidative stress. Free Radic Biol Med 75:156–166

Carmichael ST, Kathirvelu B, Schweppe CA, Nie EH (2017) Molecular, cellular and functional events in axonal sprouting after stroke. Exp Neurol 287:384–394

Caruana M, Cauchi R, Vassallo N (2016) Putative role of red wine polyphenols against brain pathology in Alzheimer's and Parkinson's disease. Front Nutr 3:31

Chen ZQ, Mou RT, Feng DX, Wang Z, Chen G (2017) The role of nitric oxide in stroke. Med Gas Res 7(3):194–203

Chen H, Guan B, Chen X, Chen X, Li C, Qiu J et al (2018) Baicalin attenuates blood-brain barrier disruption and hemorrhagic transformation and improves neurological outcome in ischemic stroke rats with delayed t-PA treatment: involvement of $ONOO^-$-MMP-9 pathway. Transl Stroke Res 9:515–529

Chen J, Zhang X, Liu X, Zhang C, Shang W, Xue J et al (2019) Ginsenoside Rg1 promotes cerebral angiogenesis via the PI3K/Akt/mTOR signaling pathway in ischemic mice. Eur J Pharmacol 856:172418

Cheng O, Li Z, Han Y, Jiang Q, Yan Y, Cheng K (2012) Baicalin improved the spatial learning ability of global ischemia/reperfusion rats by reducing hippocampal apoptosis. Brain Res 1470:111–118

Chinta SJ, Ganesan A, Reis-Rodrigues P, Lithgow GJ, Andersen JK (2013) Anti-inflammatory role of the isoflavone diadzein in lipopolysaccharide-stimulated microglia: implications for Parkinson's disease. Neurotox Res 23:145–153

Choi DK, Koppula S, Suk K (2011) Inhibitors of microglial neurotoxicity: focus on natural products. Molecules 16(2):1021–1043

Chu SF, Zhang Z, Zhou X, He WB, Chen C, Luo P et al (2019) Ginsenoside Rg1 protects against ischemic/reperfusion-induced neuronal injury through miR-144/Nrf2/ARE pathway. Acta Pharmacol Sin 40(1):13–25

Cui Y, Wan Q (2019) NKT cells in neurological diseases. Front Cell Neurosci 13:245

Dai M, Chen B, Wang X, Gao C, Yu H (2021) Icariin enhance mild hypothermia-induced neuroprotection via inhibiting the activation of NF-κB in experimental ischemic stroke. Metab Brain Dis 36(7):1779–1790

De Oliveira MR, Nabavi SF, Habtemariam S, Orhan IE, Daglia M, Nabavi SM (2015) The effects of baicalein and baicalin on mitochondrial function and dynamics: a review. Pharmacol Res 100:296–308

De Oliveira MR, Nabavi SF, Manayi A, Daglia M, Hajheydari Z, Nabavi SM (2016) Resveratrol and the mitochondria: from triggering the intrinsic apoptotic pathway to inducing mitochondrial biogenesis, a mechanistic view. Biochim Biophys Acta Gen Subj 1860(4):727–745

Dong W, Xian Y, Yuan W, Huifeng Z, Tao W, Zhiqiang L et al (2016) Catalpol stimulates VEGF production via the JAK2/STAT3 pathway to improve angiogenesis in rats' stroke model. J Ethnopharmacol 191:169–179

Dong C, Wen S, Zhao S, Sun S, Zhao S, Dong W et al (2021) Salidroside inhibits reactive astrogliosis and glial scar formation in late cerebral ischemia via the Akt/GSK-3β pathway. Neurochem Res 46:755–769

Du Q, Zhu X, Si J (2020) Angelica polysaccharide ameliorates memory impairment in Alzheimer's disease rat through activating BDNF/TrkB/CREB pathway. Exp Biol Med 245(1):1–10

Du SJ, Zhang Y, Zhao YM, Dong YJ, Tang JL, Zhou XH, Gao WJ (2021) Astragaloside IV attenuates cerebral ischemia-reperfusion injury in rats through the inhibition of calcium-sensing receptor-mediated apoptosis. Int J Mol Med 47(1):302–314

Edelbrock AN, Àlvarez Z, Simkin D, Fyrner T, Chin SM, Sato K et al (2018) Supramolecular nanostructure activates TrkB receptor signaling of neuronal cells by mimicking brain-derived neurotrophic factor. Nano Lett 18(10):6237–6247

Fan Y, Luo Q, Wei J, Lin R, Lin L, Li Y et al (2018) Mechanism of salvianolic acid B neuroprotection against ischemia/reperfusion induced cerebral injury. Brain Res 1679:125–133

Feng L, Han F, Zhou L, Wu S, Du Y, Zhang D et al (2021) Efficacy and safety of Panax notoginseng saponins (Xueshuantong) in patients with acute ischemic stroke (EXPECT) trial: rationale and design. Front Pharmacol 12:648921

Fluri F, Schuhmann MK, Kleinschnitz C (2015) Animal models of ischemic stroke and their application in clinical research. Drug Des Devel Ther 9:3445–3454

Fu J, Jin J, Cichewicz RH, Hageman SA, Ellis TK, Xiang L et al (2012) trans-(−)-ε-Viniferin increases mitochondrial sirtuin 3 (SIRT3), activates AMP-activated protein kinase (AMPK), and protects cells in models of Huntington disease. J Biol Chem 287(29):24460–24472

Gao J, Bai P, Li Y, Li J, Jia C, Wang T et al (2020) Metabolomic profiling of the synergistic effects of ginsenoside Rg1 in combination with neural stem cell transplantation in ischemic stroke rats. J Proteome Res 19(7):2676–2688

Goswami AR, Dutta G, Ghosh T (2014) Effects of vitamin C on the hypobaric hypoxia-induced immune changes in male rats. Int J Biometeorol 58:1961–1971

Gui Q, Yang Y, Ying S, Zhang M (2013) Xueshuantong improves cerebral blood perfusion in elderly patients with lacunar infarction. Neural Regen Res 8(9):792–801

Haleagrahara N, Siew CJ, Ponnusamy K (2013) Effect of quercetin and desferrioxamine on 6-hydroxydopamine (6-OHDA) induced neurotoxicity in striatum of rats. J Toxicol Sci 38(1):25–33

Harrison FE (2012) A critical review of vitamin C for the prevention of age-related cognitive decline and Alzheimer's disease. J Alzheimers Dis 29(4):711–726

Hongyun H, Tao G, Pengyue Z, Liqiang Y, Yihao D (2017) Puerarin provides a neuroprotection against transient cerebral ischemia by attenuating autophagy at the ischemic penumbra in neurons but not in astrocytes. Neurosci Lett 643:45–51

Hou Y, Wang K, Wan W, Cheng Y, Pu X, Ye X (2018) Resveratrol provides neuroprotection by regulating the JAK2/STAT3/PI3K/AKT/mTOR pathway after stroke in rats. Genes Dis 5(3):245–255

Hu Q, Liu L, Zhou L, Lu H, Wang J, Chen X, Wang Q (2020) Effect of fluoxetine on HIF-1α-Netrin/VEGF cascade, angiogenesis and neuroprotection in a rat model of transient middle cerebral artery occlusion. Exp Neurol 329:113312

Huang L, Chen C, Zhang X, Li X, Chen Z, Yang C et al (2018) Neuroprotective effect of curcumin against cerebral ischemia-reperfusion via mediating autophagy and inflammation. J Mol Neurosci 64:129–139

Huang YG, Tao W, Yang SB, Wang JF, Mei ZG, Feng ZT (2019) Autophagy: novel insights into therapeutic target of electroacupuncture against cerebral ischemia/reperfusion injury. Neural Regen Res 14(6):954–961

Huo L, Bai X, Wang Y, Wang M (2017) Betulinic acid derivative B10 inhibits glioma cell proliferation through suppression of SIRT1, acetylation of FOXO3a and upregulation of Bim/PUMA. Biomed Pharmacother 92:347–355

Hwang O (2013) Role of oxidative stress in Parkinson's disease. Exp Neurobiol 22(1):11

Jha SK, Jha NK, Kar R, Ambasta RK, Kumar P (2015) p38 MAPK and PI3K/AKT signalling cascades in Parkinson's disease. Int J Mol Cell Med 4(2):67

Jiang DP (2017) Effects of breviscapine injection on platelet activity indexes in patients with transient ischemic attack. China Med Eng 25(12):50–52

Jing YH, Chen KH, Kuo PC, Pao CC, Chen JK (2013) Neurodegeneration in streptozotocin-induced diabetic rats is attenuated by treatment with resveratrol. Neuroendocrinology 98(2):116–127

Jing X, Wei X, Ren M, Wang L, Zhang X, Lou H (2016) Neuroprotective effects of tanshinone I against 6-OHDA-induced oxidative stress in cellular and mouse model of Parkinson's disease through upregulating Nrf2. Neurochem Res 41:779–786

Kasiotis KM, Pratsinis H, Kletsas D, Haroutounian SA (2013) Resveratrol and related stilbenes: their anti-aging and anti-angiogenic properties. Food Chem Toxicol 61:112–120

Kim DW, Lee KT, Kwon J, Lee HJ, Lee D, Mar W (2015) Neuroprotection against 6-OHDA-induced oxidative stress and apoptosis in SH-SY5Y cells by 5, 7-Dihydroxychromone: activation of the Nrf2/ARE pathway. Life Sci 130:25–30

Kim SM, Park YJ, Shin MS, Kim HR, Kim MJ, Lee SH et al (2017) Acacetin inhibits neuronal cell death induced by 6-hydroxydopamine in cellular Parkinson's disease model. Bioorg Med Chem Lett 27(23):5207–5212

Lei J, Chen Q (2018) Resveratrol attenuates brain damage in permanent focal cerebral ischemia via activation of PI3K/Akt signaling pathway in rats. Neurol Res 40(12):1014–1020

Li HY, Yuan ZY, Wang YG, Wan HJ, Hu J, Chai YS et al (2012) Role of baicalin in regulating Toll-like receptor 2/4 after ischemic neuronal injury. Chin Med J 125(09):1586–1593

Li YG, Zhu W, Tao JP, Xin P, Liu MY, Li JB, Wei M (2013) Resveratrol protects cardiomyocytes from oxidative stress through SIRT1 and mitochondrial biogenesis signaling pathways. Biochem Biophys Res Commun 438(2):270–276

Li Z, Fang F, Wang Y, Wang L (2016a) Resveratrol protects CA1 neurons against focal cerebral ischemic reperfusion-induced damage via the ERK-CREB signaling pathway in rats. Pharmacol Biochem Behav 146:21–27
Li W, Suwanwela NC, Patumraj S (2016b) Curcumin by down-regulating NF-kB and elevating Nrf2, reduces brain edema and neurological dysfunction after cerebral I/R. Microvasc Res 106:117–127
Li S, Sun X, Xu L, Sun R, Ma Z, Deng X et al (2017a) Baicalin attenuates in vivo and in vitro hyperglycemia-exacerbated ischemia/reperfusion injury by regulating mitochondrial function in a manner dependent on AMPK. Eur J Pharmacol 815:118–126
Li Y, Suo L, Liu Y, Li H, Xue W (2017b) Protective effects of ginsenoside Rg1 against oxygen-glucose-deprivation-induced apoptosis in neural stem cells. J Neurol Sci 373:107–112
Li W, Suwanwela NC, Patumraj S (2017c) Curcumin prevents reperfusion injury following ischemic stroke in rats via inhibition of NF-κB, ICAM-1, MMP-9 and caspase-3 expression. Mol Med Rep 16(4):4710–4720
Li Y, Yang Y, Zhao Y, Zhang J, Liu B, Jiao S, Zhang X (2019) Astragaloside IV reduces neuronal apoptosis and parthanatos in ischemic injury by preserving mitochondrial hexokinase-II. Free Radic Biol Med 131:251–263
Li S, Lu Y, Ding D, Ma Z, Xing X, Hua X, Xu J (2020) Fibroblast growth factor 2 contributes to the effect of salidroside on dendritic and synaptic plasticity after cerebral ischemia/reperfusion injury. Aging (Albany NY) 12(11):10951
Li L, Gan H, Jin H, Fang Y, Yang Y, Zhang J et al (2021a) Astragaloside IV promotes microglia/macrophages M2 polarization and enhances neurogenesis and angiogenesis through PPARγ pathway after cerebral ischemia/reperfusion injury in rats. Int Immunopharmacol 92:107335
Li C, Sui C, Wang W, Yan J, Deng N, Du X et al (2021b) Baicalin attenuates oxygen–glucose deprivation/reoxygenation–induced injury by modulating the BDNF-TrkB/PI3K/Akt and MAPK/Erk1/2 signaling axes in neuron–astrocyte cocultures. Front Pharmacol 12:599543
Liang C, Ni GX, Shi XL, Jia L, Wang YL (2020) RETRACTED: astragaloside IV regulates the HIF/VEGF/Notch signaling pathway through miRNA-210 to promote angiogenesis after ischemic stroke. Restor Neurol Neurosci 38(3):271–282
Lin TY, Huang WJ, Wu CC, Lu CW, Wang SJ (2014) Acacetin inhibits glutamate release and prevents kainic acid-induced neurotoxicity in rats. PLoS One 9(2):e88644
Lin CH, Nicol CJ, Cheng YC, Yen C, Wang YS, Chiang MC (2020) Neuroprotective effects of resveratrol against oxygen glucose deprivation induced mitochondrial dysfunction by activation of AMPK in SH-SY5Y cells with 3D gelatin scaffold. Brain Res 1726:146492
Liu X, Huang Z, Zou X, Yang Y, Qiu Y, Wen Y (2015a) Possible mechanism of PNS protection against cisplatin-induced nephrotoxicity in rat models. Toxicol Mech Methods 25(5):347–354
Liu H, Mao P, Wang J, Wang T, Xie CH (2015b) Allicin protects PC12 cells against 6-OHDA-induced oxidative stress and mitochondrial dysfunction via regulating mitochondrial dynamics. Cell Physiol Biochem 36(3):966–979
Liu S, Cao Y, Qu M, Zhang Z, Feng L, Ye Z et al (2016) Curcumin protects against stroke and increases levels of Notch intracellular domain. Neurol Res 38(6):553–559
Liu Z, Ran Y, Huang S, Wen S, Zhang W, Liu X et al (2017) Curcumin protects against ischemic stroke by titrating microglia/macrophage polarization. Front Aging Neurosci 9:233
Liu X, Wen S, Yan F, Liu K, Liu L, Wang L et al (2018a) Salidroside provides neuroprotection by modulating microglial polarization after cerebral ischemia. J Neuroinflammation 15:1–11
Liu D, Ye Y, Xu L, Yuan W, Zhang Q (2018b) Icariin and mesenchymal stem cells synergistically promote angiogenesis and neurogenesis after cerebral ischemia via PI3K and ERK1/2 pathways. Biomed Pharmacother 108:663–669
Liu Q, Jin Z, Xu Z, Yang H, Li L, Li G et al (2019) Antioxidant effects of ginkgolides and bilobalide against cerebral ischemia injury by activating the Akt/Nrf2 pathway in vitro and in vivo. Cell Stress Chaperones 24(2):441–452
Lu Z, Liu Y, Shi Y, Shi X, Wang X, Xu C et al (2018) Curcumin protects cortical neurons against oxygen and glucose deprivation/reoxygenation injury through flotillin-1 and extracellular signal-regulated kinase1/2 pathway. Biochem Biophys Res Commun 496(2):515–522

Luo L, Kim SW, Lee HK, Kim ID, Lee H, Lee JK (2018) Gastrodin exerts robust neuroprotection in the postischemic brain via its protective effect against Zn^{2+}-toxicity and its anti-oxidative effects in astrocytes. Anim Cells Syst (Seoul) 22(6):429–437

Ma S, Fan L, Li J, Zhang B, Yan Z (2020) Resveratrol promoted the M2 polarization of microglia and reduced neuroinflammation after cerebral ischemia by inhibiting miR-155. Int J Neurosci 130(8):817–825

Magalingam KB, Radhakrishnan A, Haleagrahara N (2014) Protective effects of flavonol isoquercitrin, against 6-hydroxy dopamine (6-OHDA)-induced toxicity in PC12 cells. BMC Res Notes 7:1–8

Miao Y, Zhao S, Gao Y, Wang R, Wu Q, Wu H, Luo T (2016) Curcumin pretreatment attenuates inflammation and mitochondrial dysfunction in experimental stroke: the possible role of Sirt1 signaling. Brain Res Bull 121:9–15

Miura T, Tanno M, Sato T (2010) Mitochondrial kinase signalling pathways in myocardial protection from ischaemia/reperfusion-induced necrosis. Cardiovasc Res 88(1):7–15

Mo ZT, Liao YL, Zheng J, Li WN (2020) Icariin protects neurons from endoplasmic reticulum stress-induced apoptosis after OGD/R injury via suppressing IRE1α-XBP1 signaling pathway. Life Sci 255:117847

Mohamed HE, El-Swefy SE, Hasan RA, Hasan AA (2014) Neuroprotective effect of resveratrol in diabetic cerebral ischemic-reperfused rats through regulation of inflammatory and apoptotic events. Diabetol Metab Syndr 6:1–14

Moridi Farimani M, Namazi Sarvestani N, Ansari N, Khodagholi F (2011) Calycopterin promotes survival and outgrowth of neuron-like PC12 cells by attenuation of oxidative-and ER-stress-induced apoptosis along with inflammatory response. Chem Res Toxicol 24(12):2280–2292

Newman DJ, Cragg GM (2016) Natural products as sources of new drugs from 1981 to 2014. J Nat Prod 79(3):629–661

Olatunji OJ, Feng Y, Olatunji OO, Tang J, Ouyang Z, Su Z (2016) Cordycepin protects PC12 cells against 6-hydroxydopamine induced neurotoxicity via its antioxidant properties. Biomed Pharmacother 81:7–14

Park HJ, Zhao TT, Lee KS, Lee SH, Shin KS, Park KH et al (2015) Effects of (-)-sesamin on 6-hydroxydopamine-induced neurotoxicity in PC12 cells and dopaminergic neuronal cells of Parkinson's disease rat models. Neurochem Int 83:19–27

Pengyue Z, Tao G, Hongyun H, Liqiang Y, Yihao D (2017) Breviscapine confers a neuroprotective efficacy against transient focal cerebral ischemia by attenuating neuronal and astrocytic autophagy in the penumbra. Biomed Pharmacother 90:69–76

Qin C, Zhou LQ, Ma XT, Hu ZW, Yang S, Chen M et al (2019) Dual functions of microglia in ischemic stroke. Neurosci Bull 35:921–933

Qiu CW, Liu ZY, Zhang FL, Zhang L, Li F, Liu SY et al (2019) Post-stroke gastrodin treatment ameliorates ischemic injury and increases neurogenesis and restores the Wnt/β-Catenin signaling in focal cerebral ischemia in mice. Brain Res 1712:7–15

Ren J, Yuan L, Wang W, Zhang M, Wang Q, Li S et al (2019) Tricetin protects against 6-OHDA-induced neurotoxicity in Parkinson's disease model by activating Nrf2/HO-1 signaling pathway and preventing mitochondria-dependent apoptosis pathway. Toxicol Appl Pharmacol 378:114617

Roberts RF, Tang MY, Fon EA, Durcan TM (2016) Defending the mitochondria: the pathways of mitophagy and mitochondrial-derived vesicles. Int J Biochem Cell Biol 79:427–436

Seok JY, Jeong YJ, Hwang SK, Kim CH, Magae J, Chang YC (2018) Upregulation of AMPK by 4-O-methylascochlorin promotes autophagy via the HIF-1α expression. J Cell Mol Med 22(12):6345–6356

Seto SW, Chang D, Jenkins A, Bensoussan A, Kiat H (2016) Angiogenesis in ischemic stroke and angiogenic effects of Chinese herbal medicine. J Clin Med 5(6):56

Shah SA, Yoon GH, Kim HO, Kim MO (2015) Vitamin C neuroprotection against dose-dependent glutamate-induced neurodegeneration in the postnatal brain. Neurochem Res 40:875–884

Shang Y, Wang X, Li F, Yin T, Zhang J, Zhang T (2019) rTMS ameliorates prenatal stress–induced cognitive deficits in male-offspring rats associated with BDNF/TrkB signaling pathway. Neurorehabil Neural Repair 33(4):271–283

Shao A, Lin D, Wang L, Tu S, Lenahan C, Zhang J (2020) Oxidative stress at the crossroads of aging, stroke and depression. Aging Dis 11(6):1537

Shen C, Cheng W, Yu P, Wang L, Zhou L, Zeng L, Yang Q (2016) Resveratrol pretreatment attenuates injury and promotes proliferation of neural stem cells following oxygen-glucose deprivation/reoxygenation by upregulating the expression of Nrf2, HO-1 and NQO1 in vitro. Mol Med Rep 14(4):3646–3654

Shi C, Zou J, Li G, Ge Z, Yao Z, Xu J (2011) Bilobalide protects mitochondrial function in ovariectomized rats by up-regulation of mRNA and protein expression of cytochrome c oxidase subunit I. J Mol Neurosci 45:69–75

Shi YH, Zhang XL, Ying PJ, Wu ZQ, Lin LL, Chen W et al (2021) Neuroprotective effect of astragaloside IV on cerebral ischemia/reperfusion injury rats through Sirt1/Mapt pathway. Front Pharmacol 12:639898

Shu ZM, Shu XD, Li HQ, Sun Y, Shan H, Sun XY et al (2016) Ginkgolide B protects against ischemic stroke via modulating microglia polarization in mice. CNS Neurosci Ther 22(9):729–739

Singh S, Singh TG (2020) Role of nuclear factor kappa B (NF-κB) signalling in neurodegenerative diseases: an mechanistic approach. Curr Neuropharmacol 18(10):918–935

Sommer CJ (2017) Ischemic stroke: experimental models and reality. Acta Neuropathol 133(2):245–261

Soni D, Salh B (2012) A neutraceutical by design: the clinical application of curcumin in colonic inflammation and cancer. Scientifica (Cairo) 2012(1):757890

Su XT, Wang L, Ma SM, Cao Y, Yang NN, Lin LL et al (2020) Mechanisms of acupuncture in the regulation of oxidative stress in treating ischemic stroke. Oxidative Med Cell Longev 2020(1):7875396

Sui Y, Bian L, Ai Q, Yao Y, Yu M, Gao H et al (2019) Gastrodin inhibits inflammasome through the STAT3 signal pathways in TNA2 astrocytes and reactive astrocytes in experimentally induced cerebral ischemia in rats. NeuroMolecular Med 21:275–286

Sun L, Han R, Guo F, Chen H, Wang W, Chen Z et al (2020a) Antagonistic effects of IL-17 and Astragaloside IV on cortical neurogenesis and cognitive behavior after stroke in adult mice through Akt/GSK-3β pathway. Cell Death Discov 6(1):74

Sun L, Zhang H, Wang W, Chen Z, Wang S, Li J et al (2020b) Astragaloside IV exerts cognitive benefits and promotes hippocampal neurogenesis in stroke mice by downregulating interleukin-17 expression via Wnt pathway. Front Pharmacol 11:421

Tang F, Guo S, Liao H, Yu P, Wang L, Song X et al (2017) Resveratrol enhances neurite outgrowth and synaptogenesis via sonic hedgehog signaling following oxygen-glucose deprivation/reoxygenation injury. Cell Physiol Biochem 43(2):852–869

Tao J, Cui Y, Duan Y, Zhang N, Wang C, Zhang F (2017) Puerarin attenuates locomotor and cognitive deficits as well as hippocampal neuronal injury through the PI3K/Akt1/GSK-3β signaling pathway in an in vivo model of cerebral ischemia. Oncotarget 8(63):106283

Tian YS, Zhong D, Liu QQ, Zhao XL, Sun HX, Jin J et al (2018) Upregulation of miR-216a exerts neuroprotective effects against ischemic injury through negatively regulating JAK2/STAT3-involved apoptosis and inflammatory pathways. J Neurosurg 130(3):977–988

Uyeda A, Muramatsu R (2020) Molecular mechanisms of central nervous system axonal regeneration and remyelination: a review. Int J Mol Sci 21(21):8116

Uzdensky AB (2019) Apoptosis regulation in the penumbra after ischemic stroke: expression of pro-and antiapoptotic proteins. Apoptosis 24(9):687–702

Valenti D, de Bari L, de Rasmo D, Signorile A, Henrion-Caude A, Contestabile A, Vacca RA (2016) The polyphenols resveratrol and epigallocatechin-3-gallate restore the severe impairment of mitochondria in hippocampal progenitor cells from a Down syndrome mouse model. Biochim Biophys Acta 1862(6):1093–1104

Venkat P, Shen Y, Chopp M, Chen J (2018) Cell-based and pharmacological neurorestorative therapies for ischemic stroke. Neuropharmacology 134:310–322

Vissers MC, Bozonet SM, Pearson JF, Braithwaite LJ (2011) Dietary ascorbate intake affects steady state tissue concentrations in vitamin C–deficient mice: tissue deficiency after suboptimal intake and superior bioavailability from a food source (kiwifruit). Am J Clin Nutr 93(2):292–301

Walle T (2011) Bioavailability of resveratrol. Ann N Y Acad Sci 1215(1):9–15

Wan D, Zhou Y, Wang K, Hou Y, Hou R, Ye X (2016) Resveratrol provides neuroprotection by inhibiting phosphodiesterases and regulating the cAMP/AMPK/SIRT1 pathway after stroke in rats. Brain Res Bull 121:255–262

Wang B, Jin K (2015) Current perspectives on the link between neuroinflammation and neurogenesis. Metab Brain Dis 30:355–365

Wang B, Sun J, Ma Y, Wu G, Tian Y, Shi Y, Le G (2014a) Resveratrol preserves mitochondrial function, stimulates mitochondrial biogenesis, and attenuates oxidative stress in regulatory T cells of mice fed a high-fat diet. J Food Sci 79(9):H1823–H1831

Wang YH, Yu HT, Pu XP, Du GH (2014b) Myricitrin alleviates methylglyoxal-induced mitochondrial dysfunction and AGEs/RAGE/NF-κB pathway activation in SH-SY5Y cells. J Mol Neurosci 53:562–570

Wang H, Quirion R, Little PJ, Cheng Y, Feng ZP, Sun HS et al (2015) Forkhead box O transcription factors as possible mediators in the development of major depression. Neuropharmacology 99:527–537

Wang P, Cao Y, Yu J, Liu R, Bai B, Qi H et al (2016a) Baicalin alleviates ischemia-induced memory impairment by inhibiting the phosphorylation of CaMKII in hippocampus. Brain Res 1642:95–103

Wang Y, Chen G, Yu X, Li Y, Zhang L, He Z et al (2016b) Salvianolic acid B ameliorates cerebral ischemia/reperfusion injury through inhibiting TLR4/MyD88 signaling pathway. Inflammation 39:1503–1513

Wang JF, Mei ZG, Fu Y, Yang SB, Zhang SZ, Huang WF et al (2018a) Puerarin protects rat brain against ischemia/reperfusion injury by suppressing autophagy via the AMPK-mTOR-ULK1 signaling pathway. Neural Regen Res 13(6):989–998

Wang S, Nan Y, Zhu W, Yang T, Tong Y, Fan Y (2018b) Gastrodin improves the neurological score in MCAO rats by inhibiting inflammation and apoptosis, promoting revascularization. Int J Clin Exp Pathol 11(11):5343

Wang P, Shao BZ, Deng Z, Chen S, Yue Z, Miao CY (2018c) Autophagy in ischemic stroke. Prog Neurobiol 163:98–117

Wang J, Wan D, Wan G, Wang J, Zhang J, Zhu H (2019) Catalpol induces cell activity to promote axonal regeneration via the PI3K/AKT/mTOR pathway in vivo and in vitro stroke model. Ann Transl Med 7(23):756

Wang Y, Su Y, Lai W, Huang X, Chu K, Brown J, Hong G (2020) Salidroside restores an anti-inflammatory endothelial phenotype by selectively inhibiting endothelial complement after oxidative stress. Inflammation 43:310–325

Wang L, Zhu T, Xu HB, Pu XP, Zhao X, Tian F et al (2021) Effects of notoginseng leaf triterpenes on small molecule metabolism after cerebral ischemia/reperfusion injury assessed using MALDI–MS imaging. Ann Transl Med 9(3):246

Waseem M, Parvez S (2016) Neuroprotective activities of curcumin and quercetin with potential relevance to mitochondrial dysfunction induced by oxaliplatin. Protoplasma 253:417–430

Watanabe R, Kurose T, Morishige Y, Fujimori K (2018) Protective effects of fisetin against 6-OHDA-induced apoptosis by activation of PI3K-Akt signaling in human neuroblastoma SH-SY5Y cells. Neurochem Res 43:488–499

Wei Y, Hong H, Zhang X, Lai W, Wang Y, Chu K et al (2017a) Salidroside inhibits inflammation through PI3K/Akt/HIF signaling after focal cerebral ischemia in rats. Inflammation 40:1297–1309

Wei H, Sun T, Tian Y, Wang K (2017b) Ginkgolide B modulates BDNF expression in acute ischemic stroke. J Korean Neurosurg Soc 60(4):391–396

Wen Y, Gu Y, Tang X, Hu Z (2020) PINK1 overexpression protects against cerebral ischemia through parkin regulation. Environ Toxicol 35(2):188–193

Xie CJ, Gu AP, Cai J, Wu Y, Chen RC (2018) Curcumin protects neural cells against ischemic injury in N2a cells and mouse brain with ischemic stroke. Brain Behav 8(2):e00921

Xie W, Zhu T, Zhou P, Xu H, Meng X, Ding T et al (2020) Notoginseng leaf triterpenes ameliorates OGD/R-induced neuronal injury via SIRT1/2/3-Foxo3a-MnSOD/PGC-1α signaling pathways mediated by the NAMPT-NAD pathway. Oxidative Med Cell Longev 2020(1):7308386

Xiong D, Deng Y, Huang B, Yin C, Liu B, Shi J, Gong Q (2016) Icariin attenuates cerebral ischemia–reperfusion injury through inhibition of inflammatory response mediated by NF-κB, PPARα and PPARγ in rats. Int Immunopharmacol 30:157–162

Xu S, Zhong A, Ma H, Li D, Hu Y, Xu Y, Zhang J (2017) Neuroprotective effect of salvianolic acid B against cerebral ischemic injury in rats via the CD40/NF-κB pathway associated with suppression of platelets activation and neuroinflammation. Brain Res 1661:37–48

Xu H, Nie B, Liu L, Zhang C, Zhang Z, Xu M, Mei Y (2019) Curcumin prevents brain damage and cognitive dysfunction during ischemic-reperfusion through the regulation of miR-7-5p. Curr Neurovasc Res 16(5):441–454

Xu X, Cui Y, Li C, Wang Y, Cheng J, Chen S et al (2021) SETD3 downregulation mediates PTEN upregulation-induced ischemic neuronal death through suppression of actin polymerization and mitochondrial function. Mol Neurobiol 58(10):4906–4920

Xue B, Huang J, Ma B, Yang B, Chang D, Liu J (2019) Astragaloside IV protects primary cerebral cortical neurons from oxygen and glucose deprivation/reoxygenation by activating the PKA/CREB pathway. Neuroscience 404:326–337

Yang J, Huang J, Shen C, Cheng W, Yu P, Wang L et al (2018) Resveratrol treatment in different time-attenuated neuronal apoptosis after oxygen and glucose deprivation/reoxygenation via enhancing the activation of Nrf-2 signaling pathway in vitro. Cell Transplant 27(12):1789–1797

Yang J, Shao C, Li W, Wan H, He Y, Yang J (2021) Protective effects of Astragaloside IV against oxidative injury and apoptosis in cultured astrocytes by regulating Nrf2/JNK signaling. Exp Brain Res 239:1827–1840

Yu P, Wang L, Tang F, Zeng L, Zhou L, Song X et al (2017) Resveratrol pretreatment decreases ischemic injury and improves neurological function via sonic hedgehog signaling after stroke in rats. Mol Neurobiol 54(1):212–226

Yuste JE, Tarragon E, Campuzano CM, Ros-Bernal F (2015) Implications of glial nitric oxide in neurodegenerative diseases. Front Cell Neurosci 9:322

Zhang Z, Zheng L, Zhao Z, Shi J, Wang X, Huang J (2014) Grape seed proanthocyanidins inhibit H2O2-induced osteoblastic MC3T3-E1 cell apoptosis via ameliorating H2O2-induced mitochondrial dysfunction. J Toxicol Sci 39(5):803–813

Zhang XS, Ha S, Wang XL, Shi YL, Duan SS, Li ZA (2015a) Tanshinone IIA protects dopaminergic neurons against 6-hydroxydopamine-induced neurotoxicity through miR-153/NF-E2-related factor 2/antioxidant response element signaling pathway. Neuroscience 303:489–502

Zhang LQ, Sa F, Chong CM, Wang Y, Zhou ZY, Chang RCC et al (2015b) Schisantherin A protects against 6-OHDA-induced dopaminergic neuron damage in zebrafish and cytotoxicity in SH-SY5Y cells through the ROS/NO and AKT/GSK3β pathways. J Ethnopharmacol 170:8–15

Zhang C, Li C, Chen S, Li Z, Jia X, Wang K et al (2017a) Berberine protects against 6-OHDA-induced neurotoxicity in PC12 cells and zebrafish through hormetic mechanisms involving PI3K/AKT/Bcl-2 and Nrf2/HO-1 pathways. Redox Biol 11:1–11

Zhang C, Li C, Chen S, Li Z, Ma L, Jia X et al (2017b) Hormetic effect of panaxatriol saponins confers neuroprotection in PC12 cells and zebrafish through PI3K/AKT/mTOR and AMPK/SIRT1/FOXO3 pathways. Sci Rep 7(1):41082

Zhang Y, Yan Y, Cao Y, Yang Y, Zhao Q, Jing R et al (2017c) Potential therapeutic and protective effect of curcumin against stroke in the male albino stroke-induced model rats. Life Sci 183:45–49

Zhang X, Du Q, Yang Y, Wang J, Liu Y, Zhao Z et al (2018) Salidroside alleviates ischemic brain injury in mice with ischemic stroke through regulating BDNK mediated PI3K/Akt pathway. Biochem Pharmacol 156:99–108

Zhang HS, Liu MF, Ji XY, Jiang CR, Li ZL, OuYang B (2019a) Gastrodin combined with rhynchophylline inhibits cerebral ischaemia-induced inflammasome activation via upregulating miR-21–5p and miR-331–5p. Life Sci 239:116935

Zhang X, Lai W, Ying X, Xu L, Chu K, Brown J et al (2019b) Salidroside reduces inflammation and brain injury after permanent middle cerebral artery occlusion in rats by regulating PI3K/PKB/Nrf2/NFκB signaling rather than complement C3 activity. Inflammation 42:1830–1842

Zhang Y, Huang N, Chen M, Jin H, Nie J, Shi J, Jin F (2019c) Procyanidin protects against 6-hydroxydopamine-induced dopaminergic neuron damage via the regulation of the PI3K/Akt signalling pathway. Biomed Pharmacother 114:108789

Zhang Y, Zhang Y, Jin XF, Zhou XH, Dong XH, Yu WT, Gao WJ (2019d) The role of astragaloside IV against cerebral ischemia/reperfusion injury: suppression of apoptosis via promotion of P62-LC3-autophagy. Molecules 24(9):1838

Zhang HS, Ouyang B, Ji XY, Liu MF (2021) Gastrodin alleviates cerebral ischaemia/reperfusion injury by inhibiting pyroptosis by regulating the lncRNA NEAT1/miR-22-3p axis. Neurochem Res 46:1747–1758

Zheng T, Jiang H, Jin R, Zhao Y, Bai Y, Xu H, Chen Y (2019) Ginsenoside Rg1 attenuates protein aggregation and inflammatory response following cerebral ischemia and reperfusion injury. Eur J Pharmacol 853:65–73

Zhong ZF, Han J, Zhang JZ, Xiao Q, Chen JY, Zhang K et al (2019) Neuroprotective effects of salidroside on cerebral ischemia/reperfusion-induced behavioral impairment involves the dopaminergic system. Front Pharmacol 10:1433

Zhou X, Chen M, Zeng X, Yang J, Deng H, Yi L, Mi MT (2014) Resveratrol regulates mitochondrial reactive oxygen species homeostasis through Sirt3 signaling pathway in human vascular endothelial cells. Cell Death Dis 5(12):e1576–e1576

Zhou ZQ, Li YL, Ao ZB, Wen ZL, Chen QW, Huang ZG et al (2017a) Baicalin protects neonatal rat brains against hypoxic-ischemic injury by upregulating glutamate transporter 1 via the phosphoinositide 3-kinase/protein kinase B signaling pathway. Neural Regen Res 12(10):1625–1631

Zhou J, Yao W, Li C, Wu W, Li Q, Liu H (2017b) Administration of follicle-stimulating hormone induces autophagy via upregulation of HIF-1α in mouse granulosa cells. Cell Death Dis 8(8):e3001–e3001

Zhou Z, Lu J, Liu WW, Manaenko A, Hou X, Mei Q et al (2018) Advances in stroke pharmacology. Pharmacol Ther 191:23–42

Zhu T (2019) Polarization of microglia/macrophages and its role in ischemic stroke repair. Chin Pharm J:1046–1050

Zhu HF, Shao Y, Qin L, Wang JH, Feng S, Jiang YB, Wan D (2019a) Catalpol enhances neurogenesis and inhibits apoptosis of new neurons via BDNF, but not the BDNF/Trkb pathway. Drug Des Devel Ther 13:4145–4157

Zhu H, Wang J, Shao Y, Wan D (2019b) Catalpol may improve axonal growth via regulating miR-124 regulated PI3K/AKT/mTOR pathway in neurons after ischemia. Ann Transl Med 7(14):306

Zhu T, Wang L, Tian F, Zhao X, Pu XP, Sun GB, Sun XB (2020) Anti-ischemia/reperfusion injury effects of notoginsenoside R1 on small molecule metabolism in rat brain after ischemic stroke as visualized by MALDI–MS imaging. Biomed Pharmacother 129:110470

Zhu T, Wang L, Xie W, Meng X, Feng Y, Sun G, Sun X (2021) Notoginsenoside R1 improves cerebral ischemia/reperfusion injury by promoting neurogenesis via the BDNF/Akt/CREB pathway. Front Pharmacol 12:615998

Neurotransmitter Modulation and Cognitive Enhancement

4

Dipak Kumar, Ranvijay Singh, and Abhishek Kumar Mishra

Abstract

Cognition is complex attribute of the brain encompassing the functions of memory, decision, language, attention, gnosis, orientation, praxis, visuospatial skills, executive functions, and social cognition. It is the function of neuronal network in the central nervous system (CNS) and pertaining neurotransmitter modulation. Neurotransmitters are chemical messengers packaged in vesicles at presynaptic terminals. The communication process is regulated in the specialized areas of the plasma membrane called active zones. Neuromodulation has further evolved to allow for the flexibility in the dynamically updating the effectivity of the static anatomical connectivity and cognitive functions. Psychiatric and developmental disorders which compromise the cognitive ability have been linked with the variabilities of CNS such as synaptic plasticity, hippocampal health, epigenetic modification, epigenetic priming, hemispheric symmetry, prefrontal cortex, etc. Neuromodulators such as histone deacetylase inhibitors, retinoic acid receptor activators, and neuroplasticity enhancers can reverse the effect of neurodegeneration, aging, autism, schizophrenia, and other psychiatric disorders. There are several phytochemicals such as vitamin A, valproate, flavonoids, monoterpenes, and peppermint which works as neuromodulators or cognitive enhancers.

D. Kumar (✉)
Department of Zoology, Munger University, Munger, Bihar, India

R. Singh
Department of Psychology, Munger University, Munger, Bihar, India

A. K. Mishra
Department of Zoology, Government Shaheed Gendsingh College,
Charama Uttar Bastar Kanker, Chhattisgarh, India

S. Jahan et al. (eds.), *A Mechanistic Exploration of Natural Compounds for Neuronal Health*, https://doi.org/10.1007/978-981-95-0125-0_4

Keywords

Cognition · Neurotransmitters · Neuromodulation · Neurodegeneration · Psychiatric disorders · Phytochemicals

4.1 Introduction

Vertebrate nervous system has two main divisions. One, the central nervous system (CNS) consists of the brain and spinal cord, and second, the peripheral nervous system (PNS) in humans includes 12 pairs of cranial nerves and 31 pairs of spinal nerves. These nerve pairs work as either an afferent (sensory) system or an efferent (motor) system. The afferent system transmits impulses from the peripheral tissues to the CNS while efferent system transmits impulses from the CNS to the peripheral tissues to cause an effect or action. Further, PNS has been divided into somatic nervous system (SNS) and autonomic nervous system (ANS). SNS supplies motor impulses to skeletal muscles which are in conscious control hence are called as voluntary nervous system. In contrast, ANS supplies impulses to the smooth muscles, cardiac muscles, and glandular epithelia which are autonomic or involuntary in function so they are referred as involuntary nervous system. The ANS may be sympathetic, parasympathetic, or enteric in function. Sympathetic nerves are excitatory in function and are activated during the emergency or "fight–or–flight" situations. Parasympathetic nerves are inhibitory in nature and remain active during the "rest–and–digestion" or "feed–and–breed" activities. Enteric nerves, sometimes considered as independent system (gut brain), function to control the gastrointestinal system.

The brain is enclosed in cranium, a bony structure while the spinal cord is protected by the units of the vertebral column, i.e. vertebrae. Foramen magnum connects the brain and the spinal cord uniting them to function as the centre or hub of nervous system. Mammalian brain has cerebrum, diencephalons, brain stem, and cerebellum as its main parts. It is imperative to note here that the development of cerebral cortex including cytoarchitecture and cortical folding is conserved across the mammalian species (Molnár et al. 2014; Sun and Hevner 2014). The development has a common starting point which progresses in the guidance of conserved genetic and cellular mechanisms (Sun and Hevner 2014). Structurally, the cerebral cortex develops into the archipallium (hippocampus), paleopallium (olfactory lobe), and neopallium (dorsal cortex) which are unique to the mammals. Neopallium or neocortex greatly enlarges covering almost all other parts of the brain (Fig. 4.1). This structure has been evolutionarily attaining greater complexities assuming the higher neural activities of cognition such as learning, correlation, and association (Molnár et al. 2014). Further, neopallium in primates and humans expands extensively resulting in convoluted surface increasing the total surface area (Hill et al. 2010). Segregated yet connected brain structures perform complex functions of memory, decision, language, attention, gnosis, orientation, praxis, visuospatial skills, executive functions, and social cognition.

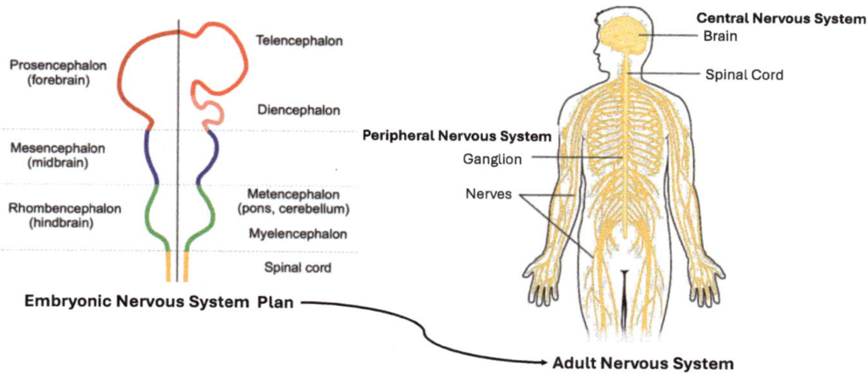

Fig. 4.1 Nervous system is ectodermal in origin. During embryonic development, neural plate folds forming neural tube which further differentiate into prosencephalon, telencephalon, mesencephalon, diencephalon, rhombencephalon, metencephalon, myelencephalon, and spinal cord. Former seven parts form brain. The brain and spinal cord are parts of central nervous system, while ganglion and nerves arising from both the brain and spinal cord forms peripheral nervous system. Cognitive and neuromotor functions are the result of cooperation and communication between these two systems

4.2 Role of Neurotransmitters in Neuronal Health and Cognitive Function

Cognition or mental activity designates the procurement, storage, transformation, and utilization of information or knowledge. These functions are performed by the brain. A human brain is an incredibly complex structure composed of nearly 100 billion neurons and 1000 billion glial cells (Yang et al. 2022). Recent techniques such as image-derived phenotypes (IDPs), single nucleotide polymorphisms (SNPs), and magnetic resonance imaging (MRI) have collected evidence of connections among brain structures and their functions, brain development, behaviour differences, and diseases (Wang et al. 2023; Zhao et al. 2023). Neurotransmitters are the chemical means of neuronal function and serve as information sharing agent among the brain structures and cells. These chemical messengers are small molecules synthesized in the presynaptic axon terminals of the neurons or recycled locally, transported into vesicles, and released to convey extracellular signals. Meanwhile, cell body of the neuron synthesizes neuropeptides which works as neuromodulators (Yeo et al. 2022) (Fig. 4.2).

Neurotransmitters and neuropeptides are transported in designated vesicles for exocytosis at localised synaptic cleft and generalised cell membrane, respectively. At the presynaptic terminal of the axon, action potential travelling down the line triggers Ca^{2+} influx, which releases neurotransmitters into the space between presynaptic and postsynaptic membranes. Receptors on the postsynaptic membrane receive the signals and alter the activities of pertaining neurons. This relay of information transfer travels up to the effector cell. Further, receptors on the postsynaptic neuron release the transmitters for their recycling or diffusion. The postsynaptic

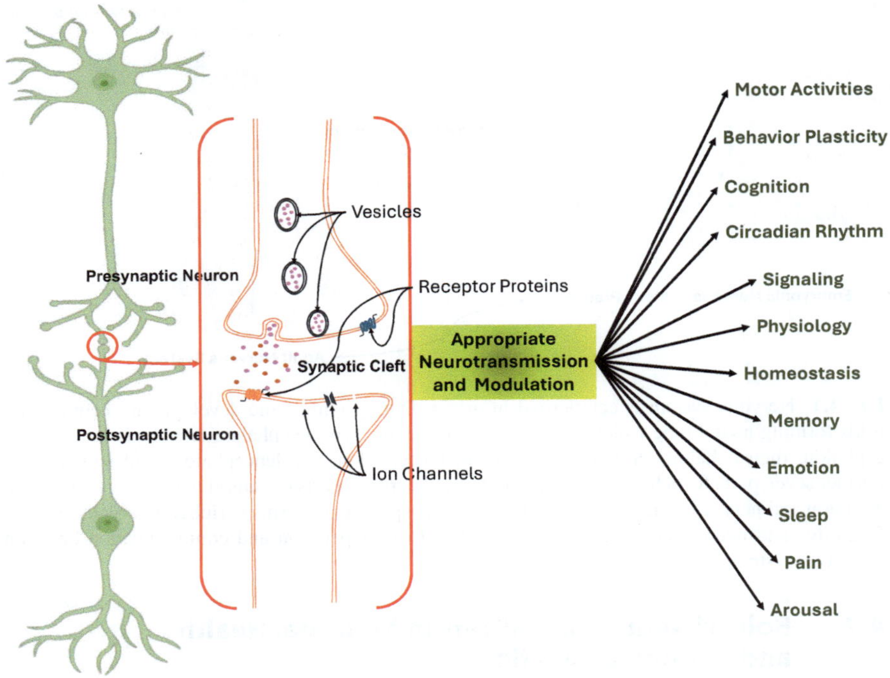

Fig. 4.2 The exchange of molecules between the presynaptic and postsynaptic neurons through the space of synaptic cleft is the center of neurotransmissions and communications among the cells of CNS and PNS. When this chemical physiology works optimally with the help of presynaptic transmitting vesicles, postsynaptic ion channels and pre- and post-synaptic receptor proteins. Then, cognitive, motor, and sensory functions are regulated and modulated properly

membrane then sends reverse messages to ensure that the presynaptic membrane stops sending the same signal again and again (Lovinger 2008). The signal in the form of a transmitter is deciphered by the receptor protein embedded in the membrane. When the active site of the receptor protein binds its ligand, the receptor can either open a channel (ionotropic effect) or activate a downstream signalling pathway (metabotropic effect) (Osimo and Stahl 2023). The ionotropic effect acts sudden (less than a millisecond) and lasts for a short duration (less than 20 millisecond) while the metabotropic effect starts slower (more than 30 millisecond) and lasts longer, maybe a few seconds, minutes, or even longer (Osimo and Stahl 2023; Roth 2019).

CNS controls the functions of cardiac, skeletal, and smooth muscles; secretion of hormones and enzymes; organ and organ system functions through the synaptic transmission of neurotransmitters; and secretion of neuropeptides (Teleanu et al. 2022). Signals from nerve to nerve and/or nerve to other cell types are transmitted, amplified, and modulated using these endogenous molecules. That is why these molecules play a crucial role in communicating motor, sensory, and integrate signals among the neurons of the CNS, ANS, and PNS. Apart from routine nervous system functions, these molecules regulate neuronal survival, growth, and differentiation (Teleanu et al. 2022). Consequently, they affect many cognitive and psychotic functions such as emotions, memories, thoughts, behaviors, movements, locomotion, sleep patterns, ideas, philosophies, moods, etc. (McCutcheon et al. 2020; Teleanu et al. 2022). Conversely, the abnormality in neurotransmitter physiology has been found to be responsible for the dysregulation of nervous system functions terminating in various physical, cognitive, psychotic, and neurodegenerative diseases (Sarter et al. 2007).

More than 200 chemical messengers have been characterized since 1921. The list is still growing for such compounds exhibiting neuroactivities. To qualify as a neuroactive compound, a molecule should have (a) been synthesized and released from the stock of presynaptic vesicles of the same neuron, (b) a particular inducing effect on the postsynaptic neuron, (c) the same effect if exogenously administered, and (d) some specific stopping mechanism for the induced effect on the postsynaptic neuron (Alavijeh and Palmer 2010). So, there are some well-defined neuroactive compounds known as canonical neurotransmitters and some are still under investigation known as noncanonical neurotransmitters.

4.2.1 Canonical Neurotransmitters

Well-characterised neuroactive compounds may be categorised into amino acids, modified amino acids, amines, peptides, purines, and gases (Table 4.1). Chemically, amino group(s) (NH_2) has been found to be predominantly present in these molecules. Among these, an amino acid has a hydrocarbon skeleton with a carboxylic acid (–COOH) and an amino (–NH_2) as functional groups. Acetylcholine (Ach) and monoamines such as indoleamines and catecholamines are modified amino acid molecules. In Ach, NH_2 has been replaced by $N(CH_3)_3$ group, while several amino

Table 4.1 List of neurotransmitters

Types of neurotransmitters	Examples of neurotransmitters
Amino acid	Glutamate, γ–aminobutyric acid (GABA), glycine, aspartate
Modified amino acid	Ach
Monoamines	Indoleamine: serotonin, catecholamines: dopamine, norepinephrine, epinephrine
Neuropeptides	Substance P, neuropeptide Y, opioids: enkephalins, endorphin, dynorphin, tachykinins: substance P, neurokinin Glucagon-related: vasoactive intestinal polypeptide (VIP), pituitary adenylate cyclase-activating peptide (PACAP), somatostatin, neurotensin, calcitonin gene-related peptide (CGRP), galanin etc.
Purines	ATP, adenosine
Gases	NO (nitric oxide), CO (carbon monoxide)
Endocannabinoids	Arachidonoyl ethanolamide (AEA), 2-archidonoyl glycerol (2–AG)

acids join by peptide bonds forming the polypeptide chain (neuropeptides) that also act as neurotransmitters. There are two types of nitrogenous bases found in the living system, purines and pyrimidines. Purine, such as adenosine, has been noted to be functioning as the neurotransmitter. Some gases such as nitric oxide (NO) and carbon monoxide (CO) are produced in the living cells and act as neurotransmitters too (Snyder et al. 1998). These unconventional neurotransmitters directly interact with the receptors on the nucleus as retinoids and vitamin D or with cytosolic enzymes as guanylyl cyclase (Leonelli et al. 2009). Moreover, some lipid-based neurotransmitters as endogenous cannabinoids or endocannabinoids have also been characterised (Lu and Mackie 2016). These recently discovered neurotransmitters have their receptors on the cells' surface throughout the body. The unconventional neurotransmitters are different from their counterpart as they are neither stored in vesicles nor released by exocytosis at presynaptic terminals and are often associated with retrograde transmission, i.e. from postsynaptic to presynaptic terminals (Leonelli et al. 2009).

Almost all excitatory ionotropic synapses of the brain use glutamate as the neurotransmitter (Reiner and Levitz 2018). Meanwhile, inhibitory ionotropic synapses use the neurotransmitter GABA to open chloride gates allowing Cl^- ions to enter the cell (Nakamura et al. 2015). Similarly, glycine is an inhibitory neurotransmitter (Bowery and Smart 2006; Nakamura et al. 2015). Ach, one of the best-studied neurotransmitters works at several ionotropic synapses, has mainly excitatory functions. Ach receptor (AchR) also referred as cholinergic receptors are found on the postsynaptic membrane of sympathetic as well as parasympathetic neurons (Tiwari et al. 2013). These receptors are transmembrane complex multimeric protein having multiple subunits: two α, one β, one γ, and one σ arranged around a central channel. Ach binds at the outer binding site opening the channel and allowing diffusion of Na^+ and K^+ ions. Nicotinic AchR stimulates sympathetic postganglionic neurons, adrenal chromaffin cells, and parasympathetic postganglionic neurons for their physiological actions. Muscarinic AchR is mainly associated with parasympathetic neurons linked with peripheral tissues such as glands and smooth muscles for their physiological functions (Tiwari et al. 2013). The ionotropic receptors include ionotropic nicotinic Ach receptor (nAcR), $GABA_A$ receptor ($GABA_AR$), and glycine receptor (GlyR).

There are several neurotransmitters which bind to metabotropic receptors or G-protein coupled receptors (GPCRs). These receptors are generally monomeric transmembrane proteins having seven hydrophobic outer domains (Osimo and Stahl 2023). GPCRs include muscarinic Ach receptors, glutamate receptors, $GABA_B$ receptors ($GABA_BR$), most serotonin receptors, and receptors for epinephrine, norepinephrine, histamine, neuropeptides, dopamine, and endocannabinoids. While the outer domains receive neurotransmitters, the inner domain which is attached to the G-protein (a trimer: one α, one β and one γ) coupled with GTP activates the secondary messengers. The second messengers, such as cyclic adenosine monophosphate (cAMP), relays information to the effector proteins inside the cell (Osimo and Stahl 2023) (Fig. 4.3).

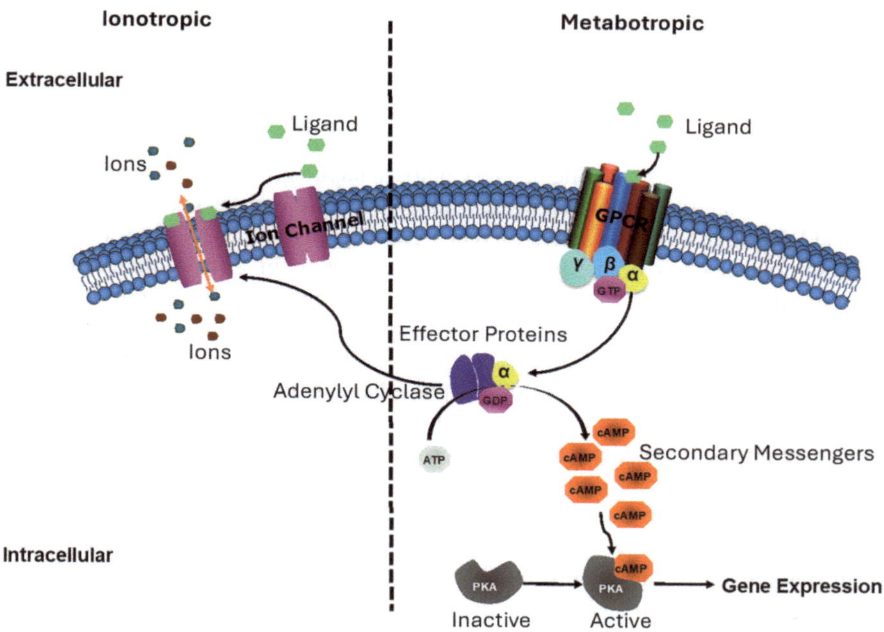

Fig. 4.3 There are two schemes of neurotransmitter functioning: ionotropic and metabotropic. In ionotropic system, ligands (neurotransmitters) bind to the ion channels allosterically. In response to the binding, ion channels open to allow the flow of ions through the cell membrane resulting in physiological modulation the neuron, while in metabotropic system, ligands bind to the GPCR triggering a cascade of events. This involves effector protein such as adenylyl cyclase which either can open ion channels or convert ATP into cAMPs. Ion channel responds same as in case of ionotropic system. cAMP further acts as secondary messenger activating PKA which initiates the procedure of gene expression

4.2.2 Noncanonical Neurotransmitters

Apart from the abovementioned well-defined molecules, there are some more entities which show neuroactivity. Neurons have been found to be communicating with the help of exosomes (Pascual et al. 2020). These are the subtype of extracellular vesicles (EVs). Exosomes are bilipid EVs resembling the synaptic vesicles. It is imperative to note here that these vesicles are released by exocytosis throughout the surface of plasma membrane of all cell types in a packaged form. The packaged cargo contains various molecules including nucleic acids, metabolites, functional proteins, growth factors, cytokines, etc. (Morel et al. 2013). Apart from nervous system, they have been reported to influence varied physiological phenomena such as immune response, tissue repair, stem cell proliferation and differentiation, pathological physiology related to neurodegeneration, cardiovascular diseases, general inflammation, cancer, etc. (De Toro et al. 2015). Specifically for neurons, they can modify presynaptic and/or postsynaptic signalling, support synapse, control production and release of neurotransmitters, enhance or suppress neurite outgrowth, initiate the removal of neurites, start axon regeneration, and modulate neurotransmitter recycling (Huo et al. 2021).

Exosomes contain synaptotagmin-4 (Syt4), α-amino-3-hydroxy-5-methyl-4-isoxazolepropionic acid (AMPA) receptor (AMPAR), and Wnt (Wingless-related integration site) which have been reported to be regulating synaptic plasticity (Xia et al. 2022). Syt4 trans synaptic delivery helps synaptic growth and augments properties of presynaptic membrane. Meanwhile, AMPAR enhances neuronal excitability and Wnt ligand binding activates Wnt signalling (Xia et al. 2022). Additionally, the resemblance of exosomes with the neurotransmitters in terms of their release in response to the action potential from the presynaptic membrane highlights their role in nerve-to-nerve communication. These exosomes contain neuropeptides and some other ligands which bind at the receptor site of GPCR activating phospholipase C (PLC) which generates inositol 1,4,5 triphosphate (IP_3) (Estelles et al. 2007; Xia et al. 2022). When GPCR is activated, the soluble N-ethylmaleimide-sensitive factor activating protein receptor (SNARE) proteins along with the neuronal exosomes induce release of Ca^{2+} from the endoplasmic reticulum via IP_3 receptor (IP_3R) and Ca^{2+} influx through the Ca^{2+} channel at the plasma membrane (Quetglas et al. 2000; Stewart et al. 2000). This increased intracellular Ca^{2+} further affects the response from postsynaptic neurons, opening a period of neuronal receptors and opening/closing positions of ion channels on the plasma membrane.

Steroids are another group of chemicals which have been recently studied as neurotransmitters. Besides being a hormone, steroids are also responsible for signalling at membrane and nucleus regulating brain functions, activating signalling cascade inside the cytosol. They also increase the intracellular Ca^{2+} release and activate Src ("Src" stand for short form of sarcoma, it is a non-receptor tyrosine kinase protein encoded by the SRC gene), mitogen–activated protein kinase pathway (MAPK) or (phosphatidylinositol–3–kinase–protein kinase B or PKB) PI3K–AKT pathways (Rostène et al. 1995; Rudolph et al. 2016; Vicencio et al. 2011). D–Aspartic acid, an amino acid has been found in the nervous system of several animals, especially during the developmental period of human brain. Their presence establishes their role in the nervous system development as well as neuronal activities (Usiello et al. 2020). In neurons D–aspartate is synthesised from L–aspartate by racemase enzyme. This amino acid molecule establish cell–cell communication and gets eliminated from postsynaptic membrane of neuron after its action (Herring et al. 2015).

4.2.3 Cognitive Physiology and Functions

Catecholamines has important functions to perform right through developmental period to the adult life of primates including human. Catecholaminergic neurons are generated during the embryonic development of telencephalic vesicles (Verney 1999). In adults, they mediate various CNS functions like motor control, emotion, endocrine modulation, memory processing, and general cognition (Kobayashi 2001). Imperatively, their dysfunction in neurotransmission is responsible for range of neurological and neuropsychiatric disorders. Particularly, norepinephrine plays crucial role in consolidation process of long-term memory. This compound modulates neuronal activities in amygdala and cerebral cortex for conditioning learning (Ranjbar-Slamloo and Fazlali 2020). Neuronal activities of catecholamines as these are possible as it works through adrenergic receptors (ARs). The receptors are heptahelical transmembrane protein, guanine nucleotide-binding linked with GPCR. ARs have two subtypes α (α_1, α_2)-receptors and β (β_1, $\beta_{2,}$ β_3)-receptors. These receptors mediate different roles of catecholamines by different mode of action mechanisms (Graham 1990). For instance, the specific receptors on endothelial surface get activated sensing the local and systemic release of catecholamines. Here, the adrenergic system is principal regulators of vascular and cardiac functions. Among various cognitive and noncognitive functions, β-receptors especially mediate memory functions. Moreover, ARs have been reported to be working with nicotinic system synergistically via postganglionic sympathetic nerve terminals where noradrenalin is released (Yoon et al. 2014) (Table 4.2). The cascade may be working to modulate the memory processes.

In mammalian brain, glutamate is the primary excitatory neurotransmitter as it is widely distributed in the brain and has wide area of metabolic functions. It plays crucial role in the regulation of neurogenesis, neurite outgrowth, synaptogenesis, and overall survival of neurons in the brain via N-methyl-dextro-aspartic acid (NMDA) receptor (NMDAR) (Reiner and Levitz 2018). NMDARs are obligate heterotetramers having mainly three distinct subunits as GluN1, GluN2, and GluN3 (Yamamoto et al. 2015). These subunits combine differently to form receptor subtypes. Extracellular ligand binding domain works with intracellular carboxy-terminal domain which activates Na^+ and Ca^{2+} channels. It is established that NMDAR's transient activation triggers the induction of long-term potentiation at synapse of neurons of CA3 and CA1 in the hippocampus (Lüscher and Malenka 2012). This leads to hippocampus-mediated learning and memory. Thus, cognitive functions are synchronised by the glutamatergic activities via synaptic plasticity restoration. Any dysfunction in glutamatergic component can cause cognitive impairment, loss of synaptic plasticity, and compromised neuronal survival.

Dopaminergic system has been extensively studied. It's been found to be working in context of various neuropsychiatric aspects such as attention deficit hyperactivity disorder (ADHD), schizophrenia, Parkinsonism, cognitive child development, aging, memory, and drug addiction. The broad area of dopamine (DA) functions in neuromodulation, cortical and subcortical network, cognition, and age-dependent changes in behavior through the life span of an individual puts it at the center of brain functioning. DA works through two types of receptors: D1-like receptors (includes D1 and D5) which is coupled with Gs proteins and D2-like receptors (includes D2, D3 and D4) coupled with the Gi proteins (Mishra et al. 2018). Both groups of D1 and D2 receptors are differentially distributed in the brain attributed to the different functions of the different regions. This tissue- and cell-specific expression of DA receptors are altered during the brain maturation, senescence stages, and diseases. Comparatively, D1 receptors are more abundant than D2 receptors when whole brain is considered. Specifically, D1 receptors are highly concentrated in

Table 4.2 List of signalling systems, receptor types, and their association for cognitive effects

System	System association	Receptor type	Receptor activation	Cognition effects
Adrenergic	Nicotine	α_1	Agonists	Cognitive stimulation
			Antagonists	Cognitive inhibition
		α_2	Agonists	Cognitive stimulation
			Antagonists	Cognitive inhibition
		β_1	Agonists	Cognitive inhibition
			Antagonists	Cognitive stimulation
		β_2	Agonists	Cognitive stimulation
			Antagonists	Cognitive inhibition
		β_3	Agonists	Cognitive stimulation
			Antagonists	Cognitive inhibition
Glutamate		NMDA	Agonists	Cognitive stimulation
			Antagonists	Cognitive inhibition
Dopaminergic		D_1	Agonists	Cognitive stimulation
			Antagonists	Cognitive inhibition
		D_5	Agonists	Cognitive stimulation
			Antagonists	Cognitive inhibition
		D_3	Agonists	Cognitive inhibition
			Antagonists	Cognitive stimulation
		D_4	Agonists	Cognitive inhibition
			Antagonists	Cognitive stimulation
		D_2	Agonists	Cognitive stimulation
			Antagonists	Cognitive inhibition

Serotoninergic	NA	5-HT$_1$	Agonists	Cognitive inhibition
			Antagonists	Cognitive stimulation
		5-HT$_{1B}$	Agonists	Cognitive inhibition
			Antagonists	Cognitive stimulation
		5-HT$_{1D}$	Agonists	No effect
			Antagonists	No effect
		5-HT$_{1E}$	Agonists	No effect
			Antagonists	No effect
		5-HT$_{1F}$	Agonists	No effect
			Antagonists	No effect
	Nicotine	5-HT$_{2A}$	Agonists	Cognitive inhibition
			Antagonists	Cognitive stimulation
	NA	5-HT$_{2B}$	Agonists	No effect
			Antagonists	No effect
		5-HT$_{2C}$	Agonists	No effect
			Antagonists	No effect
		5-HT$_3$	Agonists	NA
			Antagonists	
	Nicotine	5-HT$_4$	Agonists	Cognitive stimulation
			Antagonists	Cognitive inhibition
	NA	5-HT$_5$	Agonists	NA
			Antagonists	
		5-HT$_{5A}$	Agonists	No effect
			Antagonists	No effect
		5-HT$_{5B}$	Agonists	NA
			Antagonists	
	Nicotine	5-HT$_6$	Agonists	Cognitive inhibition
			Antagonists	Cognitive stimulation
	NA	5-HT$_7$	Agonists	NA
			Antagonists	

both striatum and neocortex while D2 receptors are only concentrated in striatum with lower expression found in brainstem and thalamus with a minute distribution in the neocortex (Mishra et al. 2018). It's been variously reported that nicotine stimulates DA functions including the cognitive one with some reports saying dopaminergic system negatively regulate nicotinic system (Benowitz 2009). Intracellularly, the receptors initiate a cascade reaction involving cyclic adenosine monophosphate (cAMP) formation and activation of cAMP-dependent protein kinase A (PKA) which results into cell physiological output.

DA neurons originating from midbrain widely innervate the various parts of cortical and subcortical regions forming nigrostriatal, mesolimbic, mesocortical, and thalamic dopaminergic systems (De Keyser et al. 1989; Islam et al. 2021). Nigrostriatal and mesolimbic pathway are subcortical DA neuron system. Cell bodies of nigrostriatal DA neurons are in the region of substantia nigra. While mesolimbic DA system originates from more diffused neuron collection in the ventral tegmental area (Islam et al. 2021). Some of the neurons from here projects to limbic system as nucleus accumbens, amygdala, hippocampus, and anterior cingulate cortex. For mesocortical DA system, the neurons originate from ventral tegmentum and projects all over the area of neocortex. Last one, the thalamic DA system projects its neuron to the area of paraventricular nucleus (PVN) and rostral ventrolateral medulla. All these four DA systems are involved in wide spectrum of brain functions such as reinforcement learning, episodic memory, working memory, cognitive control, and award/novelty-related motivational processing (De Keyser et al. 1989; Islam et al. 2021). It is imperative to note here that the PVN is an integral part of hypothalamic neuroendocrine circuit which controls appetite, fluid balance, and general metabolism of an individual (Ahima and Antwi 2008; Li et al. 2019a).

Serotonin or 5-hydroxytryptamine (5HT) has been reported to be working in across all the phyla of animal kingdom. This chemical is required for fine tuning of almost all the animal behaviors and neuronal responses. Chemically, 5HT is derived from amino acid tryptophan by the action of two enzymes, tryptophan hydroxylase and decarboxylase. 5HT can bind to seven receptor types, each having further subtypes. These receptors can either be serotonin-gated ion channels (e.g. 5HT3) or GPCR receptors (e.g. 5HT1, 5HT2, and 5HT4–7) (Tsegay et al. 2020). At intracellular level, these receptors increase the concentration of cAMP which is an important component of synaptic long-term potentiation. This neurotransmitter is found in both CNS and PNS. Functionally, this serotoninergic system is associated with various neurophysiological and neuropathological aspects. This includes, control of memory, circadian rhythm, motor activities, arousal, sleep, feeding, social interactions, anxiety, mood, aggressiveness, impulsive and compulsive dimensions, and general behavioral plasticity (Jenkins et al. 2016; Tortora et al. 2023). Moreover, 5HT receptors have ability to increase the Ach release in brain regions responsible for memory processes and learning as to cause the serotonin's positive effect on general cognition in both young and aged animals (Nair and Gudelsky 2004).

4.3 Exploration of Natural Compounds That Modulate Neurotransmitter Systems

The phytochemical system represents a captivating frontier in neuroscience and neuropharmacology, offering plenty of bioactive molecules with diverse therapeutic potential. With a growing interest in harnessing the therapeutic benefits of natural products, researchers are delving into the intricate interplay between phytochemicals and neurotransmitter pathways in the brain. These natural compounds, derived from plants, herbs, and other sources, can interact with crucial neurotransmission components, including receptors, enzymes, and transporters. By regulating the action of neurotransmitters like serotonin, dopamine, norepinephrine, Ach, GABA, and glutamate, these natural compounds show potential for treating many neuropsychiatric illnesses, cognitive abnormalities, and overall brain health. As scientific understanding deepens and innovative research methodologies emerge, the exploration of natural compounds continues to unveil novel therapeutic avenues with the potential to revolutionize the field of neuropsychopharmacology.

4.3.1 Phytochemicals

Phytochemicals are plant-derived natural compounds that have potential human health benefits. A diverse array of phytochemicals, including alkaloids, flavonoids, terpenoids, carotenoids, alkanes, fatty acids, vitamins, valproate, saponins, and monoterpenes, are obtained from the fruits, seeds, vegetables, leaves, roots, and stem of plants. These bioactive compounds are being extensively researched for cognitive enhancement and their association to resist disease onsets, including cardiovascular disease, cancer, diabetes, and neurodegenerative disorders (Asgharian et al. 2022). They possess diverse biological roles and multiple modes of action to influence cellular health through different mechanisms, such as scavenging free radicals, modulating cell signalling pathways, and altering enzyme activity (Küpeli Akkol et al. 2021; Mohd Sairazi and Sirajudeen 2020).

4.3.1.1 Types of Phytochemicals That Modulate CNS

Although all phytochemicals are commonly derived from the kingdom Plantae, each has a different originating source, biosynthesis pathway, chemical nature, action mechanism, and biological role. Thus, these bioactive compounds are classified as follows:

A. *Flavonoids*: Citrus fruits, apples, cocoa, berries, and teas are rich sources of flavonoids. In plants, flavonoids are secondary metabolites having a typical C15 benzene ring (C6–C3–C6) structure derived from the phenylpropanoid metabolic pathway (Liu et al. 2021). These compounds hold antioxidant and anti-inflammatory properties that promote synaptogenesis and neurogenesis and modulate synaptic plasticity (Cichon et al., 2020; Ramezani et al. 2022). Flavonoids also impact endothelial function, enhance the cerebrovascular system (Rees et al. 2018; Rendeiro et al. 2015), and influence the neurotransmitter system, such as norepinephrine, dopamine, and serotonin (Pannu et al. 2021). The flavonoid-induced modulation of neurotransmitters increases the formation of brain-derived neurotrophic factor (BDNF) and other neuronal factors in the brain, thereby improving cognitive function and neuroprotection (German-Ponciano et al. 2018).

B. *Phenolic acids*: Berries, nuts, and whole grains are abundant sources of phenolic acids. Phenolic acids are potent anti-inflammatory chemicals that inhibit neuronal inflammatory pathways and decrease neuronal oxidative stress (Dong et al. 2022; Fakhri et al. 2022; González de Llano et al. 2023). In mice, ferulic acid, a phenolic acid, inhibits apoptosis, preserves neuronal integrity, and improves cognitive health by reducing oxidative stress (Mei et al. 2023). In rats, gallic acid administration decreases pro-inflammatory cytokines, restores morphology and dendritic spine density of hippocampal neurons, and improves cognition (Diaz et al. 2021). Hase et al. (2019) reported that Rosmarinic acid feeding restored monoamines, including dopamine, levels in the cerebral cortex by inhibiting monoamine oxidase B (MAO–B) enzyme in mice models of AD (Hase et al. 2019). Thus, phenolic acids preserve neuronal integrity and improve cognitive health.

C. *Carotenoids*: Carotenoids are among the most common and widely distributed pigments responsible for the vibrant colors of fruits and vegetables. Lutein and zeaxanthin are the two carotenoids studied for their direct effect on cognitive health. The two carotenoids are reported to accumulate in the CNS and retina, prevent the neuronal membrane from extrinsic oxidative damage and inflammation, and thereby improve visual episodic memory and cognitive function in humans (Lopresti et al. 2022; Wang et al. 2022). However, the mechanism is yet to be determined.

D. *Polyphenols*: Polyphenols are free radical scavengers found exclusively in cocoa, grapes, and olive oil. According to an animal study, tea polyphenols decreased oxidative stress, altered the AKT/CREB/BDNF signalling cascade, increased the nuclear translocation of Nrf2, and augmented phase II detoxification enzymes essential for brain tissue survival (Qi et al. 2017). Meta-analysis of clinical studies on polyphenol supplementation reported that it enhances the circulation of pro-cognitive neurotrophic factors in the brain (Ammar et al. 2020). Research shows that polyphenols interact with the brain proteins and lipid signalling cascade, promoting synaptic plasticity and survivability, and enhancing vascularisation and neurogenesis (Vauzour 2017). Moreover, polyphenols also regulate neurotransmitter systems, including the metabolism and function of chemical messengers like Ach, GABA, glutamate, serotonin, and dopamine (Rebas et al. 2020).

E. *Terpenoids*: Terpenoids, also known as terpenes, constitute a significant class of phytochemicals responsible for the characteristic scents and flavors of many spices, oils, and herbs. Occurring less frequently in the diet, this structural group of phytochemicals plays both attractant and deterrent roles in plants. The research investigated that a single dose of monoterpenes derived from *Salvia* officinalis (Sage) and *Mentha piperita* (peppermint), diterpenes derived from *Ginkgo biloba*, and triterpenes containing plant extract obtained from *Panax ginseng* (Ginseng) and *Bacopa monnieri* improve cognitive function, alertness, and psychological state (Kennedy 2019). Terpenoids are excellent cholinesterase inhibitors and are being studied to cure AD; however, future research will determine their potency in treating AD (Min et al. 2022).

F. *Alkaloids*: Alkaloids are nitrogen-containing phytocompounds in various plant species and play a vital role in plant growth and defence. The compound also acts as a nitrogen reservoir in plants. Isoquinoline alkaloids and amide alkaloids present in plants potentially ameliorate neuronal deterioration in neurodegenerative diseases, like AD, PD, and cerebral ischemia, by suppressing enzymes involved in progressive neuronal loss, such as MAO and β-secretase (Li et al. 2023; Othman et al. 2022). Caffeine, a purine alkaloid, modulates the activities of various synaptic receptors and channels. It influences mobilization of calcium ions, inhibition of phosphodiesterase, and activities of NMDA receptor, adenosine receptor, and GABA receptor in the hippocampal and neocortex region (Martins et al. 2020; Yoshimura 2005). These actions of caffeine turn up the synaptic plasticity, activity, and transmission and restore synaptic morphology to enhance cognitive health (Martins et al. 2020; Yoshimura 2005).

G. *Saponins*: Saponins are triterpenoid or steroidal glycosides consisting of an aglycone part (non-sugar), a glycone part, and at least one glycosidic linkage to bind glycone and non-glycone part. They exhibit neuroprotective properties against many neurodegenerative diseases of CNS, including AD. In AD, they modulate neuroinflammation and oxidative stress to improve cognitive health and brain function (Abduljawad et al. 2022).

4.3.1.2 Phytochemicals Interaction with Neurotransmitters

A. *Phytochemicals modulating cholinergic neurotransmission*: Ach is a major neurotransmitter that plays a crucial role in learning and memory. During attention, Ach is released from the cholinergic neurons and facilitates spike time-dependent plasticity (Sugisaki et al. 2022). Moreover, the released Ach binds and activates muscarine Ach receptors and nicotine Ach receptors. The two receptors in the hippocampus are essential for postsynaptic changes and memory consolidation (Sugisaki et al. 2022). Several Acetylcholinesterase (AchE) inhibitors—chemicals that inhibit the catalytic function of AchE, an enzyme that breaks down Ach into acetic acid and choline—are being investigated for their therapeutic role in cognition restoration in disease and impairments. Many phytochemicals of the diverse family are being studied for their interaction with AchE, restoration of Ach, and cognitive enhancement. Koly et al. (2023) reported that berberine, anthocyanin, and diterpene alkaloid easily cross the blood–brain barrier, bind and inhibit AchE activity to preserve Ach, and restore cognition in AD (Koly et al. 2023). The in silico study investigated that either C(2) substitution of Huperzine A or its dimerization with tacrine develops a noble derivative that potentially binds with AchE and inhibits its activity (Anukanon et al. 2021; Mak et al. 2021). Also, rosmarinic acid derived from *Salvia officinalis*, galantamine derived from *Galanthus nivalis*, and turmeric's bioactive phytochemical curcumin inhibit AchE activity (Bisset et al. 2022; Kamli et al. 2022). Thus, understanding phytochemicals impeding AchE activity could be applied to develop therapeutic compounds and improve cognitive health in neurodegenerative diseases and mental impairments.

B. *Phytochemicals modulating dopaminergic neurotransmission*: Dopamine is a neurotransmitter of catecholamine and phenethylamine families synthesized from its precursor L-DOPA in the brain. Flavonoids, such as catechins and quercetin, are reported to optimize dopamine metabolism and promote dopamine release in rodent PD models (Josiah et al. 2022). Salidroside, a glucoside extracted from *Rhodiola rosea*, protects dopaminergic neurons and prevents PD; however, its role in dopamine receptor activity and its release is yet to be investigated (Li et al. 2019b). The literature demonstrating the dietary supplement of amino acid L-theanine showed that L-theanine increased relaxation and mental clarity, contributing to enhanced cognition by augmenting BDNF and dopamine levels in the brain (Xu et al. 2022). In addition, polyphenols such as resveratrol and carotenoids like lycopene are potent phytochemicals that improve dopamine levels in the brain (Khaksar et al. 2016; Li et al. 2017).

C. *Phytochemicals modulating serotonergic neurotransmission*: Serotonin, is a monoamine neurotransmitter primarily synthesized in serotonergic neurons and intestines. The neurotransmitter is biosynthesized through an enzyme-catalysed process, packed into the specialized synaptic vesicles, and released from the presynaptic terminals of serotonergic neurons when required. It regulates various physiological processes, including mood, sleep, appetite, and cognitive function. An imbalance in serotonin levels affects mental well-being and is linked to mood disorders such as depression and anxiety (Yohn et al. 2017). Kaempferol, a flavonoid extracted from *Cudrania tricuspidata*, is reported to interact with serotonin type 3A (5-HT3A) receptor by binding it in concentration-dependent and voltage-dependent manner and inhibiting the serotonergic pathway (Lee et al. 2022b). The research investigated that the inhibition of the 5–HT3A receptor and attenuation of serotonin in cholinergic pathways and dopaminergic pathways increases the expulsion of Ach and dopamine in the brain, respectively, and enhances memory and cognition (Dremencov et al. 2006; Maura et al. 1992). The in silico molecular docking technique revealed that the isoflavones (subclass of flavonoid family), L-tryptophan, peonidin, and catechin in banana and pineapple bind to serotonin transporter and inhibit its activity (Ofir et al. 2003; Purnawati et al. 2022) These phytochemicals could be a potent antidepressant agent to enhance mood (Purnawati et al. 2022). Several other phytochemicals modulating the serotonergic neurotransmission are curcumin, resveratrol, quercetin, and epigallocatechin gallate (EGCG) (Kulkarni and Dhir 2010; Li et al. 2020; Mendonça et al. 2022; Moore et al. 2018).

D. *Phytochemicals modulating GABAnergic neurotransmission*: In the CNS, GABA is an important inhibitory neurotransmitter regulating neuronal excitability. It is synthesized in the GABAnergic neurons, released from there, and binds to GABA receptors on adjacent postsynaptic neurons, leading to their hyperpolarization and neuronal firing inhibition (Excitatory et al. 1997). The inhibitory action of GABA is vital in modulating anxiety, stress, and overall brain function (Sharmin et al. 2023). Several phytochemicals interact with GABA receptors and potentially enhance cognitive function. The L-theanine in green tea reduces anxiety and improves attention and memory by strengthening GABAnergic neurotransmission; increased GABA increases the dopamine level, reduces serotonin level, and results in well-being (Higashiyama et al. 2011; Li et al. 2022). The anxiolytic effects of valerenic acid are also due to enhanced GABAnergic neurotransmission and the promotion of inhibitory signalling in the brain (Trauner et al. 2008; Yuan et al. 2004). Magnolol, a phenolic phytochemical obtained from the bark of *Magnolia officinalis*, and kavalactones, bioactive compounds in kava root, promote GABA activity and improve relaxation and sleep by modulating GABA receptors and increasing GABA levels in the hippocampus (Alexeev et al. 2012; Shi et al. 2014). Additionally, curcumol is an allosteric modulator of GABA receptors to modulate GABAnergic neurotransmission and increase GABA levels (Liu et al. 2017).

4.3.2 Neurotransmitters Imbalance and Cognitive Disorders

Various cognitive disorders arise from the inadequate release or/and biodegradation of neurotransmitters in the brain, leading to their insufficiency and disruption of brain function (Fig. 4.4). Some of these disorders include the following:

A. *Depression*: It is a heterogeneous cognitive disorder mainly characterized by agitation, anxiety, insomnia, loss of appetite, negative emotions, and suicidality. While in the subgroup of patients, decreased activity of serotonin onsets depression pathogenesis, in other subgroups, depletion of dopamine, and norepinephrine in CNS develops this disorder (Rihmer et al. 2022; Tavakoli et al. 2018).
B. *Anxiety disorders*: Imbalance in serotonin, GABA, and dopamine, particularly within the hippocampus, amygdala, locus coeruleus, and raphe nuclei, contribute to a spectrum of anxiety disorders (Nuss 2015; Pałasz et al. 2021). The anxiety disorder is diagnosed by excessive apprehension, fear, restlessness, palpitations, and sweating.
C. *Schizophrenia*: Individuals with schizophrenia develop dopamine dysregulation—hyperactive neurotransmission in the mesolimbic zones and hypoactive transmission in the prefrontal cortex; however, deviated levels of glutamate, GABA, and serotonin are also reported in different midbrain areas of people with schizophrenia (Brisch et al. 2014). The hallmarks of the disorder include hallucinations and delusion that appear due to increased subcortical release of dopamine and augmented activation of the D_2 receptor and anhedonia, impaired cognition, and disorganized thinking due to reduced activation of the D_1 receptor in the prefrontal cortex (Brisch et al. 2014).
D. *Bipolar disorder*: Bipolar disorder is a psychiatric condition distinguished by profound fluctuations in an individual's energy, activity level, and concentra-

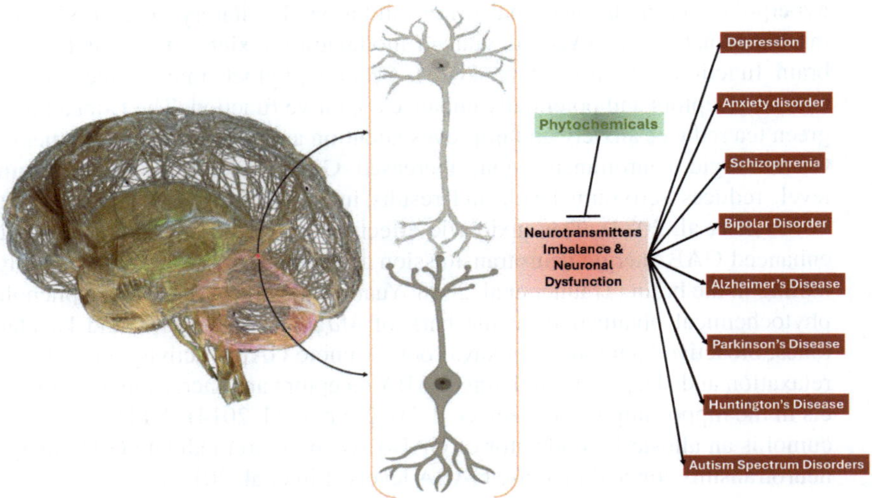

Fig. 4.4 In case of imbalance and dysfunction of neurotransmitters, the survival of neurons is at stake. The phytochemicals can act as remedy in such cases, if administered. Otherwise, the situation may aggravate causing various cognitive, motor, and sensory disorders such as depression, anxiety, schizophrenia, AD, PD, HD, ASD etc.

tion, in addition to extreme mood swings. It is often linked to neurotransmitter dysregulation, such as elevated serotonin levels and decreased norepinephrine and dopamine levels (Lee et al. 2022a).

E. *Alzheimer's disease*: It is a neurodegenerative disorder characterized by a gradual deterioration of memory function and cognition, including confusion, difficulties with language, and decision-making. Depletion of Ach is a hallmark of AD. However, significant depletion of glutamate, GABA, and serotonin levels is also marked in AD patients (Yang et al. 2023).
F. *Parkinson's disease*: The progressive and specific loss of A9 dopaminergic neurons and depletion of dopamine in the brain are the causes of the onset of PD. The condition manifests with motor signs such as resting tremors, bradykinesia, rigidity, postural instability, and non–motor symptoms such as cognitive impairment, mood disruption, and sleep issues. While the motor symptoms in PD are due to the depletion of dopamine, the non–motor symptoms in PD are linked to the insufficiency of GABA and abundance of Ach and glutamate in the extrapyramidal system (Takeda et al. 2021; Werner and Coveñas 2014).
G. *Huntington's disease*: A mutation in the huntingtin gene causes the disorder that leads to neuronal degeneration, particularly in the basal ganglia, resulting in cognitive decline, psychiatric disturbances, and hyperkinetic movements. The symptoms in HD patients appear due to imbalances in dopamine and glutamate (Chen et al. 2013).
H. *Autism spectrum disorder (ASD)*: ASD is a spectrum of neurodevelopment disorders identified by social communication challenges, repetitive behaviours,

and restricted interests. The contributing neurotransmitter systems for the onset of the disorder are monoamine neurotransmitters, including serotonin, norepinephrine, and dopamine (Kuo and Liu 2022).

4.3.3 Phytochemicals for Amelioration of Cognitive Disorders

Phytochemicals offer promising restoration of cognitive health due to their neuroprotective properties. The flavonoids increase cognitive function by promoting neurogenesis, improving blood flow, and reducing cellular oxidative stress (Cichon et al., 2020; Ramezani et al. 2022; Rees et al. 2018). Additionally, phenolic acids resist neuronal loss and cognitive decline through their anti-inflammatory properties (Dong et al. 2022; Fakhri et al. 2022; González de Llano et al. 2023). The antioxidant and anti-inflammatory properties of curcumin protect against neurodegeneration in PD and mitigate amyloid plaque formation and cognitive decline in AD (Bisset et al. 2022). Curcumin binds and significantly inhibits Achsterase and butyrylcholinesterase, critical enzymes involved in AD (Bisset et al. 2022). Clinical investigations indicated that the Achsterase inhibitory effect of Huperzine A, a natural alkaloid produced from the Chinese club moss, enhances cognitive function and memory in individuals with AD and vascular dementia (Wang et al. 2009; Zhang 2012). Data from clinical trials suggest that bacosides extracted from Brahmi improve attention, memory consolidation, and cognitive performance (Fatima et al. 2022; Kumar et al. 2016). Similarly, Panax ginseng, crocin, and safranal improve attention, cognition function, and age-related cognitive decline (Abbaszadeh-Mashkani et al. 2021; D'Onofrio et al. 2021; Park et al. 2019), refer to Table 4.3 for elaboration.

4.4 Discussion of Their Impact on Cognitive Enhancement and Neuroprotection

Phytochemicals, which are bioactive compounds found in plants, have garnered significant attention for their potential in enhancing cognitive functions and providing neuroprotection. Research in modern psychology and neuroscience indicates that these compounds can influence brain function by modulating neurotransmitter systems, enhancing synaptic plasticity, and protecting against neural damage (Ahmad and Pal 2022; Dubey and Pandey 2023; Kamli et al. 2022; Kennedy 2019; Kumar and Khanum 2012). Cognitive functions such as memory, attention, and executive function are crucial for daily living and are often impaired in neurodegenerative diseases like AD and PD. Flavonoids, present in fruits such as blueberries and grapes, have been shown to enhance memory and learning by increasing cerebral blood flow and promoting synaptic plasticity through the upregulation of BDNF (Ammar et al. 2020). BDNF is crucial for the survival and growth of neurons, and its increased expression is associated with improved cognitive performance (Lu et al. 2014). As discussed, Huperzine A derived from the Chinese club moss,

Table 4.3 Chemical nature and functions of phytochemicals

Sr. no.	Phytochemicals	Source	Chemical family	Function	References
1.	Huperzine A	Club moss (*Huperzia serrata*)	Alkaloid	Improves cognition function and memory in patients with AD and vascular dementia	Wang et al. (2009), Zhang (2012)
2.	Rosmarinic acid	Lamiaceae herbs	Polyphenolic acid	Potentially suppresses amyloid β aggregation and cognitive deficit in AD	Yamamoto et al. (2021)
3.	Salvianolic acid	*Salvia miltiorrhiza*	Polyphenolic acid	Salvianolic acid exposure in rats shows significant improvement in cognitive deficits associated with vascular dementia	Ma et al. (2017)
4.	Galantamine	Amaryllidaceae family	Alkaloid	It inhibits Achsterase and amplifies the action of Ach on nicotinic receptors. Its short-term dose is effective in ameliorating dementia and cognitive function in mild to moderate AD	Razay and Wilcock (2008)
5.	Curcumin	Turmeric	Polyphenol	Curcumin improves the physiological and behavioral cognition in AD and its animal models. However, the human studies are limited and inconsistent	Bisset et al. (2022), Voulgaropoulou et al. (2019)
6.	Epigallocatechin gallate (EGCG)	Green tea	Catechin	In animals, it remediates cognitive impairment by augmenting glutamate and GABA levels	Mostafa et al. (2022)

7.	Apigenin	Chamomile, Parsley, and Celery	Flavonoid	It corrects induced cognitive dysfunction in mice. Few clinical trials have shown its therapeutic potency for cognitive impairments in humans	Kim et al. (2021), Nabavi et al. (2018)
8.	Quercetin	Citrus fruits, parsley, tea, onions, and berries	Flavonoid	It promotes dopamine release and serotonergic neurotransmission, but its role in neurocognitive function in humans is obscure	Broman-Fulks et al. (2012), Chakraborty et al. (2014), Josiah et al. (2022)
9.	Salidroside	*Rhodiola rosea*	Glucoside	It improves the cognitive dysfunction associated with neurological disorders in model animals. Further research is needed to test its clinical efficacy	Lai et al. (2015), Zhong et al. (2018)
10.	L-Theanine	Green tea	Amino acid derivative	L-Theanine improves attention and enhances memory and executive functions	Baba et al. (2021)
11.	Resveratrol	Grapes, peanuts, and mulberries	Polyphenol	Randomized clinical trials in humans showed improved cerebral blood flow, perceived performances, and some parameters of cognitive decline	Cicero et al. (2019)
12.	Oleuropein	Oleaceae family	Polyphenol	It has multiple action modes to combat several neurological disorders, but its role in mitigating cognitive dysfunction needs further clinical trials to gain knowledge	Grubić Kezele and Ćurko-Cofek (2022)

(continued)

Table 4.3 (continued)

Sr. no.	Phytochemicals	Source	Chemical family	Function	References
13.	Berberine	Goldenseal, Barberry, and Oregon grapes	Alkaloid	Berberine improves cognitive impairments in rats and mice models of several diseases. Its effect on humans is not known	Wu et al. (2021), Yao et al. (2023)
14.	Lycopene	Tomatoes, watermelon, etc.	Carotenoid	Evidence of clinical potency to increase cognitive longevity and dementia in humans is insufficient	Crowe-White et al. (2019), Imran et al. (2020)
15.	Astaxanthin	Marine algae	Pigment	It is a promising therapeutic agent for multifactorial neurodegenerative diseases and synaptic plasticity, but its clinical role in improving cognitive capabilities needs further research	Grimmig et al. (2017), Queen et al. (2024)
16.	Limonene	Rutaceae family and coniferous trees	Monocyclic terpene	Limonene protects dopaminergic and GABAnergic neurons and inhibits anxiety–related behavior in rodents. Its efficacy in humans needs clinical validation	Song et al. (2021)
17.	Caffeine	Coffee beans, tea leaves, kola nuts, etc.	Alkaloid	Low to moderate doses of caffeine enhance alertness, attention, and reaction time but have inconsistent effects on memory, judgment, and decision-making	McLellan et al. (2016)

functions as an Achsterase inhibitor. This action has been shown to enhance cognitive performance, particularly in individuals with AD, where Ach levels are typically diminished (Wang et al. 2009). Clinical studies have confirmed that Huperzine A can improve memory and cognitive functions, making it a promising candidate for managing cognitive impairments (Anukanon et al. 2021; Zhang 2012).

Neuroprotection involves safeguarding neurons from damage and death caused by oxidative stress, inflammation, and excitotoxicity. Curcumin, the active compound in turmeric, has potent antioxidant and anti-inflammatory properties. It inhibits the aggregation of amyloid-beta plaques, which are implicated in AD and reduces oxidative damage to neural cells (Bisset et al. 2022; Voulgaropoulou et al. 2019). Similarly, resveratrol, found in grapes and red wine, activates sirtuins and enhances mitochondrial function, thereby offering protection against cognitive decline and neurodegeneration (Cicero et al. 2019). Overall, phytochemicals offer a multifaceted approach to enhancing cognitive function and protecting against neurodegenerative diseases (Table 4.3). They achieve these benefits through mechanisms such as antioxidant activity, anti-inflammatory effects, and modulation of neurotrophic factors, underscoring their potential as therapeutic agents in cognitive health. Polyphenols, another group of phytochemicals found in a variety of fruits, vegetables, and beverages like tea and red wine, also contribute to cognitive enhancement. Resveratrol, a well-known polyphenol in grapes and red wine, has been shown to improve cerebral blood flow and support neuronal health through the activation of sirtuins and enhancement of mitochondrial function. Further, research suggests that resveratrol supplementation can lead to improvements in memory and overall cognitive performance in older adults, thereby helping to mitigate age-related cognitive decline (Cicero et al. 2019; Li et al. 2017; Moore et al. 2018).

Food sources rich in neurotransmitter precursors and cofactors can influence neurotransmitter synthesis and function in the brain. For example, amino acids such as tryptophan (a precursor of serotonin) and tyrosine (a precursor of dopamine and norepinephrine) are found in protein-rich foods like meat, dairy products, nuts, and seeds. These amino acids cross the blood-brain barrier and participate in neurotransmitter synthesis when available in sufficient quantities. Similarly, cofactors such as vitamins (e.g. vitamin B6, folate) and minerals (e.g. magnesium, zinc) are essential for the enzymatic processes involved in neurotransmitter metabolism and function. Foods rich in these nutrients, including leafy greens, whole grains, and legumes, support optimal neurotransmitter activity and overall brain health (Table 4.4).

The balanced synthesis and function of neurotransmitters are crucial for maintaining cognitive functions such as memory, attention, and mood regulation. Dysregulation of neurotransmitter levels or function has been implicated in various neurological and psychiatric disorders, including depression, anxiety disorders, schizophrenia, and PD. Therefore, dietary strategies that promote the intake of neurotransmitter precursors, cofactors, and supportive nutrients play a significant role in supporting brain function and mental well-being. Further research into the interplay between diet, neurotransmitter systems, and brain health continues to provide insights into potential therapeutic approaches for optimizing cognitive function and managing neurological conditions.

Table 4.4 Neurotransmitter structure, function, and food sources

Neurotransmitter	Chemical structure	Function name	Functions	Food source
Dopamine	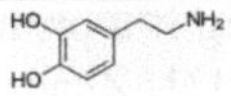	Pleasure neurotransmitter	Enthusiasm, recollection, cognition, awareness, decision-making, sleep control	Poultry, fish, sesame, banana, soya product, legumes, wheat, watermelon, tomato, olive oil
Serotonin	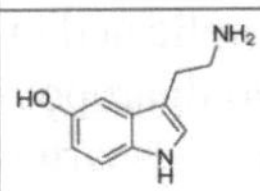	Mood neurotransmitter	Mood, appetite, social behavior, sleep, sexual drive, reminiscence, learning, gastrointestinal mobility	Banana, cherry, chicory, cabbage, pineapple, papaya, hazelnuts, coffee beans, brinjal, green grapes, legumes, onions
Ach	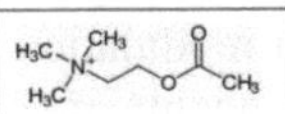	Learning neurotransmitter	Recollection, concentration, attention, muscular activities, thinking, mood	Egg yolk, turkey, beef, green split peas, liver, salmon, soya beans, mung beans, lentils, orange, strawberry, radish
Epinephrine	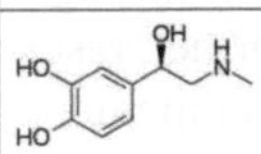	Fight or flight neurotransmitter	Increase heartbeat, tear production, blood pressure, trigger fight or flight response	Derived from phenylalanine and/or tyrosine
Norepinephrine	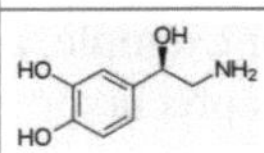	Fight or flight neurotransmitter	Increase in heart rate and blood pressure, fast breathing, strength, and physical performance elevation	Derived from phenylalanine and/or tyrosine
GABA	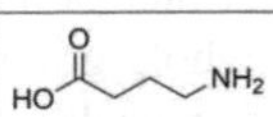	Calming neurotransmitter	Relaxation, stress diminution, CNS relaxation, mood stabilization	Lupin, adzuki beans, soya beans, common beans, pea, oat, wheat, barley, rice, potato, spinach

4.5 Conclusion

Cognition refers to the processes of acquiring, storing, transforming, and using information, managed by the brain—a complex structure housing nearly 100 billion neurons and 1000 billion glial cells (Yang et al. 2022). Advanced techniques like image-derived phenotypes (IDPs), single nucleotide polymorphisms (SNPs), and magnetic resonance imaging (MRI) have elucidated brain structure-function relationships, developmental influences, behavioral variances, and disease mechanisms (Wang et al. 2023; Zhao et al. 2023).

Neurotransmitters and neuropeptides act as chemical messengers facilitating communication among brain structures and cells. Synthesized in presynaptic axon terminals or locally recycled, neurotransmitters are stored in vesicles and released into the synaptic cleft upon calcium influx triggered by action potentials. This release activates receptors on postsynaptic membranes, initiating ionotropic (rapid and short-lived) or metabotropic (slower and longer-lasting) effects (Osimo and Stahl 2023; Roth 2019). These molecules play pivotal roles in CNS, ANS, and PNS communication, regulating motor, sensory, and integrative signals, as well as influencing cognitive and psychological functions such as emotions, memories, and behaviors (McCutcheon et al. 2020; Teleanu et al. 2022). Dysregulation of neurotransmitter physiology contributes to various neurological and psychiatric disorders, underscoring their importance in nervous system function and health (Sarter et al. 2007). Over 200 neuroactive compounds have been identified, classified into canonical neurotransmitters and those still under investigation (Alavijeh and Palmer 2010). Research continues to expand our understanding of these compounds and their roles in brain function, with implications for treating cognitive decline and neurodegenerative diseases.

The integration of findings from psychology and physiology highlights phytochemicals' potential in cognitive health by enhancing neurotransmission, reducing inflammation, and protecting brain cells from oxidative damage. Their ability to cross the blood-brain barrier underscores their effectiveness as natural agents for cognitive enhancement and neuroprotection. Continued research is essential to fully understand their mechanisms and optimize their therapeutic applications in promoting brain health and resilience against cognitive decline. Phytochemicals like flavonoids and phenolic acids, found in fruits like blueberries, grapes, and coffee, enhance cognitive functions by improving synaptic plasticity, increasing cerebral blood flow, and upregulating BDNF—a crucial factor supporting neuronal growth and synaptic connections (Lu et al. 2014). Regular consumption of these compounds correlates with improved memory, learning, and executive functions, suggesting their role in optimizing cognitive processes and protecting against neurodegenerative diseases. These phytochemicals operate through diverse mechanisms including modulation of neurotransmitter systems, neurotrophic factor upregulation, and antioxidant properties that mitigate oxidative stress and neuroinflammation. Compounds such as Huperzine A, polyphenols, curcumin, resveratrol, and ginsenosides contribute significantly by enhancing synaptic plasticity, supporting mitochondrial function, and inhibiting amyloid-beta plaque formation, critical in managing Alzheimer's and Parkinson's diseases.

References

Abbaszadeh-Mashkani S, Hoque SS, Banafshe HR, Ghaderi A (2021) The effect of crocin (the main active saffron constituent) on the cognitive functions, craving, and withdrawal syndrome in opioid patients under methadone maintenance treatment. Phytother Res 35:1486–1494. https://doi.org/10.1002/PTR.6913

Abduljawad AA, Elawad MA, Elkhalifa MEM, Ahmed A, Hamdoon AAE, Salim LHM, Ashraf M, Ayaz M, Hassan SS u, Bungau S (2022) Alzheimer's disease as a major public health concern: role of dietary saponins in mitigating neurodegenerative disorders and their underlying mechanisms. Molecules 27:6804–6804. https://doi.org/10.3390/MOLECULES27206804

Ahima RS, Antwi DA (2008) Brain regulation of appetite and satiety. Endocrinol Metab Clin N Am 37:811–823. https://doi.org/10.1016/j.ecl.2008.08.005

Ahmad A, Pal K (2022) Therapeutic efficacy of natural phytochemicals as acetylcholinesterase inhibitors against Alzheimer's disease. J Ecophysiol Occup Health 22:39. https://doi.org/10.18311/JEOH/2022/29511

Alavijeh MS, Palmer AM (2010) Measurement of the pharmacokinetics and pharmacodynamics of neuroactive compounds. Neurobiol Dis 37:38–47. https://doi.org/10.1016/j.nbd.2009.09.025

Alexeev M, Grosenbaugh DK, Mott DD, Fisher JL (2012) The natural products magnolol and honokiol are positive allosteric modulators of both synaptic and extra-synaptic GABA(A) receptors. Neuropharmacology 62:2507–2514. https://doi.org/10.1016/J.NEUROPHARM.2012.03.002

Ammar A, Trabelsi K, Boukhris O, Bouaziz B, Müller P, Glenn JM, Bott NT, Müller N, Chtourou H, Driss T, Hökelmann A (2020) Effects of polyphenol-rich interventions on cognition and brain health in healthy young and middle-aged adults: systematic review and meta-analysis. J Clin Med 9:1598. https://doi.org/10.3390/JCM9051598

Anukanon S, Pongpamorn P, Tiyabhorn W, Chatwichien J, Niwetmarin W, Sessions RB, Ruchirawat S, Thasana N (2021) In silico-guided rational drug design and semi-synthesis of C(2)-functionalized huperzine A derivatives as acetylcholinesterase inhibitors. ACS Omega 6:19924–19939. https://doi.org/10.1021/ACSOMEGA.1C02875

Asgharian P, Quispe C, Herrera-Bravo J, Sabernavaei M, Hosseini K, Forouhandeh H, Ebrahimi T, Sharafi-Badr P, Tarhriz V, Soofiyani SR, Helon P, Rajkovic J, Durna Daştan S, Docea AO, Sharifi-Rad J, Calina D, Koch W, Cho WC (2022) Pharmacological effects and therapeutic potential of natural compounds in neuropsychiatric disorders: an update. Front Pharmacol. https://doi.org/10.3389/fphar.2022.926607

Baba Y, Inagaki S, Nakagawa S, Kaneko T, Kobayashi M, Takihara T (2021) Effects of l-theanine on cognitive function in middle-aged and older subjects: a randomized placebo-controlled study. J Med Food 24:333. https://doi.org/10.1089/JMF.2020.4803

Benowitz NL (2009) Pharmacology of nicotine: addiction, smoking-induced disease, and therapeutics. Annu Rev Pharmacol Toxicol 49:57–71. https://doi.org/10.1146/annurev.pharmtox.48.113006.094742

Bisset S, Sobhi W, Bensouici C, Khenchouche A (2022) Antioxidant activity and inhibitory effect of curcumin on some enzymes involved in several diseases: acetylcholinesterase, butyrylcholinesterase, α-glucosidase and tyrosinase. Curr Enzym Inhib 18:172–179. https://doi.org/10.2174/1573408018666220602091615

Bowery NG, Smart TG (2006) GABA and glycine as neurotransmitters: a brief history. Br J Pharmacol 147 Suppl:S109–S119. https://doi.org/10.1038/sj.bjp.0706443

Brisch R, Saniotis A, Wolf R, Bielau H, Bernstein HG, Steiner J, Bogerts B, Braun K, Kumaratilake J, Henneberg M, Gos T (2014) The role of dopamine in schizophrenia from a neurobiological and evolutionary perspective: old fashioned, but still in vogue. Front Psych 5. https://doi.org/10.3389/FPSYT.2014.00047

Broman-Fulks JJ, Canu WH, Trout KL, Nieman DC (2012) The effects of quercetin supplementation on cognitive functioning in a community sample: a randomized, placebo-controlled trial. Ther Adv Psychopharmacol 2:131. https://doi.org/10.1177/2045125312445894

Chakraborty J, Singh R, Dutta D, Naskar A, Rajamma U, Mohanakumar KP (2014) Quercetin improves behavioral deficiencies, restores astrocytes and microglia, and reduces serotonin metabolism in 3-nitropropionic acid-induced rat model of Huntington's disease. CNS Neurosci Ther 20:10. https://doi.org/10.1111/CNS.12189

Chen JY, Wang EA, Cepeda C, Levine MS (2013) Dopamine imbalance in Huntington's disease: a mechanism for the lack of behavioral flexibility. Front Neurosci 7:114. https://doi.org/10.3389/fnins.2013.00114

Cicero AFG, Ruscica M, Banach M (2019) Resveratrol and cognitive decline: a clinician perspective. Arch Med Sci 15:936. https://doi.org/10.5114/AOMS.2019.85463

Cichon N, Saluk-Bijak J, Gorniak L, Przyslo L, Bijak M (2020) Flavonoids as a natural enhancer of neuroplasticity-an overview of the mechanism of neurorestorative action. Antioxidants (Basel) 9:1–19. https://doi.org/10.3390/ANTIOX9111035

Crowe-White KM, Phillips TA, Ellis AC (2019) Lycopene and cognitive function. J Nutr Sci 8:e20. https://doi.org/10.1017/jns.2019.16

D'Onofrio G, Nabavi SM, Sancarlo D, Greco A, Pieretti S (2021) Crocus sativus L. (Saffron) in Alzheimer's disease treatment: bioactive effects on cognitive impairment. Curr Neuropharmacol 19:1606. https://doi.org/10.2174/1570159X19666210113144703

De Keyser J, Ebinger G, Vauquelin G (1989) Evidence for a widespread dopaminergic innervation of the human cerebral neocortex. Neurosci Lett 104:281–285. https://doi.org/10.1016/0304-3940(89)90589-2

De Toro J, Herschlik L, Waldner C, Mongini C (2015) Emerging roles of exosomes in normal and pathological conditions: new insights for diagnosis and therapeutic applications. Front Immunol 6. https://doi.org/10.3389/fimmu.2015.00203

Diaz A, Muñoz-Arenas G, Caporal-Hernandez K, Vázquez-Roque R, Lopez-Lopez G, Kozina A, Espinosa B, Flores G, Treviño S, Guevara J (2021) Gallic acid improves recognition memory and decreases oxidative-inflammatory damage in the rat hippocampus with metabolic syndrome. Synapse 75. https://doi.org/10.1002/SYN.22186

Dong X, Luo S, Hu D, Cao R, Wang Q, Meng Z, Feng Z, Zhou W, Song W (2022) Gallic acid inhibits neuroinflammation and reduces neonatal hypoxic-ischemic brain damages. Front Pediatr 10. https://doi.org/10.3389/FPED.2022.973256/PDF

Dremencov E, Weizmann Y, Kinor N, Gispan-Herman I, Yadid G (2006) Modulation of dopamine transmission by 5HT2C and 5HT3 receptors: a role in the antidepressant response. Curr Drug Targets 7:165–175. https://doi.org/10.2174/138945006775515491

Dubey P, Pandey KB (2023) Neurotransmitter modulation by phytochemicals. In: Phytochemical drug discovery for central nervous system disorders, pp 311–325. https://doi.org/10.1002/9781119794127.CH13

Estelles A, Sperinde J, Roulon T, Aguilar B, Bonner C, LePecq J-B, Delcayre A (2007) Exosome nanovesicles displaying G protein-coupled receptors for drug discovery. Int J Nanomedicine 2:751–760

Excitatory H, John Mihic S, Adron Harris R (1997) GABA and the GABAA receptor. Alcohol Health Res World 21:127

Fakhri S, Abbaszadeh F, Moradi SZ, Cao H, Khan H, Xiao J (2022) Effects of polyphenols on oxidative stress, inflammation, and interconnected pathways during spinal cord injury. Oxidative Med Cell Longev 2022. https://doi.org/10.1155/2022/8100195

Fatima U, Roy S, Ahmad S, Ali S, Elkady WM, Khan I, Alsaffar RM, Adnan M, Islam A, Hassan MI (2022) Pharmacological attributes of Bacopa monnieri extract: current updates and clinical manifestation. Front Nutr 9. https://doi.org/10.3389/FNUT.2022.972379

German-Ponciano LJ, Rosas-Sánchez GU, Rivadeneyra-Domínguez E, Rodríguez-Landa JF (2018) Advances in the preclinical study of some flavonoids as potential antidepressant agents. Scientifica (Cairo) 2018:2963565–2963565. https://doi.org/10.1155/2018/2963565

González de Llano D, Roldán M, Parro L, Bartolomé B, Moreno-Arribas MV (2023) Activity of microbial-derived phenolic acids and their conjugates against LPS-induced damage in neuroblastoma cells and macrophages. Meta 13. https://doi.org/10.3390/METABO13010108

Graham RM (1990) Adrenergic receptors: structure and function. Cleve Clin J Med 57:481–491. https://doi.org/10.3949/ccjm.57.5.481

Grimmig B, Kim SH, Nash K, Bickford PC, Douglas Shytle R (2017) Neuroprotective mechanisms of astaxanthin: a potential therapeutic role in preserving cognitive function in age and neurodegeneration. GeroScience 39:19. https://doi.org/10.1007/S11357-017-9958-X

Grubić Kezele T, Ćurko-Cofek B (2022) Neuroprotective panel of olive polyphenols: mechanisms of action, anti-demyelination, and anti-stroke properties. Nutrients 14. https://doi.org/10.3390/NU14214533

Hase T, Shishido S, Yamamoto S, Yamashita R, Nukima H, Taira S, Toyoda T, Abe K, Hamaguchi T, Ono K, Noguchi-Shinohara M, Yamada M, Kobayashi S (2019) Rosmarinic acid suppresses Alzheimer's disease development by reducing amyloid β aggregation by increasing monoamine secretion. Sci Rep 9:1–13. https://doi.org/10.1038/s41598-019-45168-1

Herring BE, Silm K, Edwards RH, Nicoll RA (2015) Is aspartate an excitatory neurotransmitter? J Neurosci 35:10168–10171. https://doi.org/10.1523/JNEUROSCI.0524-15.2015

Higashiyama A, Htay HH, Ozeki M, Juneja LR, Kapoor MP (2011) Effects of l-theanine on attention and reaction time response. J Funct Foods 3:171–178. https://doi.org/10.1016/J.JFF.2011.03.009

Hill J, Inder T, Neil J, Dierker D, Harwell J, Van Essen D (2010) Similar patterns of cortical expansion during human development and evolution. Proc Natl Acad Sci USA 107:13135–13140. https://doi.org/10.1073/pnas.1001229107

Huo L, Du X, Li X, Liu S, Xu Y (2021) The emerging role of neural cell-derived exosomes in intercellular communication in health and neurodegenerative diseases. Front Neurosci 15:738442. https://doi.org/10.3389/fnins.2021.738442

Imran M, Ghorat F, Ul-haq I, Ur-rehman H, Aslam F, Heydari M, Shariati MA, Okuskhanova E, Yessimbekov Z, Thiruvengadam M, Hashempur MH, Rebezov M (2020) Lycopene as a natural antioxidant used to prevent human health disorders. Antioxidants (Basel) 9:1–27. https://doi.org/10.3390/ANTIOX9080706

Islam KUS, Meli N, Blaess S (2021) The development of the mesoprefrontal dopaminergic system in health and disease. Front Neural Circuits 15:746582. https://doi.org/10.3389/fncir.2021.746582

Jenkins TA, Nguyen JCD, Polglaze KE, Bertrand PP (2016) Influence of tryptophan and serotonin on mood and cognition with a possible role of the gut-brain Axis. Nutrients 8. https://doi.org/10.3390/nu8010056

Josiah SS, Famusiwa CD, Crown OO, Lawal AO, Olaleye MT, Akindahunsi AA, Akinmoladun AC (2022) Neuroprotective effects of catechin and quercetin in experimental Parkinsonism through modulation of dopamine metabolism and expression of IL-1β, TNF-α, NF-κB, IκKB, and p53 genes in male Wistar rats. Neurotoxicology 90:158–171. https://doi.org/10.1016/J.NEURO.2022.03.004

Kamli MR, Sharaf AAM, Sabir JSM, Rather IA (2022) Phytochemical screening of Rosmarinus officinalis L. as a potential anticholinesterase and antioxidant–medicinal plant for cognitive decline disorders. Plants (Basel) 11:514–514. https://doi.org/10.3390/PLANTS11040514

Kennedy DO (2019) Phytochemicals for improving aspects of cognitive function and psychological state potentially relevant to sports performance. Sports Med 49:39–58. https://doi.org/10.1007/S40279-018-1007-0

Khaksar Z, Azhdari S, Taherianfard M (2016) Histomorphometric study of lycopene effect on the number of neurons containing D1, D2 and GABA receptors in hippocampus and amygdala in offsprings rat from Parkinson's mothers. Biomed Pharmacol J 9:143–149. https://doi.org/10.13005/BPJ/921

Kim Y, Kim J, He M, Lee A, Cho E (2021) Apigenin ameliorates scopolamine-induced cognitive dysfunction and neuronal damage in mice. Molecules 26. https://doi.org/10.3390/molecules26175192

Kobayashi K (2001) Role of catecholamine signaling in brain and nervous system functions: new insights from mouse molecular genetic study. J Investig Dermatol Symp Proc 6:115–121. https://doi.org/10.1046/j.0022-202x.2001.00011.x

Koly HK, Sutradhar K, Rahman MS (2023) Acetylcholinesterase inhibition of Alzheimer's disease: identification of potential phytochemicals and designing more effective derivatives to manage disease condition. J Biomol Struct Dyn 41:1–13. https://doi.org/10.1080/07391102.2023.2166992

Kulkarni SK, Dhir A (2010) An overview of curcumin in neurological disorders. Indian J Pharm Sci 72:149. https://doi.org/10.4103/0250-474X.65012

Kumar GP, Khanum F (2012) Neuroprotective potential of phytochemicals. Pharmacogn Rev 6:81. https://doi.org/10.4103/0973-7847.99898

Kumar N, Abichandani LG, Thawani V, Gharpure KJ, Naidu MUR, Venkat Ramana G (2016) Efficacy of standardized extract of Bacopa monnieri (Bacognize®) on cognitive functions of medical students: a six-week, randomized placebo-controlled trial. Evid Based Complement Alternat Med 2016. https://doi.org/10.1155/2016/4103423

Kuo HY, Liu FC (2022) Pathophysiological studies of monoaminergic neurotransmission systems in valproic acid-induced model of autism spectrum disorder. Biomedicines 10:560–560. https://doi.org/10.3390/BIOMEDICINES10030560

Küpeli Akkol E, Tatlı Çankaya I, Şeker Karatoprak G, Carpar E, Sobarzo-Sánchez E, Capasso R (2021) Natural compounds as medical strategies in the prevention and treatment of psychiatric disorders seen in neurological diseases. Front Pharmacol. https://doi.org/10.3389/fphar.2021.669638

Lai W, Zheng Z, Zhang X, Wei Y, Chu K, Brown J, Hong G, Chen L (2015) Salidroside-mediated neuroprotection is associated with induction of early growth response genes (Egrs) across a wide therapeutic window. Neurotox Res 28:108–121. https://doi.org/10.1007/S12640-015-9529-9

Lee JG, Woo YS, Park SW, Seog DH, Seo MK, Bahk WM (2022a) Neuromolecular etiology of bipolar disorder: possible therapeutic targets of mood stabilizers. Clin Psychopharmacol Neurosci 20:228. https://doi.org/10.9758/CPN.2022.20.2.228

Lee S, Seol HS, Eom S, Lee J, Kim C, Park JH, Kim TH, Lee JH (2022b) Hydroxy pentacyclic triterpene acid, kaempferol, inhibits the human 5-Hydroxytryptamine type 3A receptor activity. Int J Mol Sci 23:544–544. https://doi.org/10.3390/IJMS23010544

Leonelli M, Torrão AS, Britto LRG (2009) Unconventional neurotransmitters, neurodegeneration and neuroprotection. Braz J Med Biol Res 42:68–75. https://doi.org/10.1590/s0100-879x2009000100011

Li Y, Yu L, Zhao L, Zeng F, Liu QS (2017) Resveratrol modulates cocaine-induced inhibitory synaptic plasticity in VTA dopamine neurons by inhibiting phosphodiesterases (PDEs). Sci Rep 7:15657–15657. https://doi.org/10.1038/S41598-017-16034-9

Li MM, Madara JC, Steger JS, Krashes MJ, Balthasar N, Campbell JN, Resch JM, Conley NJ, Garfield AS, Lowell BB (2019a) The paraventricular hypothalamus regulates satiety and prevents obesity via two genetically distinct circuits. Neuron 102:653–667.e6. https://doi.org/10.1016/j.neuron.2019.02.028

Li R, Wang S, Li T, Wu L, Fang Y, Feng Y, Zhang L, Chen J, Wang X (2019b) Salidroside protects dopaminergic neurons by preserving complex I activity via DJ-1/Nrf2-mediated antioxidant pathway. Parkinsons Dis 2019:6073496–6073496. https://doi.org/10.1155/2019/6073496

Li G, Yang J, Wang X, Zhou C, Zheng X, Lin W (2020) Effects of EGCG on depression-related behavior and serotonin concentration in a rat model of chronic unpredictable mild stress. Food Funct 11:8780–8787. https://doi.org/10.1039/D0FO00524J

Li MY, Liu HY, Wu DT, Kenaan A, Geng F, Li HB, Gunaratne A, Li H, Gan RY (2022) L-theanine: a unique functional amino acid in tea (Camellia sinensis L.) with multiple health benefits and food applications. Front Nutr 9. https://doi.org/10.3389/FNUT.2022.853846/FULL

Li J, Wu Y, Dong S, Yu Y, Wu Y, Xiang B, Li Q (2023) Research progress on neuroprotective effects of isoquinoline alkaloids. Molecules 28:4797–4797. https://doi.org/10.3390/MOLECULES28124797

Liu YM, Fan HR, Ding J, Huang C, Deng S, Zhu T, Le Xu T, Ge WH, Li WG, Li F (2017) Curcumol allosterically modulates GABA(A) receptors in a manner distinct from benzodiazepines. Sci Rep 7. https://doi.org/10.1038/SREP46654

Liu W, Feng Y, Yu S, Fan Z, Li X, Li J, Yin H (2021) The flavonoid biosynthesis network in plants. Int J Mol Sci 22. https://doi.org/10.3390/IJMS222312824

Lopresti AL, Smith SJ, Drummond PD (2022) The effects of lutein and zeaxanthin supplementation on cognitive function in adults with self-reported mild cognitive complaints: a randomized, double-blind, placebo-controlled study. Front Nutr 9. https://doi.org/10.3389/FNUT.2022.843512

Lovinger DM (2008) Communication networks in the brain: neurons, receptors, neurotransmitters, and alcohol. Alcohol Res Health

Lu H-C, Mackie K (2016) An introduction to the endogenous cannabinoid system. Biol Psychiatry 79:516–525. https://doi.org/10.1016/j.biopsych.2015.07.028

Lu B, Nagappan G, Lu Y (2014) BDNF and synaptic plasticity, cognitive function, and dysfunction. Handb Exp Pharmacol 220:223–250. https://doi.org/10.1007/978-3-642-45106-5_9

Lüscher C, Malenka RC (2012) NMDA receptor-dependent long-term potentiation and long-term depression (LTP/LTD). Cold Spring Harb Perspect Biol 4. https://doi.org/10.1101/cshperspect.a005710

Ma X, Xu W, Zhang Z, Liu N, Yang J, Wang M, Wang Y (2017) Salvianolic acid B ameliorates cognitive deficits through IGF-1/Akt pathway in rats with vascular dementia. Cell Physiol Biochem 43:1381–1391. https://doi.org/10.1159/000481849

Mak S, Li W, Fu H, Luo J, Cui W, Hu S, Pang Y, Carlier PR, Tsim KW, Pi R, Han Y (2021) Promising tacrine/huperzine A-based dimeric acetylcholinesterase inhibitors for neurodegenerative disorders: from relieving symptoms to modifying diseases through multitarget. J Neurochem 158:1381–1393. https://doi.org/10.1111/JNC.15379

Martins RS, Rombo DM, Gonçalves-Ribeiro J, Meneses C, Borges-Martins VPP, Ribeiro JA, Vaz SH, Kubrusly RCC, Sebastião AM (2020) Caffeine has a dual influence on NMDA receptor–mediated glutamatergic transmission at the hippocampus. Purinergic Signal 16:503–518. https://doi.org/10.1007/S11302-020-09724-Z

Maura G, Andrioli GC, Cavazzani P, Raiteri M (1992) 5-Hydroxytryptamine3 receptors sited on cholinergic axon terminals of human cerebral cortex mediate inhibition of acetylcholine release. J Neurochem 58:2334–2337. https://doi.org/10.1111/J.1471-4159.1992.TB10983.X

McCutcheon RA, Krystal JH, Howes OD (2020) Dopamine and glutamate in schizophrenia: biology, symptoms and treatment. World Psychiatry 19:15–33. https://doi.org/10.1002/wps.20693

McLellan TM, Caldwell JA, Lieberman HR (2016) A review of caffeine's effects on cognitive, physical and occupational performance. Neurosci Biobehav Rev 71:294–312. https://doi.org/10.1016/J.NEUBIOREV.2016.09.001

Mei Z, Hong Y, Yang H, Cai S, Hu Y, Chen Q, Yuan Z, Liu X (2023) Ferulic acid alleviates high fat diet-induced cognitive impairment by inhibiting oxidative stress and apoptosis. Eur J Pharmacol 946:175642. https://doi.org/10.1016/J.EJPHAR.2023.175642

Mendonça CR, Manhães-de-Castro R, de Santana BJRC, Olegário da Silva L, Toscano AE, Guzmán-Quevedo O, Galindo LCM (2022) Effects of flavonols on emotional behavior and compounds of the serotonergic system: a preclinical systematic review. Eur J Pharmacol 916:174697. https://doi.org/10.1016/J.EJPHAR.2021.174697

Min LS, Liew S, Chear SY, Goh NJY, Tan BH, Khaw W-N, Lai S, Min S, Liew SY, Jeng N, Chear Y, Goh H, Tan W-N, Khaw KY (2022) Plant terpenoids as the promising source of cholinesterase inhibitors for anti-AD therapy. Biology (Basel) 11:307–307. https://doi.org/10.3390/BIOLOGY11020307

Mishra A, Singh S, Shukla S (2018) Physiological and functional basis of dopamine receptors and their role in neurogenesis: possible implication for Parkinson's disease. J Exp Neurosci 12:1179069518779829. https://doi.org/10.1177/1179069518779829

Mohd Sairazi NS, Sirajudeen KNS (2020) Natural products and their bioactive compounds: neuroprotective potentials against neurodegenerative diseases. Evid Based Complement Alternat Med. https://doi.org/10.1155/2020/6565396

Molnár Z, Kaas JH, de Carlos JA, Hevner RF, Lein E, Němec P (2014) Evolution and development of the mammalian cerebral cortex. Brain Behav Evol 83:126–139. https://doi.org/10.1159/000357753

Moore A, Beidler J, Hong MY (2018) Resveratrol and depression in animal models: a systematic review of the biological mechanisms. Molecules 23. https://doi.org/10.3390/MOLECULES23092197

Morel L, Regan M, Higashimori H, Ng SK, Esau C, Vidensky S, Rothstein J, Yang Y (2013) Neuronal exosomal miRNA-dependent translational regulation of astroglial glutamate transporter GLT1. J Biol Chem 288:7105–7116. https://doi.org/10.1074/jbc.M112.410944

Mostafa MD, ElKomy MA, Othman AI, Amer ME, El-Missiry MA (2022) Epigallocatechin-3-gallate enhances cognitive and memory performance and protects against brain injury in methionine-induced hyperhomocysteinemia through interdependent molecular pathways. Neurotox Res 40:2103. https://doi.org/10.1007/S12640-022-00605-4

Nabavi SF, Khan H, D'onofrio G, Šamec D, Shirooie S, Dehpour AR, Argüelles S, Habtemariam S, Sobarzo-Sanchez E (2018) Apigenin as neuroprotective agent: of mice and men. Pharmacol Res 128:359–365. https://doi.org/10.1016/J.PHRS.2017.10.008

Nair SG, Gudelsky GA (2004) Activation of 5-HT2 receptors enhances the release of acetylcholine in the prefrontal cortex and hippocampus of the rat. Synapse 53:202–207. https://doi.org/10.1002/syn.20054

Nakamura Y, Darnieder LM, Deeb TZ, Moss SJ (2015) Regulation of GABAARs by phosphorylation. Adv Pharmacol 72:97–146. https://doi.org/10.1016/bs.apha.2014.11.008

Nuss P (2015) Anxiety disorders and GABA neurotransmission: a disturbance of modulation. Neuropsychiatr Dis Treat 11:165–175. https://doi.org/10.2147/NDT.S58841

Ofir R, Tamir S, Khatib S, Vaya J (2003) Inhibition of serotonin re-uptake by licorice constituents. J Mol Neurosci 20:135–140. https://doi.org/10.1385/JMN:20:2:135

Osimo EF, Stahl SM (2023) Classification of receptors: metabotropic and ionotropic receptors. In: Lynall M-E, Jones PB, Stahl SME (eds) Cambridge textbook of neuroscience for psychiatrists. Cambridge University Press, United Kingdom and RCPsych Publications operates from the Royal College of Psychiatrists London headquarters at 21 Prescot Street, London, England, pp 22–23

Othman A, Sayed AM, Amen Y, Shimizu K (2022) Possible neuroprotective effects of amide alkaloids from Bassia indica and Agathophora alopecuroides: in vitro and in silico investigations. RSC Adv 12:18746–18758. https://doi.org/10.1039/D2RA02275C

Pałasz A, Menezes IC, Worthington JJ (2021) The role of brain gaseous neurotransmitters in anxiety. Pharmacol Rep 73(2):357–371. https://doi.org/10.1007/S43440-021-00242-2

Pannu A, Sharma PC, Thakur VK, Goyal RK (2021) Emerging role of flavonoids as the treatment of depression. Biomolecules 11. https://doi.org/10.3390/BIOM11121825

Park KC, Jin H, Zheng R, Kim S, Lee SE, Kim BH, Yim SV (2019) Cognition enhancing effect of panax ginseng in Korean volunteers with mild cognitive impairment: a randomized, double-blind, placebo-controlled clinical trial. Transl Clin Pharmacol 27:92. https://doi.org/10.12793/TCP.2019.27.3.92

Pascual M, Ibáñez F, Guerri C (2020) Exosomes as mediators of neuron-glia communication in neuroinflammation. Neural Regen Res 15:796–801. https://doi.org/10.4103/1673-5374.268893

Purnawati S, Wrasiati LP, Lesmana CBJ, Megantara S, Lesmana R (2022) A study of molecular docking of l-tryptophan ligand as a compound in pineapples and bananas binding with the human serotonin transporter (SERT). Bali Med J 11:1243–1249. https://doi.org/10.15562/BMJ.V11I3.3526

Qi G, Mi Y, Wang Y, Li R, Huang S, Li X, Liu X (2017) Neuroprotective action of tea polyphenols on oxidative stress-induced apoptosis through the activation of the TrkB/CREB/BDNF pathway and Keap1/Nrf2 signaling pathway in SH-SY5Y cells and mice brain. Food Funct 8:4421–4432. https://doi.org/10.1039/C7FO00991G

Queen CJJ, Sparks SA, Marchant DC, Mcnaughton LR, James C, Queen J, Sparks SA, Marchant DC, Mcnaughton LR (2024) The effects of astaxanthin on cognitive function and neurodegeneration in humans: a critical review. Nutrients 16:826. https://doi.org/10.3390/NU16060826

Quetglas S, Leveque C, Miquelis R, Sato K, Seagar M (2000) Ca2+-dependent regulation of synaptic SNARE complex assembly via a calmodulin- and phospholipid-binding domain of synaptobrevin. Proc Natl Acad Sci USA 97:9695–9700. https://doi.org/10.1073/pnas.97.17.9695

Ramezani M, Meymand AZ, Khodagholi F, Kamsorkh HM, Asadi E, Noori M, Rahimian K, Shahrbabaki AS, Talebi A, Parsaiyan H, Shiravand S, Darbandi N (2022) A role for flavonoids in the prevention and/or treatment of cognitive dysfunction, learning, and memory deficits: a review of preclinical and clinical studies. Nutr Neurosci 26:156–172. https://doi.org/10.1080/1028415X.2022.2028058

Ranjbar-Slamloo Y, Fazlali Z (2020) Dopamine and noradrenaline in the brain; overlapping or dissociate functions? Front Mol Neurosci 12. https://doi.org/10.3389/fnmol.2019.00334

Razay G, Wilcock GK (2008) Galantamine in Alzheimer's disease. Expert Rev Neurother 8:9–17. https://doi.org/10.1586/14737175.8.1.9

Rebas E, Rzajew J, Radzik T, Zylinska L (2020) Neuroprotective polyphenols: a modulatory action on neurotransmitter pathways. Curr Neuropharmacol 18:431–445. https://doi.org/10.2174/1570159X18666200106155127

Rees A, Dodd GF, Spencer JPE (2018) The effects of flavonoids on cardiovascular health: a review of human intervention trials and implications for cerebrovascular function. Nutrients 10:1852. https://doi.org/10.3390/NU10121852

Reiner A, Levitz J (2018) Glutamatergic signaling in the central nervous system: ionotropic and metabotropic receptors in concert. Neuron 98:1080–1098. https://doi.org/10.1016/j.neuron.2018.05.018

Rendeiro C, Rhodes JS, Spencer JPE (2015) The mechanisms of action of flavonoids in the brain: direct versus indirect effects. Neurochem Int. https://doi.org/10.1016/j.neuint.2015.08.002

Rihmer Z, Dome P, Katona C (2022) Serotonin and depression - a riposte to Moncrieff et al. (2022). Neuropsychopharmacol Hung 24:120–125

Rostène W, Sarrieau A, Nicot A, Scarceriaux V, Betancur C, Gully D, Meaney M, Rowe W, De Kloet R, Pelaprat D (1995) Steroid effects on brain functions: an example of the action of glucocorticoids on central dopaminergic and neurotensinergic systems. J Psychiatry Neurosci 20:349–356

Roth BL (2019) Molecular pharmacology of metabotropic receptors targeted by neuropsychiatric drugs. Nat Struct Mol Biol 26:535–544. https://doi.org/10.1038/s41594-019-0252-8

Rudolph LM, Cornil CA, Mittelman-Smith MA, Rainville JR, Remage-Healey L, Sinchak K, Micevych PE (2016) Actions of steroids: new neurotransmitters. J Neurosci 36:11449–11458. https://doi.org/10.1523/JNEUROSCI.2473-16.2016

Sarter M, Bruno JP, Parikh V (2007) Abnormal neurotransmitter release underlying behavioral and cognitive disorders: toward concepts of dynamic and function-specific dysregulation. Neuropsychopharmacology 32:1452–1461. https://doi.org/10.1038/sj.npp.1301285

Sharmin D, Mian MY, Marcotte M, Prevot TD, Sibille E, Witkin JM, Cook JM (2023) Synthesis and receptor binding studies of α5 GABAAR selective novel imidazodiazepines targeted for psychiatric and cognitive disorders. Molecules 28:4771. https://doi.org/10.3390/MOLECULES28124771/S1

Shi Y, Dong J-W, Zhao J-H, Tang L-N, Zhang J-J (2014) Herbal insomnia medications that target GABAergic systems: a review of the psychopharmacological evidence. Curr Neuropharmacol 12:289. https://doi.org/10.2174/1570159X11666131227001243

Snyder SH, Jaffrey SR, Zakhary R (1998) Nitric oxide and carbon monoxide: parallel roles as neural messengers. Brain Res Rev 26:167–175. https://doi.org/10.1016/S0165-0173(97)00032-5

Song Y, Seo S, Lamichhane S, Seo J, Hong JT, Cha HJ, Yun J (2021) Limonene has anti-anxiety activity via adenosine A2A receptor-mediated regulation of dopaminergic and GABAergic neuronal function in the striatum. Phytomedicine 83:153474. https://doi.org/10.1016/J.PHYMED.2021.153474

Stewart BA, Mohtashami M, Trimble WS, Boulianne GL (2000) SNARE proteins contribute to calcium cooperativity of synaptic transmission. Proc Natl Acad Sci USA 97:13955–13960. https://doi.org/10.1073/pnas.250491397

Sugisaki E, Fukushima Y, Nakajima N, Aihara T (2022) The dependence of acetylcholine on dynamic changes in the membrane potential and an action potential during spike timing-dependent plasticity induction in the hippocampus. Eur J Neurosci 56:5972–5986. https://doi.org/10.1111/EJN.15832

Sun T, Hevner RF (2014) Growth and folding of the mammalian cerebral cortex: from molecules to malformations. Nat Rev Neurosci 15:217–232. https://doi.org/10.1038/nrn3707

Takeda A, Tomiyama M, Hanajima R (2021) The relationship between pathophysiology and neurotransmitters in Parkinson's disease. Brain Nerve 73:829–837. https://doi.org/10.11477/MF.1416201843

Tavakoli R, Nouri A, Masoudi R, Mirhoseini M, Ahmadpour F, Masoudi R (2018) Serum markers related to depression: a systematic review. J Med Biomed Sci 6:30–42. https://doi.org/10.4314/JMBS.V6I3.5

Teleanu RI, Niculescu A-G, Roza E, Vladâcenco O, Grumezescu AM, Teleanu DM (2022) Neurotransmitters-key factors in neurological and neurodegenerative disorders of the central nervous system. Int J Mol Sci 23. https://doi.org/10.3390/ijms23115954

Tiwari P, Dwivedi S, Singh MP, Mishra R, Chandy A (2013) Basic and modern concepts on cholinergic receptor: a review. Asian Pac J Trop Dis. https://doi.org/10.1016/S2222-1808(13)60094-8

Tortora F, Hadipour AL, Battaglia S, Falzone A, Avenanti A, Vicario CM (2023) The role of serotonin in fear learning and memory: a systematic review of human studies. Brain Sci 13. https://doi.org/10.3390/brainsci13081197

Trauner G, Khom S, Baburin I, Benedek B, Hering S, Kopp B (2008) Modulation of GABAA receptors by valerian extracts is related to the content of valerenic acid. Planta Med 74:19–24. https://doi.org/10.1055/S-2007-993761

Tsegay EW, Demise DG, Hailu NA, Gufue ZH (2020) Serotonin type 6 and 7 receptors as a novel therapeutic target for the treatment of schizophrenia. Neuropsychiatr Dis Treat 16:2499–2509. https://doi.org/10.2147/NDT.S263424

Usiello A, Di Fiore MM, De Rosa A, Falvo S, Errico F, Santillo A, Nuzzo T, Chieffi Baccari G (2020) New evidence on the role of D-aspartate metabolism in regulating brain and endocrine system physiology: from preclinical observations to clinical applications. Int J Mol Sci 21. https://doi.org/10.3390/ijms21228718

Vauzour D (2017) Polyphenols and brain health. In: OCL - oilseeds and fats, crops and lipids, EDP Sciences 24: France. 17, Avenue du Hoggar, Parc d'Activité de Courtaboeuf BP 112, 91944 Les Ulis Cedex A, France. https://doi.org/10.1051/OCL/2017008

Verney C (1999) Distribution of the catecholaminergic neurons in the central nervous system of human embryos and fetuses. Microsc Res Tech 46:24–47. https://doi.org/10.1002/(SICI)1097-0029(19990701)46:1<24::AID-JEMT3>3.0.CO;2-E

Vicencio JM, Estrada M, Galvis D, Bravo R, Contreras AE, Rotter D, Szabadkai G, Hill JA, Rothermel BA, Jaimovich E, Lavandero S (2011) Anabolic androgenic steroids and intracellular calcium signaling: a mini review on mechanisms and physiological implications. Mini Rev Med Chem 11:390–398. https://doi.org/10.2174/138955711795445880

Voulgaropoulou SD, van Amelsvoort TAMJ, Prickaerts J, Vingerhoets C (2019) The effect of curcumin on cognition in Alzheimer's disease and healthy aging: a systematic review of pre-clinical and clinical studies. Brain Res 1725:146476. https://doi.org/10.1016/J.BRAINRES.2019.146476

Wang BS, Wang H, Wei ZH, Song YY, Zhang L, Chen HZ (2009) Efficacy and safety of natural acetylcholinesterase inhibitor huperzine A in the treatment of Alzheimer's disease: an updated meta-analysis. J Neural Transm (Vienna) 116:457–465. https://doi.org/10.1007/S00702-009-0189-X

Wang H, Wang G, Billings R, Li D, Haase SR, Wheeler PF, Vance DE, Li W (2022) Can diet supplements of macular pigment of lutein, zeaxanthin, and meso-zeaxanthin affect cognition? J Alzheimers Dis 87:1079–1087. https://doi.org/10.3233/JAD-215736

Wang C, Shi Z, Li Y, Xia X, Kuang X, Qian S, Xue L, Lan L, Wu Y, Zhang N, Tao J, Hu X, Cao W, He N, Guo Y, Chen W, Zhang J, Luo J, Wang H, Tian M (2023) Protocol for brain magnetic resonance imaging and extraction of imaging-derived phenotypes from the China Phenobank project. Phenomics 3:642–656. https://doi.org/10.1007/s43657-022-00083-w

Werner F-M, Coveñas R (2014) Classical neurotransmitters and neuropeptides involved in Parkinson's disease: a multi-neurotransmitter system. J Cytol Histol 5. https://doi.org/10.4172/2157-7099.1000266

Wu Y, Chen Q, Wen B, Wu N, He B, Chen J (2021) Berberine reduces Aβ42 deposition and tau hyperphosphorylation via ameliorating endoplasmic reticulum stress. Front Pharmacol 12. https://doi.org/10.3389/FPHAR.2021.640758

Xia X, Wang Y, Qin Y, Zhao S, Zheng JC (2022) Exosome: a novel neurotransmission modulator or non-canonical neurotransmitter? Ageing Res Rev 74:101558. https://doi.org/10.1016/j.arr.2021.101558

Xu W, Song Y, Xiao W, Gong Z (2022) Regulatory effects and mechanisms of L-theanine on neurotransmitters via liver-brain axis under a high protein diet. https://doi.org/10.21203/RS.3.RS-2119382/V1

Yamamoto H, Hagino Y, Kasai S, Ikeda K (2015) Specific roles of NMDA receptor subunits in mental disorders. Curr Mol Med 15:193–205. https://doi.org/10.2174/1566524015666150330142807

Yamamoto S, Kayama T, Noguchi-Shinohara M, Hamaguchi T, Yamada M, Abe K, Kobayashi S (2021) Rosmarinic acid suppresses tau phosphorylation and cognitive decline by downregulating the JNK signaling pathway. NPJ Sci Food 5:1–11. https://doi.org/10.1038/s41538-021-00084-5

Yang J, Sui S-F, Liu Z (2022) Brain structure and structural basis of neurodegenerative diseases. Biophys Rep 8:170–181. https://doi.org/10.52601/bpr.2022.220013

Yang Z, Zou Y, Wang L (2023) Neurotransmitters in prevention and treatment of Alzheimer's disease. Int J Mol Sci 24. https://doi.org/10.3390/IJMS24043841

Yao J, Wei W, Wen J, Cao Y, Li H (2023) The efficacy and mechanism of berberine in improving aging-related cognitive dysfunction: a study based on network pharmacology. Front Neurosci 17. https://doi.org/10.3389/FNINS.2023.1093180/FULL

Yeo XY, Cunliffe G, Ho RC, Lee SS, Jung S (2022) Potentials of neuropeptides as therapeutic agents for neurological diseases. Biomedicine 10. https://doi.org/10.3390/biomedicines10020343

Yohn CN, Gergues MM, Samuels BA (2017) The role of 5-HT receptors in depression. Mol Brain 10. https://doi.org/10.1186/S13041-017-0306-Y

Yoon J-Y, Jung S-R, Hille B, Koh D-S (2014) Modulation of nicotinic receptor channels by adrenergic stimulation in rat pinealocytes. Am J Physiol Cell Physiol 306:C726–C735. https://doi.org/10.1152/ajpcell.00354.2013

Yoshimura H (2005) The potential of caffeine for functional modification from cortical synapses to neuron networks in the brain. Curr Neuropharmacol 3:309–316. https://doi.org/10.2174/157015905774322543

Yuan CS, Mehendale S, Xiao Y, Aung HH, Xie JT, Ang-Lee MK (2004) The gamma-aminobutyric acidergic effects of valerian and valerenic acid on rat brainstem neuronal activity. Anesth Analg 98:353–358. https://doi.org/10.1213/01.ANE.0000096189.70405.A5

Zhang HY (2012) New insights into huperzine A for the treatment of Alzheimer's disease. Acta Pharmacol Sin 33(9):1170–1175. https://doi.org/10.1038/aps.2012.128

Zhao X, Yang A, Zhang Z-C, Yang YT, Zhao X-M (2023) Deciphering the genetic architecture of human brain structure and function: a brief survey on recent advances of neuroimaging genomics. Brief Bioinform 24:bbad060. https://doi.org/10.1093/bib/bbad060

Zhong Z, Han J, Zhang J, Xiao Q, Hu J, Chen L (2018) Pharmacological activities, mechanisms of action, and safety of salidroside in the central nervous system. Drug Des Devel Ther 12:1479. https://doi.org/10.2147/DDDT.S160776

Natural Compounds and Neuroinflammation

5

Vishal Mathur, Shaheen Ali, Iqra Zai, and Anam Ilyas

Abstract

Neurodegenerative disorders (NDs) are considered as the most prevalent disorder in the World with limited cure. Many factors are responsible as causative agent like age, life style, intrinsic factor and extrinsic factor. Inflammation is also one of the responsible factors for the onset of NDs. Commercial therapies are available but less effective and high cost. Therefore, researchers are working to dig out the easiest and budget friendly cure. Natural products retrieved from plants along with their primary and secondary metabolites have been utterly studied over the years to explore their therapeutic efficiency against neurodegenerative diseases such as Parkinson's disease, amyotrophic lateral sclerosis, Alzheimer's disease, and Huntington's disease. Identification of major mechanisms of phytomolecules exerting their neuroprotective action are such as antioxidant, neuroinflammation, anti-apoptotic, antithrombotic, aging, and monoamine oxidase inhibition. This chapter focuses to investigate in-depth study of selection of phytoconstituents based on their strong antioxidant and anti-

V. Mathur (✉)
Amity Institute of Pharmacy, Amity University, Noida, UP, India

Department of Pharmaceutical Chemistry, School of Pharmaceutical Education and Research, Jamia Hamdard, New Delhi, India

S. Ali
Department of Molecular Medicine, SIST, Jamia Hamdard, New Delhi, India

I. Zai
Department of Pharmaceutics, School of Pharmaceutical Education and Research, Delhi, India

A. Ilyas
Department of Pharmaceutical Chemistry, School of Pharmaceutical Education and Research, Jamia Hamdard, New Delhi, India

S. Jahan et al. (eds.), *A Mechanistic Exploration of Natural Compounds for Neuronal Health*, https://doi.org/10.1007/978-981-95-0125-0_5

inflammatory potential, suggesting high efficacy in preclinical alongside clinical studies. We further explained an overview of mechanism of action enrolled in NDs and their therapeutic potential.

Keywords

Neurodegeneration · Parkinson's disease · Alzheimer's disease · Neuroinflammation · Natural compounds

5.1 Introduction

Neurological disorders (NDs) are ailments, majorly affecting peripheral and central nervous system. NDs could be initiated through various factors such as ischemia, autoimmune-mediated neuronal loss, ER cellular stress, injury of nervous system, viral or prion infections, and abnormal protein aggregation in neural tissue. Depending upon site infected or affected, neuronal loss, demyelination, or gliosis could further lead to behavioral changes, motor deficits, and cognitive decline (Sharifi et al. 2020a). Investigation incorporating 28 countries around the globe includes data from 1985 to 2012 depicting prevalence of psychiatric disorder recorded to be 13.4% in adults and children. As previously reported, aging of higher population is extremely contributing in NDs count and age-related diseases, as NDs are found to be irreversible in nature. Huntingon's disease (HD) is well characterized by loss of medium-sized spiny striatal neurons; Parkinson's disease (PD) causes loss of dopaminergic nigrostriatal neurons, whereas Alzheimer's disease (AD) is produced via diffused brain atrophy. Iron dyshomeostasis plays crucial role in strengthening neuroinflammatory phenotype, accentuating importance of Fe+ balance where human transferrin plays key role in this aspect. Patients suffering with NDs develop various variable symptoms such as overlapping cognitive and memory impairment, ability to communicate, breathe, and walk. Moreover, mitochondrial dysfunction, oxidative stress, protein misfolding dysfunction and agglomeration, and neuroinflammation are some biological processes linked to NDs. To date, wide research leads us to explain therapeutic potential and process of natural compounds in neuroprotection. Phytomolecules are well known for their medicinal properties and safer use since ancient times. Phytoconstitutents and their bioactive compounds have been widely evaluated with emphasis on nutritional principles, therapeutic benefits, biological processes, and potential health (Rahman et al. 2021).

AD is a chronic neurodegenerative disorder, distorting human brain's hippocampus and cerebral cortex regions. This neurological complication is associated with numerous symptoms including cognitive decline and mental impairment, which predominantly affects elder patients above the age 65. Majorly, it is distinguished via two prominent features (i) deposition and extracellular development of amyloid-β plaques and (ii) intracellular accretion of hyperphosphorylated tau (τ) protein, which helps in forming neurofibrillary tangles (NFTs) in brain.

PD is one of the most common neurodegenerative disorders which leads to impairing eminence of life and dependency, leading to menace of premature decease in affected patients. Parkinson is instigated by a substantial damage to dopaminergic neurons, pushing to reduce motor functions and produce symptoms like resting tremor, muscular rigidity, and bradykinesia. These diseases are distinguished via Lewy neurites, accretion of protein segregates, and Lewy bodies with misfolded presynaptic protein α-synuclein (Fig. 5.1).

ALS is pathologically characterized via substantial degeneration and ending of both upper and lower neurons, miscarrying respiratory system leading to paralysis and further death. Several aspects underlying mechanism of action of ALS including autoimmune response, excitotoxicity, oxidative stress, genetic factors, mitochondrial dysfunction, and neurofilament aggregation are considered to be potential contributor to progression and development of ALS. At present, therapeutic and preventive treatments for neurological complications remain limited. However, improvements in life expectancy have been associated with a decreased inflammatory response to such diseases. (Goyal et al. 2024).

Despite our extensive knowledge on this topic, the underlying question of causality of this issue remain unanswered. In this chapter, we focus in providing steep

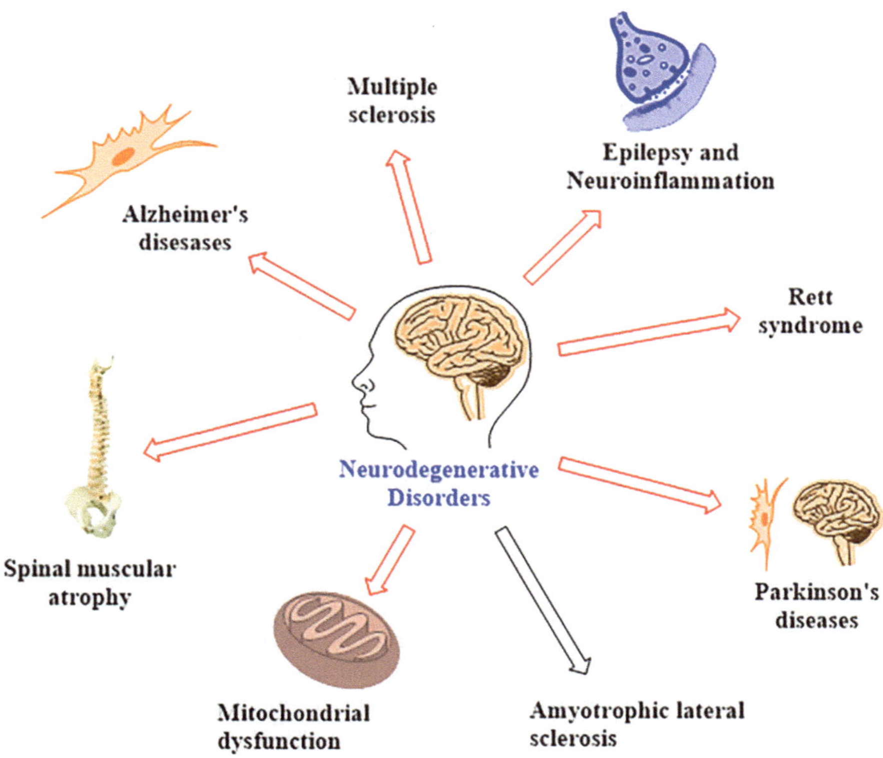

Fig. 5.1 Various neurological disorders. (https://revvitysignals.com/)

insight to the relationship between neuroinflammation and oxidative stress via Alzheimer, Parkinson, epilepsy, and several neuro-oriented diseases leading to various impairments in synaptic functions, as well as the role of G protein-coupled receptor 55 (GPR55) in oxidative and inflammatory mechanisms (Parsons et al. 2022).

5.2 In-Depth Analysis of Anti-inflammatory Properties of Natural Compounds

Phytochemicals have garnered recognition as promising active agents due to the increasing demand for innovative and effective treatments. Recent studies have illustrated that animal-derived compounds like omega-3 fatty acids, along with plant-based compounds, can reduce cellular damage and exhibit anti-inflammatory effects through neuroprotective therapies (Wollen 2010). Various natural compounds have shown anti-inflammatory properties via different pathways. These strategies involve inhibiting the activation of microglia, decreasing the amount of pro-inflammatory cytokines released by activated microglia, and limiting the activation of p38 MAPK and NF-κB. Furthermore, a pathway that contributes to the anti-neuroinflammatory actions of certain natural products is markedly activated by Nrf2. Phytochemicals with antioxidant, anti-inflammatory, and anti-amyloid properties can engage with neuro-inflammatory mediators. Although conclusive evidence for the blood brain barrier (BBB) permeability of various phytocompounds is lacking, their extensive traditional use and substantial data from animal studies suggest their capacity to cross the BBB (Kam et al. 2012).

5.2.1 Polyphenols

Polyphenols are an important part of the diet in many regions around the world. Polyphenols are mainly found in foods like berries, wine, almonds, cloves, turmeric, and various other spices. They can be generally divided into four main categories: lignans, stilbenes, flavonoids, and phenolic acids. These groups of bioactive compounds are being researched for their potential to treat various malignancies, high blood pressure, heart conditions, type II diabetes, and neurological diseases. Free radicals and oxidants tend to accumulate with aging, causing irreparable damage to cells, particularly neurons, which accelerates the aging process. Various diets worldwide include polyphenols that are found to be helpful for age-related dementia and other neurodegenerative diseases. Since neurodegenerative diseases are defined by oxidative stress and neuroinflammation, polyphenols are being extensively researched for their anti-inflammatory and antioxidant properties. Recent studies highlight resveratrol, catechins, and tannins as promising agents for neurodegenerative disorders. However, improving their bioavailability and ability to cross the blood-brain barrier is crucial (Grabska et al. 2023).

5.2.1.1 Resveratrol

In 1940, Takaoka first extracted resveratrol via roots of *Veratrum grandiflorum O. Loes*. Over the following decades, key discoveries identified resveratrol as a potential therapeutic agent for neurological, cardiovascular, and cancer diseases (Pezzuto 2019). Recently, researchers have explored the mechanisms and chemical constituents responsible for its anti-inflammatory and antioxidant properties. Grapes contain a vast array of bioactive phenolic derivatives that offer a broad range of health welfare. Recent research has investigated grape-derived polyphenols for their potential to delay or prevent the onset of given neurodegenerative diseases via various pathways. Resveratrol, a prominent grape-derived polyphenol, has garnered significant interest due to its propitious biological activity. It is the richest stilbene found in grapes, existing in both cis and trans forms, with more biologically active in trans form. Resveratrol is also existing in other fruits such as nuts, blueberries, bananas, raspberries, and citrus fruits (Chan et al. 2019).

Resveratrol induces microglial polarization from a proinflammatory to an anti-inflammatory form in lipopolysaccharide-induced BV2 cells (termed as mouse microglial cells). It blocks the MAPK and NFκB inflammatory pathway through an mTOR-dependent process. Pretreatment with resveratrol stops MAPK activation, thereby inhibiting the activation of CREB and NFκB. Microglial cells are essential for synapse formation and neural circuit pruning. They switch between proinflammatory and anti-inflammatory states, providing neuroprotection under normal conditions (Zhong et al. 2012). After CNS injuries like stroke or trauma, microglia activate into a proinflammatory state, crucial for addressing threats. Ideally, they revert to an anti-inflammatory state post-damage to aid tissue repair and maintain CNS balance (Shi and Holtzman 2018). However, chronic activation can lead to neuroinflammation, significant in diseases like Parkinson's, Alzheimer's, ALS, and multiple sclerosis (MS). Polyphenols, known for their ability to shift microglia towards anti-inflammatory states, hold promising treatment in neurodegenerative diseases and mitigating neuroinflammation.

Alzheimer's disease (AD) is a major disorder of the central nervous system characterized by the aggregation of amyloid β peptide and hyper phosphorylated tau protein. Normally, tau protein maintains stability and microtubule assembly when phosphorylated appropriately. Besides GSK-3β and Cyclin-dependent kinases, protein phosphatase 2A (PP2A) plays a crucial role in regulating tau protein phosphorylation levels to maintain homeostasis (Yadikar et al. 2020). In Alzheimer's disease (AD), PP2A downregulation, mediated by MID1, leads to tau protein hyper phosphorylation. Resveratrol reduces MID1 expression in vitro and decreases tau phosphorylation and MID1 levels in mice. However, oral resveratrol in mild AD patients did not reduce phosphorylated tau; it improved cognitive function and reduced MMP9 levels, impacting blood-brain barrier integrity and inflammation (Winter et al. 2016). Amyloid β-aggregation triggers neuro-inflammation through the NFκB pathway, activating proinflammatory genes. Thioredoxin-interacting protein (TXNIP) is also activated, inhibiting thioredoxin's anti-inflammatory role and promoting neuroinflammation. Resveratrol treatment in amyloid β-42-activated BV2 cells reduces NFκB and IκB phosphorylation, cleaved caspase 1 levels, and proinflammatory cytokines IL-1β, TNF-α, and IL-6. It also suppresses TXNIP expression, suggesting its potential in mitigating neuroinflammation associated with amyloid β aggregation (Feng and Zhang 2019).

5.2.1.2 Curcumin

Curcumin is a primary bioactive molecule in turmeric rhizome (*Curcuma longa*). Belonging to the *Zingiberaceae* family and predominantly extracted from root tuber of Curcuma aromatica Salisb, it is well-known for its antioxidant, anti-inflammatory, and antitumor properties (Sharifi et al. 2020b). Traditional Chinese medicines, including turmeric, are known for promoting blood circulation and removing blood stasis and have been utilized in China for centuries to treat inflammation, pain, and various ailments. In India, turmeric is most commonly used spice and is recognized in Ayurveda as a remedy in inflammation condition. In Western herbal medicine, turmeric is valued for its anti-inflammatory properties. The active compounds in turmeric, known as curcumin and curcuminoids, have proven to be effective therapies. The chief curcuminoids are demethoxycurcumin, curcumin, and bisdemethoxycurcumin. Curcumin-containing dietary supplements are highly popular, with numerous antioxidant and anti-inflammatory options available in the market (Aggarwal and Harikumar 2009).

Numerous studies have shown that curcumin has neuroprotective effects in animal models of neurodegenerative diseases, such as AD, PD, multiple sclerosis, and depression. Inflammation involves effectors, mediators, sensors, and inducers. Various inflammatory stimuli trigger distinct physiological and pathological pathways. Anti-inflammatory drugs commonly work by modifying tissue responses, counteracting mediator effects, targeting receptors and signaling pathways, and promoting the synthesis of anti-inflammatory mediators. Curcumin specifically reduces inflammation by regulating inflammatory signaling pathways and lowering the synthesis of inflammatory mediators (Gao et al. 2019). Curcumin links to Toll-like receptors (TLRs) and modulates downstream signaling pathways like mitogen-activated protein kinases (MAPK), nuclear factor kappa-B (NF-κB), and activator protein 1 (AP-1). These actions regulate inflammatory mediators and are beneficial in treating inflammatory ailments. Additionally, curcumin downregulates NF-κB by interacting with peroxisome proliferator-activated receptor gamma (PPARγ), (Yu et al. 2018). In cellular and animal models of neuroinflammation, curcumin has been extensively researched. It reduces the synthesis of TNFα, PGE2, and nitric oxide (NO) and inhibits NF-κB and MAPK activation. These effects were observed in experiments using BV-2 microglial cells activated by LTA, demonstrating curcumin's ability to mitigate neuroinflammation (Sorrenti et al. 2018). The same findings were observed in BV-2 microglia stimulated with LPS. These results were confirmed by animal experimental models of AD, where curcumin treatment improved cognitive function and reduced neuroinflammation induced by LPS exposure and in p25 transgenic mice. Curcumin is found to also inhibit the formation of amyloid β-oligomers and fibrils and decrease amyloid deposition in animal models of AD (Porro et al. 2019). Another study has highlighted curcumin's protective effects against ischemic stroke in experimental rodent models. However, its specific impact on inhibiting the polarization of LPS-induced microglial phenotypes and promoting M2 polarization remains unclear. Additionally, the precise molecular mechanisms through which curcumin exerts its anti-neuroinflammatory effects are not fully understood. Key signaling molecules like TLR4/NF-κB and TREM2 play crucial roles in modulating neuroinflammation and microglial activation. TLR4 is predominantly classified in microglia within the brain and responds to ligands like lipopolysaccharide (LPS), activating the NF-κB pathway and promoting M1 microglial activation, leading to the transcription of proinflammatory cytokines like IL-6, IL-1β, and TNF-α. In contrast, TREM2, uniquely expressed on microglia, helps regulate neuroinflammation in the central nervous system (CNS) (Payne et al. 2022).

5.2.1.3 Epigallocatechin Gallate (EGCG)

Green tea extract contains a substantial amount of epigallocatechin 3-gallate (EGCG), a polyphenol with diverse bioactivities that have been shown to treat various diseases (Unno et al. 2017). Green tea contains six primary catechin components: epicatechin (EC), epicatechin gallate (ECG), gallocatechin, epigallocatechin (EGC), and epigallocatechin gallate (EGCG). Among these, EGCG is the key active and extensively studied polyphenol. Each 250 mL (1.250% w/v) of green tea contains approximately 177.0 mg of EGCG. Epigallocatechin gallate (EGCG; $C_{22}H_{18}O_{11}$) is a type of catechin and major flavanol found in green tea. Its structure comprises of three aromatic rings (A, B, and C), along with a gallate group (sometimes referred to as the D ring). The B and D rings, rich in hydroxyl groups, are primarily responsible for its antioxidant activity by scavenging reactive oxygen species (Zhang et al. 2020). The D ring, in particular, has demonstrated anti-inflammatory and anticancer properties. Because of the presence of seven hydroxyl groups spaced out over three aromatic rings, EGCG is water soluble and has high permeability across the BBB, typically penetrating it within 0.5 hours. However, only 57.54% of EGCG can pass through the BBB. Despite its significant biological and pharmacological activity, oral EGCG has relatively low bioavailability. A study indicated that consuming two cups of green tea resulted in maximum plasma levels of EGCG at just 0.15 μM. This low bioavailability is due to incomplete absorption and instability in the intestinal and blood environments. Enhancing the bioavailability of oral EGCG could be achieved through structural modification or nanomaterial-based preservation and delivery systems (Kumazoe et al. 2013).

EGCG binds to various surface receptors, initiating cell signaling pathways and regulating their activities. Specifically, EGCG interacts to the 67-kDa laminin receptor (67LR), which is essential for its antitumor effects. Activation of 67LR by EGCG triggers the apoptotic signaling Akt/eNOS/PKCδ/NO/cGMP pathway, with up-regulation of cGMP being a crucial step in 67LR-dependent apoptosis (Hong et al. 2010). EGCG also inhibits TLR4 signaling via 67LR, exerting anti-inflammatory effects by binding with pathways like NF-κB, AP-1, and MAPK. EGCG's inhibition of TLR4 has been depicted to deregulate inflammation & improve insulin signaling in adipose tissue and alleviate hepatic insulin resistance in non-alcoholic fatty liver ailment. EGCG also targets cell surface growth factor receptors, specifically receptor tyrosine kinases, which are involved in cell proliferation, survival, and angiogenesis (Minnelli et al. 2021). For example, EGCG inhibits PDGFR-induced mitogenesis, EGFR-mediated tumor cell proliferation and migration, VEGFR-stimulated angiogenesis, and IGFR-1 associated with various cancers. These actions result in antitumor, antioxidative, anti-inflammatory, and neuroprotective effects (Cheng et al. 2021).

Number of studies have highlighted the significance of neuroinflammation in the pathophysiology of PD since McGeer and their colleagues first discovered active microglial infiltration in the substantia nigra of postmortem PD brains in early 1980s. Research has shown that cytokine-induced inflammatory responses are crucial in this disease, with peripheral immune cell infiltration and astrocyte/microglia activation leading to chronic neuroinflammation and neuronal injury (Ozduran et al. 2022). Targeting chronic inflammation is now considered a viable therapeutic approach for PD, with pharmacological treatments aiming to reduce neuroinflammation to prevent or slow disease progression. EGCG, a polyphenol founded in green tea, displays anti-inflammatory effects in vitro. It inhibits microglial secretion of tumor necrosis factor-α (TNF-α) and nitric oxide (NO) triggered by LPS which downregulates iNOS and TNF-α expression. EGCG also protects against dopaminergic neuron damage in PD models and preserves neuronal viability in various cell cultures, including SH-SY5Y and first rat mesencephalic cells. Additionally, EGCG reduces TNF-α and IL-1β production induced by 6-OHDA in SK-N-AS cells, thereby preventing apoptosis and promoting cell survival. In PD, anti-inflammatory effects of EGCG have been supported by biological studies (Prasanth et al. 2019). Rats treated with EGCG prior to rotenone exposure exhibited reduced levels of NO and neuroinflammatory markers in the striatum, along with lower TNF-α and NO. This treatment also enhances the number of TH-immunoreactive neurons. Furthermore, EGCG reduced pro-inflammatory cytokine levels, improved motor function in PD models, and EGCG-loaded liposomes lowers NO and TNF-α production in LPS evoked BV-2 microglia (Yang et al. 2020).

EGCG has shown multifaceted effects on potential targets related to AD. In hippocampal neuron cultures, it protects against beta-amyloid-induced neurotoxicity. EGCG supports the conversion of amyloid precursor protein to soluble APP, through activation of protein kinase C (PKC), thereby reducing the synthesis of neurotoxic Aβ. Furthermore, EGCG blocks β-secretase (BACE1), which changes sAPP to Aβ. Additionally, EGCG induces glycogen synthase kinase 3 (GSK3) and inhibits cAbl/FE65, a tyrosine kinase involved in neurodevelopment and nuclear translocation, thereby mitigating Aβ-induced neurotoxicity (Kou et al. 2022).

5.3 Flavonoids

Flavonoids are naturally occurring compounds with a core structure of 2-phenylchromone. They are abundant in various dietary sources such as fruits, vegetables, cereals, tea, and herbs. Flavonoids can be found either in their free form or as glycosides. An essential class of naturally occurring polyphenolic compounds, flavonoids have been found to be beneficial in problems related to the heart and metabolism, cancer, autoimmune diseases, and neurological disorders. They have immunomodulatory, anti-inflammatory, anti-apoptotic, and antioxidant qualities. The recognition of the use of flavonoids in neurological disorder is growing because of their excellent pharmacokinetics, cost-effectiveness, BBB crossing ability, safety, and clinical application readiness. It also has been shown to mitigate neuroinflammation by inhibiting the synthesis of proinflammatory mediators, boosting the release of anti-inflammatory factors, and altering the polarization of microglia and astrocytes. This effect is mainly achieved by inhibiting the activation of the NLRP3 inflammasome and the MAPK, NF-κB, and STAT/JAK pathways. Additionally, flavonoids enhance Nrf2, CREB/BDNF, AMPK, Akt/PI3k signaling, Wnt/β-Catenin, and SIRT1-mediated HMGB1 deacetylation (Zhang et al. 2021).

5.3.1 Quercetin

Quercetin belongs to the class of flavonols, and it can be available in a severity of foods, including tomatoes, onions, shallots, and apples, as well as in medicinal plants like *Ginkgo biloba*, *Sambucus canadensis*, and *Hypericum perforatum*. Quercetin is a significant bioflavonoid that has been shown to have anti-inflammatory, anti-hypertensive, vasodilatory, anti-obesity, anti-hyper-cholesterolemic, and anti-atherosclerotic properties (Batiha et al. 2020). Anti-inflammatory drugs can inhibit neuroinflammation by modulating the production of nitric oxide (NO), glial activation cascades, and inflammatory cytokines, as well as by preventing neuronal death (Zhang et al. 2011). Extensive in vivo and in vitro studies have demonstrated that quercetin possesses potent anti-inflammatory properties. Quercetin mitigates inflammation induced by 6-hydroxydopamine (6-OHDA) toxicity in zebrafish models and PC12 cells by lowering the overproduction of NO, the iNOS enzyme, and other inflammatory genes (Bureau et al. 2008).

Quercetin also inhibits the expression of iNOS and downregulated various signaling pathways, including extracellular signal-regulated kinase, Janus kinase-1, Akt, Src, c-Jun N-terminal kinase, and activating protein-1 (AP-1), while upregulating heme oxygenase (HO)-1 to prevent inflammation caused by lipopolysaccharide (LPS) and interferon γ. In microglial N9-neuronal (PC12) co-culture, quercetin reduced microglial-driven neuronal cell death and the production of pro-inflammatory cytokines such as TNF-α and IL-1α in astrocytes. Additionally, β-cyclodextrin-dodecylcarbonate nanoparticles with quercetin showed potent anti-inflammatory effects by downregulating TLR4 and COX2, enhancing their pharmacokinetics and targeting within the brain. In brain tissue, quercetin decreased the expression of IL-1β, monocyte chemoattractant protein (MCP)-1, and inflammation-induced neuron death triggered by 1-methyl-4-phenylpyridinium. Proinflammatory cytokines (IL-1, TNF-α, and IL-6) and neurotoxic substances released by astrocytes contribute to neuronal damage. Quercetin protects astrocytes from the harmful effects of tert-butyl hydroperoxide and H_2O_2 by regulating ROS synthesis, reducing apoptosis, and increasing HO-1 transcription. Additionally, quercetin vitally increases the expression of GSH synthetase and glutamate cysteine ligase (GCL) in astrocytes, reducing inflammation by suppressing IL-1β, IL-8, IL-6, ROS, and release of monocyte chemoattractant protein-1 (Sharma et al. 2013). Quercetin also inhibits TNF-α by enhancing the NF-κB signaling pathway, showcasing significant anti-inflammatory activity due to its antioxidant properties. Quercetin effectively reduces inflammatory responses in LPS-induced primary microglia or BV2 cells by inhibiting phagocytic activity and lowering levels of NO, IL-1β, TNF-α, MCP-1, CXCL10, IL-6, iNOS, lipocalin-2, and COX-2. It promotes IL-10 secretion through AMPK and Nrf2 signaling pathways while suppressing NF-κB activation. In animal models, quercetin mitigates age or LPS-induced behavioral abnormalities, suppresses microglia and astrocyte activation, and decreases IL-1β levels by upregulating SIRT1 and reducing NLRP3 and cleaved-caspase-1 expression in the brain. Moreover, quercetin reduces pro-inflammatory mediators in LPS-stimulated or traumatic brain injury rats by inhibiting NF-κB and activating the Nrf2/HO-1 pathway. Therefore, quercetin serves as a safe and effective dietary supplement for attenuating neuroinflammation. It achieves this by modulating the NF-κB, Akt, AMPK, and Nrf2/HO-1 pathways, increasing SIRT1 protein expression, and blocking the activation of NLRP3 inflammasomes (Kim et al. 2005).

5.3.2 Baicalein

Baicalein is a major flavonoid found in the roots of *Scutellaria baicalensis Georgi*, also known as the traditional Chinese herbal medicine Huangqin. It has been widely used to treat cancer, hypertension, inflammation, and cardiovascular diseases. In LPS plus IFN-γ-induced BV2 cells, baicalein inhibits microglial activation and polarization by reducing the production of TNF-α, iNOS, IL-1β, IL-6, CD16, and CD86, while increasing Arg-1 and CD206 levels. This effect is mediated through the activation of STAT1 expression and inhibition of the TLR4/NF-κB pathway. In rats induced with middle cerebral artery occlusion (MCAO), baicalein mitigates microglial activation and polarization by inactivating JNK, ERK, IκBα, and p38, as well as preventing the nuclear translocation of p65. Furthermore, baicalein has been shown to downregulate cleaved GSDMD, cleaved caspase-1, and NLRP3, while promoting the expression of PSD95, p-Akt, BDNF, p-TrkB, CREB, p-CaMK II, SYP, and p-PI3k. In the substantia nigra and midbrain of MPTP/rotenone induced Parkinson's disease mice, baicalein effectively suppresses the expression of GFAP and Iba-1 (Jung et al. 2010).

Baicalein could degrade the synthesis of proinflammatory cytokines such as TNF-α, IL-1β, and IL-6, which are known to increase in experimental models of PD induced by MPTP. Additionally, strong experimental evidence supports the neuroprotective effects of baicalein in mice treated with MPTP. In the rat model of 6-hydroxydopamine-induced muscle tremor, baicalein therapy alleviated tremors and preserved tyrosine hydroxylase (TH)-positive neurons. Furthermore, baicalein reduced neuroinflammatory astrocyte activation in the MPTP animal model and prevented rotenone-induced apoptosis in PC12 cells. These findings strongly offer that baicalein may be a reliable treatment to prevent neurodegeneration associated with PD. Recent research has shown that the MPTP-induced mouse PD model activates caspase-1, and baicalein inhibits this activation, indicating that inflammasome pathway plays its role in the pathophysiology of experimental Parkinson's disease. However, the precise effects of baicalein on the canonical caspase 1-associated inflammasome pathway in MPTP-induced PD have not been thoroughly investigated (Chen et al. 2006). Another study used rotenone induced PD mice and BV2 microglia to investigate the in vivo and in vitro mechanisms of baicalein's anti-neuroinflammatory effects. In the rotenone-induced PD mouse model, baicalein greatly enhanced motor performance and reduced the number of inflammatory responses. Additionally, baicalein prevented caspase-1 from activating NLRP3 from being expressed in brain tissues. Likewise, baicalein successfully reduced the rotenone-induced inflammatory response in BV2 microglia. Baicalein was shown to decrease ROS levels and block the TLR4/NF-κB signaling pathway, which in turn lowered the production of NLRP3 and activated caspase-1 in vitro studies. These findings imply that baicalein inhibits the activation of the NLRP3 inflammasome, which is one way in which it reduces neuroinflammatory responses (Tsai et al. 2021).

5.4 Alkaloids

Alkaloids belong to a varied group of phytochemicals characterized by the presence of basic nitrogen atoms, predominantly sourced from plants. These compounds have been utilized for centuries in orthodox medicine owing to their wide range of pharmacological properties, including analgesic, anti-inflammatory, antimicrobial, and antitumor effects (Lee et al. 2020). Alkaloids play a significant role in human health by serving as the basis for many modern medications, such as morphine for pain relief, quinine for treating malaria, and atropine for various medical conditions. Their therapeutic potential extends to neuroprotection, making them significant in the management of neurodegenerative diseases. For instance, alkaloids like galantamine, extracted from the snowdrop plant, are used in treating Alzheimer's disease (AD). Galantamine acts as an acetylcholinesterase inhibitor, thereby enhancing cholinergic function and improving cognitive symptoms in AD patients. Additionally, alkaloids like nicotine and caffeine have been studied for their neuroprotective effects. Nicotine, despite its addictive properties, has shown potential in protecting against Parkinson's disease (PD) by modulating dopaminergic pathways and reducing neuroinflammation. Similarly, caffeine, an alkaloid, known for its widespread consumption, correlated with a lower risk of developing PD and AD due to its antioxidant effect and its role in blocking adenosine receptors, which helps to reduce neuronal damage (Lofrumento et al. 2011). The multifaceted mechanisms of alkaloids in modulating neurotransmission, reducing oxidative stress, and inhibiting inflammatory pathways underscore their importance in neuroprotection. As research advances, the potential of alkaloids in developing therapeutic strategies for neurodegenerative diseases continues to grow, highlighting their significant role in enhancing human health and addressing complex neurological disorders (Rui et al. 2020).

5.4.1 Berberine

Berberine, an isoquinoline alkaloid, is predominantly present in variety of plants, notably within Berberis species, including *Berberis vulgaris* (barberry), *Berberis aquifolium* (Oregon grape), and *Berberis aristata* (tree turmeric). This alkaloid has been traditionally used in Chinese and Ayurvedic medicine due to its extensive antimicrobial properties, notably for treating gastrointestinal infections (Xin et al. 2021).

The neuroprotective mechanisms of berberine are multifaceted and involve a range of biochemical pathways. One of the primary mechanisms is its anti-inflammatory action, which is crucial in combating neurodegenerative diseases marked by chronic inflammation, like AD and PD (Gutierrez et al. 2020). Berberine inhibits the NF-κB signaling pathway, a crucial regulator of the inflammatory response. By lowering NF-κB activation, berberine diminishes the synthesis of pro-inflammatory cytokines like IL-1β, TNF-α, and IL-6, thereby mitigating inflammation and neuronal damage. Apart from its role in anti-inflammatory properties, berberine also exerts antioxidant properties. It increases the activity of antioxidant enzymes like superoxide dismutase and catalase and increases the levels of glutathione, a major intracellular antioxidant. These actions help in neutralizing ROS and reducing oxidative stress, which is a key factor in neuronal degeneration. Berberine also stimulates the AMPK pathway, essential for cellular energy regulation, activation of AMPK via berberine leads to enhancement of mitochondrial function, improvement in autophagy, and reduction in apoptosis, all of which contribute to neuronal survival and protection against neurodegenerative conditions (Babashpour et al. 2023).

As it pertains to Alzheimer's disease, berberine has shown promising effects by targeting multiple pathological aspects of the disease. It reduces amyloid-β (Aβ) production and deposition, which are hallmarks of AD. Berberine inhibits the activity of beta-secretase (BACE1), an enzyme crucial for the formation of Aβ peptides. Moreover, it has been noted to decrease the hyper phosphorylation of tau protein, which is another hallmark of AD, by controlling the activity of glycogen synthase kinase-3β. These actions help in maintaining synaptic function and cognitive abilities in AD models (Postuma et al. 2012).

In Parkinson's disease, berberine has demonstrated protective effects on dopaminergic neurons, which are progressively lost in PD. It reduces oxidative damage and mitochondrial dysfunction in these neurons. Berberine has also been found to inhibit the accumulation of α-synuclein, a protein that forms toxic aggregates in PD, and modulates autophagic pathways to clear these aggregates. Additionally, by reducing neuroinflammation, berberine helps in preserving the integrity of dopaminergic pathways and preventing further neuronal loss (Tian et al. 2023).

Beyond AD and PD, berberine's neuroprotective effects extend to other neurodegenerative diseases like Huntington's disease and ALS. Its ability to regulate multiple pathways involved in oxidative stress, mitochondrial dysfunction, protein aggregation, and inflammation underscores its potential as a therapeutic agent (Li et al. 2014).

5.5 Sulfur-Containing Compounds

Sulfur-consisting compounds are a varied cluster of bioactive molecules which play an essential role in various biological methods and have profound implications for human health. These compounds, including glutathione, taurine, and allicin, are integral to maintaining cellular redox balance, detoxification, and immune function. Glutathione, a tripeptide consists of glutamine, cysteine, and glycine, is one of the most critical antioxidants in the body, protecting cells from oxidative stress and maintaining the balance of other antioxidants. Taurine, an amino sulfonic acid, supports cardiovascular health, bile salt formation, and supports the development and function of skeletal muscles, CVS, and the cardiovascular system. Allicin, derived from garlic, exhibits potent anti-inflammatory, antimicrobial, and antioxidant properties, contributing to CVH by reducing cholesterol levels and blood pressure (Yuan et al. 2019).

Sulfur-containing compounds have attracted considerable interest in neuroprotection for their capacity to combat neurodegenerative diseases. These compounds aid in reducing OS and neuroinflammation, which play crucial roles in the advancement of conditions like AD, PD, and ALS (Jia et al. 2012).

5.5.1 Sulforaphane

Sulforaphane is a naturally occurring isothiocyanate present in cruciferous vegetables, particularly in high concentrations in broccoli sprouts. It was first isolated in the early 1990s by Paul Talalay and his colleagues at Johns Hopkins University during their research on cancer prevention. The discovery originated from research aimed to identify the compounds behind broccoli's chemoprotective properties, which led to recognizing sulforaphane as a potent inducer of phase II detoxification enzymes. This marked the beginning of extensive research into its broad-spectrum health benefits, including its key role in neuroprotection (Francioso et al. 2020).

Sulforaphane exhibits a variety of powerful neuroprotective mechanisms, mainly due to its activation of the nuclear factor erythroid 2-related factor 2 (Nrf2) pathway. As a transcription factor, Nrf2 governs the expression of various cytoprotective and antioxidant genes. Sulforaphane enhances the cellular defense against inflammation and oxidative stress by binding to the antioxidant response element (ARE) in the promoter regions of these genes. This activation leads to the increased production of glutathione, superoxide dismutase, and catalase, which are essential for neutralizing ROS and safeguarding neuronal cells from oxidative damage (Miekus et al. 2020).

Furthermore, sulforaphane inhibits the NF-κB signaling pathway, which is significant in mediating inflammatory responses. By blocking NF-κB activation, sulforaphane decreases the synthesis of proinflammatory cytokines, like TNF-α, IL-1β, and IL-6. This anti-inflammatory action is particularly beneficial in neurodegenerative diseases, where chronic inflammation exacerbates neuronal damage and disease progression (Santulli et al. 2023).

Apart from its anti-inflammatory and antioxidant effects, sulforaphane also influences the expression of heat shock proteins (HSPs) and genes related to autography. HSP assist in the proper folding and functioning of other proteins, protecting cells from stress-induced damage. Sulforaphane-induced expression of HSPs helps in maintaining protein homeostasis and preventing the aggregation of misfolded proteins, which are characteristic of neurodegenerative conditions like PD and AD. Sulforaphane also boosts autophagy, a cellular process accountable for breaking down and recycling damaged proteins and organelles. The increased autophagy facilitates the removal of toxic protein aggregates, like amyloid-β plaques in AD and α-synuclein in PD, thereby reducing their neurotoxic effects (Zhang et al. 1992).

5.6 Other Compounds

5.6.1 Alpha-Lipoic Acid (ALA)

Alpha-lipoic acid (ALA), also termed as thioctic acid, which is a naturally occurring substance found in the mitochondria of almost all prokaryotic and eukaryotic cells (Zhang et al. 1994). It was first isolated from a liver sample by the biochemist Lester Reed in the 1951 (Kim 2021). ALA is both synthesized endogenously and obtained through dietary sources such as spinach, broccoli, tomatoes, peas, and organ meats like liver and heart. Its discovery marked a significant advancement in understanding cellular metabolism, particularly the role of ALA as a cofactor for mitochondrial enzyme complexes involved in energy production, specifically in the oxidative decarboxylation of alpha-keto acids (Ruhee and Suzuki 2020).

The neuroprotective mechanisms of ALA are multifaceted, primarily attributed to its potent antioxidant properties. ALA functions as a powerful antioxidant both in its reduced and oxidized form, dihydrolipoic acid (DHLA). This dual antioxidant capabilities allows ALA to scavenge a broad spectrum of ROS), and reactive nitrogen species (RNS), thereby protecting neurons from OS induced damage. Additionally, ALA can regenerate other antioxidants such as vit. E, vit. C, and glutathione, further enhancing its neuroprotective effects by maintaining the cellular redox balance (Botanska et al. 2023).

Another critical mechanism through which ALA exerts its neuroprotective effects is by chelating metal ions. Metals including iron and copper can catalyze free radical production via Fenton reactions, resulting in oxidative stress and neuronal injury. ALA's metal chelating properties help in reducing metal-induced oxidative damage, particularly in neurodegenerative conditions where metal dysregulation is often observed (Hu et al. 2022).

ALA also modulates several crucial signaling pathways participating in inflammation. It prevents the activation of NF-κB pathway, which shows a vital role in inflammatory response. By lowering NF-κB activation, alpha-lipoic acid (ALA) reduces the production of proinflammatory cytokines such as IL-1β, TNF-α, and IL-6. This anti-inflammatory action is particularly beneficial in neurodegenerative diseases like PD and AD, where chronic inflammation plays a significant role in disease progression (Tripathi et al. 2023).

In AD models, ALA has been shown to reduce amyloid-beta deposition, improve cognitive function, and decrease oxidative stress markers. Its ability to cross the BBB allows it to employ direct effects on brain tissue, making it a valuable therapeutic agent. In PD models, ALA protects dopaminergic neurons from degeneration, reduces oxidative stress, and improves motor function. These neuroprotective effects are further supported by ALA's ability to enhance synaptic plasticity and neurogenesis, which are vital for maintaining cognitive and motor functions (Capece et al. 2022).

5.6.2 Melatonin

Melatonin, also classified as 5-methoxy-N-acetyltryptamine, is a hormone naturally synthesized by the pineal gland present in the brain. It was first isolated and identified in 1958 by Dr. Aaron B. Lerner and by his team at Yale University, who were investigating compounds involved in skin pigmentation. Melatonin synthesis and release are controlled by the light/dark cycle, with peak levels occurring at night, thereby playing an evidential role in regulation of sleep-wake cycles and circadian rhythms. Beyond its role in regulating sleep, melatonin has garnered substantial attention for its broad-spectrum neuroprotective properties, owing to its potent antioxidant, anti-inflammatory, and mitochondrial-supporting actions (Takic et al. 2022).

The neuroprotective mechanisms of melatonin are diverse and robust. As a powerful antioxidant, melatonin directly scavenges a broad range of ROS and RNS, thereby decreasing OS, which is a chief contributor to the neuronal damage in NDD. Melatonin enhances its antioxidative capacity by enhancing the expression of various antioxidant enzymes like glutathione peroxidase, superoxide dismutase, and catalase. This assists in maintaining cellular redox balance and helps in protecting neurons from oxidative damage.

Another critical neuroprotective mechanism of melatonin is its anti-inflammatory action. Melatonin prevents the activation of the NF-κB pathway, an essential regulator of inflammatory responses. Melatonin decreases the synthesis of proinflammatory cytokines like TNF-α, IL-1β, and IL-6 by inhibiting NF-κB activation. This suppression of inflammation is particularly beneficial in neurodegenerative diseases such as AD and PD, where chronic inflammation exacerbates neuronal damage and disease progression.

Melatonin also supports mitochondrial function, which is vital for neuronal health and survival. It enhances mitochondrial respiration, reduces mitochondrial ROS synthesis, and prevents mitochondrial permeability transition pore opening, thereby maintaining mitochondrial integrity and function. These actions help in preserving energy production and reducing apoptosis (programmed cell death) in neurons, which is essential in preventing neurodegenerative processes.

In addition to these mechanisms, melatonin influences neurogenesis and synaptic plasticity. It promotes the survival and distinction of neural stem cells, supports synaptic formation, and enhances cognitive function. In models of AD, melatonin has been depicted to reduce amyloid-β (Aβ) plaque formation, a hallmark of the disease, and inhibit the hyperphosphorylation of tau protein, thereby preventing the neurofibrillary tangles characteristic of AD. Melatonin can directly exert these protective effects on brain tissues due to its efficacy to cross the BBB, allowing its potential as a promising therapeutic drug for AD.

In models of PD, melatonin protects dopaminergic neurons, which are gradually lost in the disease, by diminishing inflammation and OS. It also mitigates the toxicity of α-synuclein, a protein that helps in aggregation in PD, contributing to neuronal death. Additionally, melatonin has shown potential benefits in other neurodegenerative diseases like ALS and HG, where its multifaceted neuroprotective properties can help slow disease progression and improve neurological function (Toklu et al. 2010).

5.7 Mechanisms of Action in Reducing Neuroinflammation and Oxidative Stress

Neurodegenerative disorders are found to be the dominant cause of morbidity and disability worldwide, gathering attention because of enlarge impact on aging society. Such ailments sheds light on constant impairment in neuronal functions directing to brain atrophy, among mentioned diseases: Parkinson's, Alzheimer's, epilepsy, and ALS are some most common. Pathological mechanism and etiology of neurodiseases remains poorly understood. Furthermore, at present, treatment on basis of modulation of monoamine system doesn't seem to have psychiatric symptoms on all patients. Neuroinflammatory processes have received increasing attention in recent years, as elevated levels of proinflammatory cytokines are more frequently found in patients with neuropsychiatric disorders. It is vastly acknowledged that seizure hinders several biophysical and chemical processes in CNS depicting synthesis of ROS and OS. The brain is the largest energy consumer in the human body, accounting for about 20% of the total oxygen. Of this, around 70–80% is consumed by neurons, highlighting the brain's vulnerability to meatbolic and oxidative stress. OS and epilepsy seem to be linked with neuroinflammation, involving interleukins like interleukin-6 and interleukin 1-beta and tumor necrosis factor alpha.

Oxidative stress occurs via imbalance of antioxidants and oxidants in biological fluid. This imbalance leads to surplus level of ROS or impairment of antioxidant system. The role of oxygen is undebatable in biology, i.e. helps in proper cellular functioning, gene transduction, signal transduction, and other extracellular activities. Adverse event of O_2 is its univalent metabolic stepdown, which is rationale for ROS generation. Nitric oxide [NO] helps in regulating proliferation and relaxation of vascular smooth muscle cells, angiogenesis, leukocytes adhesion, thrombosis, platelet aggregation, hemodynamics, and vascular tone, whereas nitric monoxide [NO] is observed to be harmful for biomolecules. Several free radical oxygen-based chemical species are hydroxyl radical [HO●], peroxyl radicals [●O_2-], hydroperoxyl radical [HO_2●], and superoxide radical anion [ROO●] (Wang et al. 2018) (Fig. 5.2).

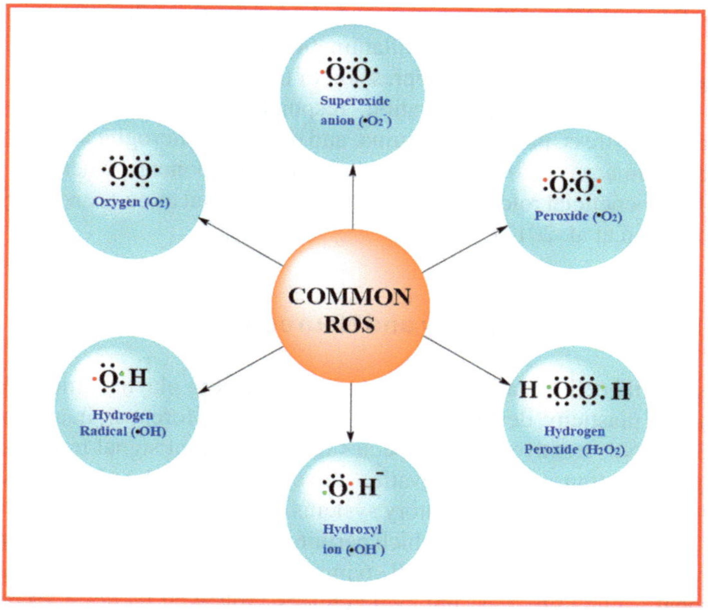

Fig. 5.2 Commonly known reactive known species (ROS). (https://revvitysignals.com/)

In mitochondrion, O_2 plays a sturdy role in oxidative phosphorylation, implicating glucose breakdown and generating ATP (cellular energy) in mitochondria. Protein and enzyme are necessary for oxidative phosphorylation (OP) encoded via mitochondrial DNA, and any mutation directly obstructs OP. Free radicals and oxygen produced exogenously (chemical and radiation, etc.) and endogenously (mitochondria, NADPH oxidase, etc.) are found to be constantly added in cellular environment. Various antioxidants like vitamin A, vitamin E, vitamin C, glutathione, taurine, creatine, and polyphenols (from tea extract) are synthesized against ROS. For instance, resveratrol (found in berries, peanuts, wine, and grapes) attenuates interleukin-1 beta, TNF-alpha, NO, and mRNA by inhibiting mPGES-1 and COX-1 via decreasing MMP9 expression, whereas in case of curcumin, it acts via inhibiting the activation of NF-kN through PI3K/Akt phosphorylation or JAK-STAT pathway inhibition (Tamagno et al. 2021).

5.7.1 Neuroinflammation

ROS and RNS help in promoting intracellular signaling leading in enhancement of pro-inflammatory genes. Neuroinflammation referred as the inflammation response of CNS factors acting against homeostasis including several cells within CNS such as microglia and astrocytes. This event could be seen in traumatic, ischemic, and infectious neurological events playing a vital role in progression of diseases. Mitochondrial ETC cycle is observed to be playing a crucial role for intracellular

ROS; NADPH NOX2 is the over-activated phagocytic inflammatory oxidative enzyme. Oxidative stress and neuroinflammation are two pivotal pathological case, inhibiting one of them results in suppression of the other. Microglia is the main element of neuro-inflammation activation, responding rapidly to imbalance brain homeostasis via stress, pathology, trauma, and diseases. Microglia activation results in release of numerous cytotoxic and inflammatory components, whereas it was also found to have beneficial roles such as remodeling neuronal circuits and CNS homeostasis (Tordjman et al. 2017).

5.7.2 Role of OS in Epilepsy and Neuroinflammation

ROS acts as vital cell signaling and potentially damaging playing chief role in neuronal synaptic plasticity, development, and excitability. Moreover, in enhancing levels of OS in CNS, reactive oxygen species impels aberrant signal transduction and redox regulation transcription activation factors like MAPKs, NF-κB, and AP-1 each playing key roles in inflammatory systems. However, hydrogen peroxide is the most efficient activator of NF-κB which permits nuclear transcription and translocation activation of IL-1β, TNF-α, and IL-6 implicating promotion of ictogenic events and neuronal excitability. Post-traumatic epilepsy followed by brain injury associated with cellular and tissue damage unleash pathophysiological cascade enhancing ROS production. Enhanced "danger signals" are frequently released via neurons at epileptogenic site, instigating pro-inflammatory response detection via pattern recognition receptors (PRRs) like nod-like (NLRs) and toll-like (TLRs) immune receptors. Higher mobility group box protein-1 (HMGB1), i.e. released via oxidative neuronal injury, initiates inflammasome activation releasing it to extracellular space. To this, CNS immune cells like microglia indicates PPRs host (for instance, RAGE/TLR4 receptors) detecting ROS/pro-inflammatory cytokines successive infiltration and activation at given site of injury. TLR4/RAGE/HMGB1 signaling pathway plays a major role in numerous neurological disorders that does not lead to seizures. However, it contributes to selective vulnerability of neuronal subpopulations (such as granule cell and pyramidal neurons) to synaptic dysfunction and hyperexcitability within an inflammatory surrounding further leading to seizure activity. HMGB1 secretion from astrocytes and microglia helps enhance the pool of DAMP, hence developing mossy fiber spouting and enhancing BBB permeability and synaptic plasticity (Potes et al. 2023) (Fig. 5.3).

A series of biomarkers for neuroinflammation are identified, hosting epilepsy disorders. Plasma level of proinflammatory cytokines incorporating IL-1B, TNF-α, and IL-6 are crucial in temporal lobe epilepsy candidates. OS plays a key role in initiation of neuroinflammation that could be seen in pilocarpine treatment rat model of SE. Administration of MnIIITDE-2ImP^{5+}, a catalytic antioxidant, has been shown to scavenge reactive oxygen species (ROS), resulting in marked reductions in proinflammatory cytokines IL-1β and TNF-α within the hippocampus. In pentylenetetrazol (PTZ)-based rat model, suppression of TLR4/RAGE/HMGB1

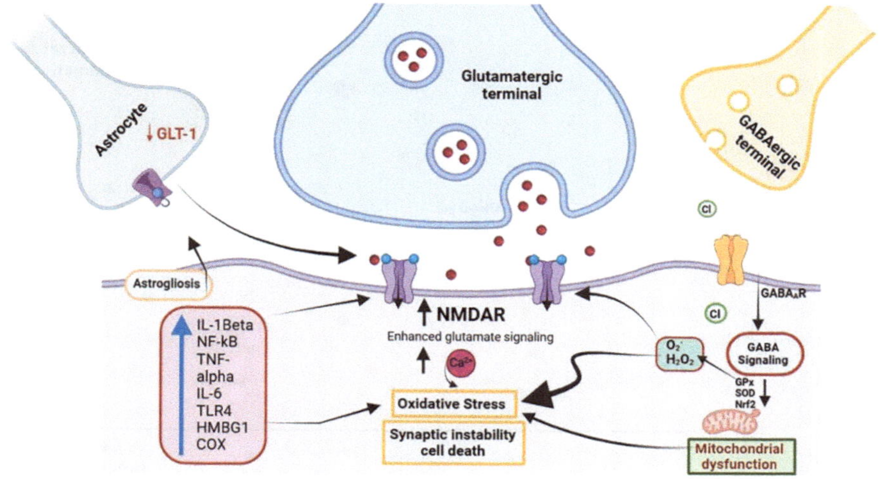

Fig. 5.3 Mechanism of epilepsy incorporating OS. Decreased GABA-mediated inhibition responses connected with enhanced glutamatergic tonus point towards higher intracellular calcium concentration which is found to be the lead source of hyperexcitability alongside with OS associated survey. (https://BioRender.com)

inflammatory cascades via pentoxifylline administration results in ROS reduction and improved hippocampal neuronal survival and cognition and memory.

Compiling these findings shed light on the fact that there is an intricate relation between neuroinflammatory response and OS associated with seizure activity, marking them as key targets for treatment of epilepsy (Ahmad et al. 2023).

5.7.3 Oxidative Stress and Parkinson's Disease

Function of Microglia: Microglia are found to be CNS-resident macrophages having homeostatic functions, constituting the immune defense system. These cells are observed to remove toxic substances, release neurotropic factors, and participate in neuronal repairment, synaptic pruning, and remodeling. To grasp the pathophysiology of PD, investigation of postmortem samples was used, implicating microglia in Parkinson's disease (Markus et al. 2013) (Fig. 5.4).

5.7.3.1 Role of Peripheral Inflammation: Novel Hypothesis for Spreading and Origin of PD

Oxidative modification of α-Syn nitration protein induces new antigenic epitopes with capability of commencing peripherally driven CD8+ and CD4+ cell responses. Activated microglia have been shown to upregulate MHC class I expression in human catecholaminergic neurons, thereby increasing their vulnerability to cytotoxic T lymphocyte-induced apoptosis. These findings underscore the immunological contribution to neurodegenerative processes. Moreover, it is depicted that

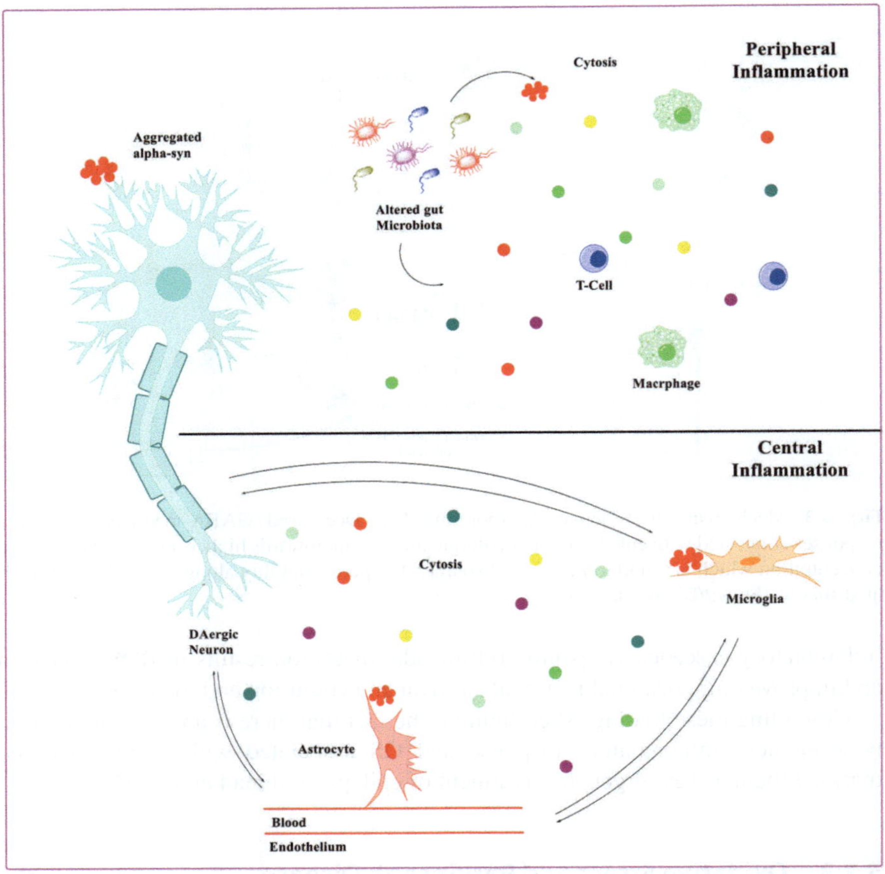

Fig. 5.4 Inflammation in PD encircling peripheral and central inflammation. Aggregated alpha-Syn suggests spreading of periphery to the brain (proteinopathy altogether with triggering factors such as ROS and impairment, etc.). (https://revvitysignals.com/)

overexpression of a sole pro-inflammatory cytokine in SN can induce a major characteristic of PD, akinesia, increasing glial activation and cell death. Enhanced cytokine level including IL-6, IFNγ, IL1β, TNF-α, and IL-2 with larger CD4+ lymphocyte counts was observed in the CSF and serum from PD patients. Peripheral inflammation (PI) is assisted via lowering CD8+ and CD4+ lymphocyte ratio and no. of Treg lymphocyte in control vs patients. PI holds an echo at brain level through autonomic nervous system or BBB and switches "primed" into "active" microglia, triggering strong action for encouraging neurodegenerative processes. The entire immune system and nervous system are found to be strongly affixed to and intrigued by the gut microbiota; the intestine is weakened via parasympathetic and sympathetic nerve fibers and correspondingly affects brain activity via intestinal immune system and vagus nerve (autonomous nervous system). Various events involving activation of ANS, regulation of sympathetic nerve activity, and increased intestinal microorganism (microbiota brain axis) impact the brain and intestine. Gastrointestinal pathophysiology is affected mostly not only via brain signals as well as gut microbiota, affecting neuropeptides, hormones, neurotransmitters, and growth factors in intestine, but also via bacterial compounds that tend to activate NOD-like receptor (NLRs), bridging between neuroinflammation and intestinal permeability. Gut microbiota regulates neuroinflammation via two ways: (i) structures mimicking self-antigens, triggering T-cells activation, and (ii) by inducing milieu mediators employing direct effect on hosts such as neurotransmitters, bacterial peptidoglycans, and short-chain fatty acids (Manchester et al. 2016).

5.7.4 Oxidative Stress and Alzheimer's Disease

Alzheimer's referred via diffusion of loss of neurons in entorhinal cortex, hippocampus, amygdala & various section of neocortex with extracellular accumulation of oligomerized amyloid β-peptide (Aβ42) termed as amyloid plaques, whereas, intraneuronal neurofibrillary tangles consists of phosphorylated τ-protein. This leads to failure of cognitive domains in a disastrous form leading to death in upcoming 3–9 years. of diagnosis and progression of dementia. Aggregation of lipid-oxidative damaged markers, nucleic acids, and proteins has been shown in several studies of postmortem brain tissue from PD patients and in CSF. Moreover, gathering of metals such as Cu and Fe in brain of AD patients as shown in postmortem studies performed via utilizing NMR and histochemical methods, alongside meta-analysis seems to challenge this impression and suggest vital citation bias. ROS could interact with any biomolecules present within neuron, damaging proteins leading to AD pathogenesis via destructing neuronal energy metabolism, redox signalling via activation of stress-based protein kinases (ERK 1/2, JNK, and p38), and proteostasis. Usually, metal-catalyzed transition leads to ROS formation in AD brain, particularly with metal-ligand with Aβ. Enhanced ROS formation leads to dysfunctional mitochondria termed as AD. Meanwhile, ROS secretion from microglia (activated) as a part of neuroinflammatory response in Alzheimer's brain is also vital where NOX plays a crucial role in above mentioned process (Srinivasan et al. 2011).

5.7.4.1 NADPH Oxidase and Alzheimer's Disease

NADPH oxidase activation is highly implicated in AD pathogenesis as apparent from postmortem studies depicting NOX2 subunits (p47phox and p67phox) translocation via cytosol to membrane. NOX enzyme activity was increased in MCI and several grades of AD compared to NOX2, NCI, and subunit (p40phox, p67phox and p47phox) levels remained elevated in various grades of AD (Fig. 5.5).

In addition, elevated levels of NOX3 and NOX1 mRNA have been spotted in the frontal lobe of early stage AD, highlighting contributions of various NOX isoforms in Alzheimer's neuropathology. Expression of NOX4 and activity of NOX3 are constantly increased in brain with consequential linear correlation between

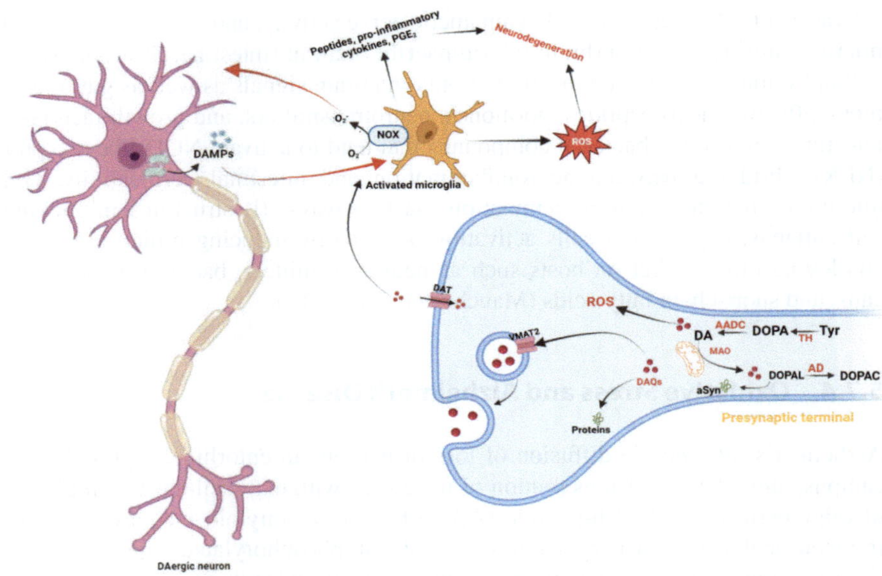

Fig. 5.5 NOX activation in Alzheimer's disease via brain microglia. Chemokines, NOX2-derived ROS, and pro-inflammatory cytokines cause activation of microglia, oxidative damage, and neuroinflammation in AD. (https://BioRender.com)

age-dependent aggregation of Aβ carrying cognitive dysfunction and NOX activity. Thus, elevated body of affirmation depicts NOX activation and ROS release in AD brain, implicating microglial response and triggering ROS synthesis. Inflammatory triggers and soluble oligomers of β-peptide (amyloid) activate microglia to M1 stage with variation in morphology, exhibiting various inflammatory markers. Microglia proliferate and congregate near β-amyloid plaques secreting proinflammatory cytokines, prostaglandins such as PGE2, chemokines, peptides, ROS and components of complement system, generating strong inflammatory response accountable for synaptic and neuronal degeneration. This further leads to release and formation of pro-inflammatory cytokines via microglia, suggesting an overlapping oxidative pathway destruction, inflammatory response in AD brain, and redox signaling mechanism (Tiong et al. 2019).

5.8 Implications for Neurodegenerative Disorders and Potential Therapeutic Applications

5.8.1 Spinocerebellar Ataxia

MSCs are found to be suitable candidates for SCA treatment as per their cognition to distinguish between immunomodulatory properties and lines. Human MSCs were introduced intracranially and i.v. to transgenic mice with polyglutamine mutation in ataxin 2 gene before and after onset loss of motor function. Intravenous transplantation leads to increase rotarod function in SCA2 mice, postponing neuronal decay; however, intracranial transplantation fails to secure a neuroprotective effect. In a clinical phase I/IIa performed in Taiwan, six volunteers with type-3 SCA & 1 with multiple atrophy cerebellar system were introduced mesenchymal stem cells (MSCs) i.v. derived from allogenic adipose tissues obtained from healthy volunteers. The follow-up depicts well tolerated with no adverse effects, concluding secure tolerability and safety of allogenic MSCs via i.v. injection (Leung et al. 2020).

5.8.2 Huntington's Disease

Transplantation of median spiny neurons (MSN) derived from ESCs and iPSCs depicts thriving neural circuit development and integration, though majorly confined to animal models. Recovered data illustrates that every third patient out of five receives merit in Creteil pilot experiment, a non-randomized, open-label, and monocentric cell transplant trial employing human's fetal donor cells retrieved from ganglionic eminence (GE). Consequently, two volunteers reported HD deterioration; however, one patient reported to have effective graft vascularization (Asemi et al. 2022).

5.8.3 Amyotrophic Lateral Sclerosis

Treating ALS patients via utilizing stem cell transplantation approach is being widely used in clinical trials. Performed six clinical studies have proven the positive effect of stem cell therapy lowering progression of ALS. Several studies have shown the efficacy of NSC therapy on amyotrophic lateral sclerosis rats. In conclusion, transplanted NSCs have the power to distinguish into neurons and form synaptic connection, delaying progression and onset of diseases therefore enhancing survival of animal models. On the basis of preclinial trials, the FDA approves 2009 clinical trials as per safety and tolerability of surgical administration of any cell toxicity or stem cells. Scientists administered fetal NSCs (Human clinical trials-I) into lumbar spinal cord of 12 ALS patients. After clinical transplantation, evaluation performed for about 6–18 months depicted no enhancement in disease progression. Intraspinal injection studies were also performed, aiming to prolong life of ALS patients via defending motor neuron groups affecting respiratory system (Mack et al. 2016).

5.8.4 Preventing Protein Aggregation and Misfolding

Major key of pathological aspect of NDs like Parkinson's disease, Huntington's disease, and AD. In Alzheimer's disease, hyperphosphorylation of tau (τ) protein leads to formation of intracellular NFT and protein accumulation resulting in degeneration of axons and dendrites. However, in AD, the accumulation of extracellular senile plaque primarily because of aggregated amyloid beta peptides and intracellular neurofibrillary tangles (NFTs) formed by hyperphosphorylated tau (τ) protein are key pathological features. These abnormalities are associated with neuroinflammation, neuronal loss, and synaptic dysfunction. Under suitable physiological conditions τ-protein associated in stabilizing microtubules. Similarly, PD is defined via proximity of α-synuclein accumulation (Lewy neurites and Lewy bodies), mostly of them finds within substantia nigra pars compacta (SNc) domain of midbrain.

α-Synuclein accumulates neurotoxic properties and accords neuronal function with survival like lysosome dysfunction, mitochondrial-dysfunction, and macroglial activation (helping neuroinflammation), disrupting axonal transportation.

Numerous plant-derived secondary metabolites hold the ability to prevent accumulation of Aβ and deregulate Aβ burden in experimental animal models of AD. These include hesperidin, apigenin, morin, isoquercetin, baicalein, and quercetin; berberine (alkaloid) found in plants of genus Berberis; curcumin catenoid from Curcuma longa; and epigallocatechin gallate catechin (EGCG) actively present in *Camellia sinesis*. Curcumin destabilizes Aβ accumulation and helps in promoting degradation of present Aβ deposits, with reduced Aβ aggregation generally mediated via γ-secretase and β-secretase inhibition, enzymes involved in the APP process; this is similar to quercetin, berberine, hesperidin, and narirutin (Bald et al. 2021).

5.8.5 Antioxidant Activity

Brain is found to be highly susceptible towards oxidative stress because of higher metabolic activity, presence of elevated redox-active metal levels, higher oxygen requirement, and oxidizable lipids. As discussed earlier, OS pathogenesis mechanism involves progression of several NDs such as disrupted energy metabolism during reperfusion of NO generation, subjecting the brain to acute OS and reactive oxygen species. Numerous compounds have been reported holding antioxidant activity such as curcumin, crocin, baicalein, quercetin, rutin, resveratrol, ginsenosides, apigenin, and berberine. Resveratrol is reported to reduce iNOS and NO expression produced via Aβ in glial cells, whereas baicalein attenuates 6-hydroxydopamine (6-OHDA) inducing neurotoxicity via lowering OS and mitochondrial dysfunction, preventing NO synthesis via inhibiting NO synthase (iNOS) present in microglia. Crocin and berberine lowers the expression of (COX2 and iNOS in vitro and in vivo and protects neurons from oxidative injury. However, quercetin promotes mitochondrial biogenesis, hence protecting neurons from ROS production and mitochondrial dysfunction leading to neuronal survival (Apweiler et al. 2024).

5.8.6 Neuroprotective Activities of Various Natural Products

Numerous phytomolecules are suggested as traditional pharmacological agents to treat NDs. Literature reported data of natural products to treat NDs, depicting

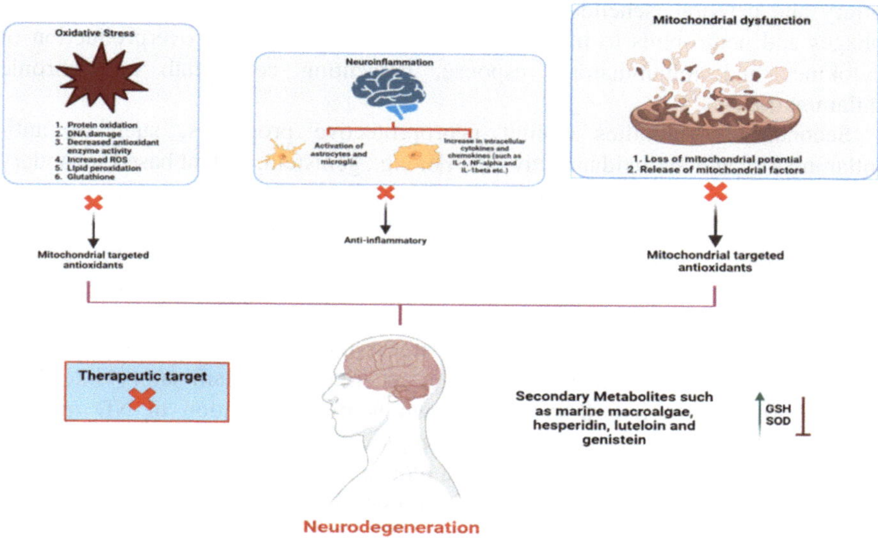

Fig. 5.6 Various phytomolecules combating NDs. (https://revvitysignals.com/)

neuroprotective activities. Figure 5.6 illustrates possible therapeutic effects of phytomolecules in treating NDs.

Luteolin, quercetin, resveratrol, apigenin, genistein, hesperidin, marine macroalgae, cyanobacteria, and *Uncaria rhynchophylla* are some common characteristic drugs specified for clinical studies targeting various common features of NDs despite their etiopathogenic and clinical differences such as abnormal cellular transports, inflammation, irregular protein deposition, intracellular Ca^{2+} overload, excitotoxicity, and unregulated ROS generation. Majorly reactive astroglia or microglia are found to be essential for pathogenesis of NDs, and various phytomolecules have been reported to treat NDs. Marine macroalgae are plant-based organisms classified as seaweed that are found around coastal areas. Green algae (Chlorophyceae), red (Rhodophyceae), and brown (Phaeophyceae) are categorized into three groups. Pigments, phenolic compounds, amino acids, peptides, proteins, and phenols are present in various bioactive materials. Various studies summarize that bioactive material and algae hold health impact, whereas carotenoids tend to have higher radical scavenging function found in marine algae as chief antioxidant (Teleanu et al. 2022).

5.8.7 Anti-Inflammatory Activity

Neuroinflammation is a major factor playing a key role in enhancement of brain damage in cerebral ischemia and neuronal tissue loss. For instance, accumulation of alpha-synuclein, tau protein, and Aβ initiates activation and recruitment of microglia leading to neuronal damage further initiating inflammatory response. Reperfusion injury and cerebral ischemia stimulate recruitment of microglia helping macrophages and neutrophils to infiltrate following enhancement of overproduction of cytokines and inflammatory response, preventing cell death and chronic inflammation.

Secondary metabolites exhibit neuroprotective properties, such as anti-inflammatory and antioxidant activities. For e.g., genistein, a plant based secondary metabolite shares structural and functional similarities with the primary female sex hormone, 17-betaestradiol. Genistein binds itself to receptive proteins such as 17-betaestradiol, stimulating estrogen-specific hormonal effect holding the capability to substitute them. Research data incorporates the anti-inflammatory activity of genistein, naringenin, curcumin, and resveratrol arbitrated via downregulation or blocking of nuclear factor-κB, regulating transcription of cytokine. Sulforaphane, ginsenosides, quercetin, and linalool inhibit cytokines (proinflammatory) like TNF-α, IL-8, IL-6, and (IL)-1β to lower down neuroinflammation in ND models. Kolaviron and curcumin (bioflavonoid) complex were extracted via Garcinia kola, depicting neuroprotection from RI/CI via reduced inflammation. However, quercetin helps in reducing GFAP in brain of PD rodent models (Singh et al. 2019).

5.8.8 Phytochemicals in Clinical Phase for Neurodegenerative Diseases

5.8.8.1 Phytomolecules in Clinical Trials for Alzheimer's Disease

Various phytochemicals were tested against their efficacy and safety towards clinical studies for Alzheimer's disease. Table 5.1 summarizes phytomolecules analyzed against AD treatment in clinical trials.

Vitamins are found to be majorly studied phytomolecules in clinical trials for Alzheimer's disease. Safety and efficacy parameter evaluation was initiated in the 1990s and is still working. In fact, vit. E has intrigued researchers as it holds soluble properties in various foods such as cereals, eggs, poultry, vegetable oil, and fruits. Animal studies illustrated anti-inflammatory and antioxidant properties of this compound with ability to minimize amyloid aggregation in transgenic mice model of AD and Aβ levels. A randomized, placebo-controlled, double blind, phase III trial was funded by the New York State Institute for Basic Research (USA), lasting 36 months with 349 participants for determining the effectiveness and safety of vit. E on AD patients (cognitive function) with Down syndrome (Pajares et al. 2020).

5.8.8.2 Phytomolecules in Clinical Trials for Parkinson's Disease

In the case of Parkinson's disease, various clinical trials were explored against phytomolecules in PD treatment as mentioned in Table 5.2.

Four different studies reported nicotine as the highest attention seeker among all phytoconstituents in clinical trials for treating PD. It clutches neuroprotective effect improving cognitive and motor symptoms, e.g. analyzed treatment for motor deficits in Parkinson's disease patients via utilizing nicotine in placebo-controlled phase I/II trial, single-center, randomized study. Volunteers with idiopathic PD (n = 65)

Table 5.1 Phytomolecules in clinical trials for Parkinson's disease (give references for clinical trials)

Natural compound	No. of subjects	Duration	Phase	Condition	Starting year	Outcomes	Status
EGCG	21	18 months	III	Early stage of AD	2009	No result released	Completed
Homotaurine	1052	78 weeks	III	Mild to moderate	2004	Slowing hippocampal atrophy	N. D
Melatonin	80	24 weeks	II	Mild to moderate	2009	Improvement in cognitive performance	Completed
Curcumin and ginkgo extract	34	24 weeks	II	AD	2004	Lesser oxidative stress (in brain)	Concluded
Huperzine A	177	16 weeks	II	Mild to moderate	2004	Safe and well tolerated at both doses; cognitive enhancement with higher dose	Concluded
Curcumin	36	24 weeks	II	Mild to moderate AD	2003	Well tolerated; no therapeutic action	Completed
Resveratrol	119	52 weeks	II	Mild to moderate AD	2012	No therapeutic effects	Completed
Resveratrol, malate and glucose	39	1 year	III	Mild to moderate AD	2007	No therapeutic action; safe and well tolerated	Concluded
Bryostatin	150	12 weeks	II	Severe AD	2015	Lesser dose; safe and well tolerated	Completed
Scyllo-inositol	353	78 weeks	II	Mild to moderate AD	2007	Acceptable safety; no therapeutic effects	Completed

(continued)

Table 5.1 (continued)

Natural compound	No. of subjects	Duration	Phase	Condition	Starting year	Outcomes	Status
Vit. E, vit. C, alpha-lipoic acid, or co-enzyme Q	75	16 weeks	I	AD	2006	Vit. C, vit. E, and alpha-lipoic acid group had shown lower oxidative stress (in brain)	Concluded
Vit. E and selegiline	341	2 years	N.D	Moderate AD	N.D	Delayed symptoms and patients deceased	Concluded
Vit. E and memantine	613	5 years	III	Mild to moderate AD	2007	Vit. E hampers functional reduction. Memantine alone and vit. E with memantine without effect	Concluded
Vit. B9, vit. B6, and vit. B12	340	18 weeks	II	Mild to moderate AD	2003	No effect	Completed
Vit. D3 and memantine	43	6 months	Pilot study	Moderate AD	2011	Improvement of patients' cognition	Concluded
Vit. E	349	36 months	III	AD with down syndrome	2000	No results posted	Completed
Vit. B3	50	24 weeks	II	Mild to moderate AD	2008	No results posted	Completed
DHA	295	18 months	III	Mild to moderate	2007	No effects	Completed

Table 5.2 Phytomolecules in clinical trials for Parkinson's disease

Natural compound	No. of subjects	Duration	Phase	Condition	Starting year	Main outcomes	Status
Vitamin B3	47	6 months	N.D	Early and late stage of PD	2016	Advancement of motor impairment; decrease of neuroinflammation	Concluded
Cannabidiol	13	5 weeks	II	Idiopathic PD	2016	Advancement in non-motor and several motor symptoms	Concluded
Vitamins B6, B9, and B12	40	6 weeks	ND	PD	2008	Advancement of homocysteine and glutathione levels	Concluded
Caffeine	119	18 months	III	Idiopathic PD	2012	Lesser value in cognitive functions and enhancement of alertness	Concluded
Nicotine	65	14 weeks	I and II	Idiopathic PD	2009	Advancement of motor symptoms	Concluded
EGCG	480	1 year	II	De-novo DP	2006	Slight clinical benefit; well tolerated	Concluded
Vitamin D	101	16 weeks	II	PD	2010	PD patient's balance	Concluded
DHA	33	1.5 years	I	PD	2018	Lowers dyskinesia	Concluded
Vitamin E with co-enzyme Q10	600	16 months	III	Early PD	2008	No therapeutic action	Concluded

were isolated into two groups of 6 mg and 1 mg of nicotine, four times/day continuously for 10 weeks. Nicotine seems to lower gait freezing and falls, two motor symptoms of PD when compared with placebo group. No fluctuation in dyskinesias was observed in groups (Ganguly et al. 2021).

5.9 Conclusion

Natural compounds exhibit potent effects in neurodegenerative ailment partly because of their anti-inflammatory properties. Rising research data based on therapeutic drug effects as neuroprotective agents is enlarging as population with AD is enhancing. In previous years, certain neuroprotective diseases have been elaborated as a dysfunction of cellular response to neuroinflammation and oxidative stress. Several natural compounds were examined depicting in vitro and in vivo results protecting neurons. The observed bioactivities of phytomolecules such as antioxidant activity, scavenging of reactive oxygen species, anticancer, and antibacterial and antiproliferative activities along with their neuroprotective effects were established. Various natural compounds including hesperidin, genistein, luteolin, and resveratrol demonstrated neurodegenerative efficacy. However, their therapeutic potential is limited by efficacy issues, solubility, and stability that impede clinical translation. This chapter focuses on neuroinflammation driven in various diseases and their importance in curing and preventing via typical molecular pathways in regulating neurodegenerative diseases.

Conflict of Interest The authors declare no conflict of interest.

Data Availability Statement Not applicable.

References

Aggarwal BB, Harikumar KB (2009) Potential therapeutic effects of curcumin, the anti-inflammatory agent, against neurodegenerative, cardiovascular, pulmonary, metabolic, autoimmune and neoplastic diseases. Int J Biochem Cell Biol 41(1):40–59. https://doi.org/10.1016/j.biocel.2008.06.010

Ahmad SB, Ali A, Bilal M, Rashid SM, Wani AB, Bhat RR, Rehman MU (2023) Melatonin and health: insights of melatonin action, biological functions, and associated disorders. Cell Mol Neurobiol 43(6):2437–2458. https://doi.org/10.1007/s10571-023-01324-w

Apweiler M, Saliba SW, Sun L, Streyczek J, Normann C, Hellwig S, Bräse S, Fiebich BL (2024) Modulation of neuroinflammation and oxidative stress by targeting GPR55 – new approaches in the treatment of psychiatric disorders. Mol Psychiatry 29(12):3779–3788. https://doi.org/10.1038/s41380-024-02614-5

Asemi-Rad A, Moafi M, Aliaghaei A, Abbaszadeh HA, Abdollahifar MA, Ebrahimi MJ, Heidari MH, Sadeghi Y (2022) The effect of dopaminergic neuron transplantation and melatonin co-administration on oxidative stress-induced cell death in Parkinson's disease. Metab Brain Dis 37(8):2677–2685. https://doi.org/10.1007/s11011-022-01021-5

Babashpour-Asl M, Kaboudi PS, Barez SR (2023) Therapeutic and medicinal effects of snowdrop (*Galanthus* spp.) in Alzheimer's disease: a review. J Educ Health Promot 12:128. https://doi.org/10.4103/jehp.jehp_451_22

Bald EM, Nance CS, Schultz JL (2021) Melatonin may slow disease progression in amyotrophic lateral sclerosis: findings from the pooled resource open-access ALS clinic trials database. Muscle Nerve 63(4):572–576. https://doi.org/10.1002/mus.27168

Batiha GE, Beshbishy AM, Ikram M, Mulla ZS, El-Hack MEA, Taha AE, Algammal AM, Elewa YHA (2020) The pharmacological activity, biochemical properties, and pharmacokinetics of the major natural polyphenolic flavonoid: quercetin. Foods 9(3):374. https://doi.org/10.3390/foods9030374

Boťanská B, Pecníková V, Fogarassyová M, Barančík M (2023) The role of heat shock proteins and autophagy in mechanisms underlying effects of sulforaphane on doxorubicin-induced toxicity in HEK293 cells. Physiol Res 72(S1):S47–S59. https://doi.org/10.33549/physiolres.935107

Bureau G, Longpré F, Martinoli MG (2008) Resveratrol and quercetin, two natural polyphenols, reduce apoptotic neuronal cell death induced by neuroinflammation. J Neurosci Res 86(2):403–410. https://doi.org/10.1002/jnr.21503

Capece U, Moffa S, Improta I, Di Giuseppe G, Nista EC, Cefalo CMA, Cinti F, Pontecorvi A, Gasbarrini A, Giaccari A, Mezza T (2022) Alpha-lipoic acid and glucose metabolism: a comprehensive update on biochemical and therapeutic features. Nutrients 15(1):18. https://doi.org/10.3390/nu15010018

Chan EWC, Wong CW, Tan YH, Foo JPY, Wong SK, Chan HT (2019) Resveratrol and pterostilbene: a comparative overview of their chemistry, biosynthesis, plant sources and pharmacological properties. J Appl Pharm Sci 9(7):124–129. https://doi.org/10.7324/JAPS.2019.90717

Chen TJ, Jeng JY, Lin CW, Wu CY, Chen YC (2006) Quercetin inhibition of ROS-dependent and -independent apoptosis in rat glioma C6 cells. Toxicology 223(1–2):113–126. https://doi.org/10.1016/j.tox.2006.03.007

Cheng CY, Barro L, Tsai ST, Feng TW, Wu XY, Chao CW, Yu RS, Chin TY, Hsieh MF (2021) Epigallocatechin-3-gallate-loaded liposomes favor anti-inflammation of microglia cells and promote neuroprotection. Int J Mol Sci 22(6):3037. https://doi.org/10.3390/ijms22063037

Feng L, Zhang L (2019) Resveratrol suppresses aβ-induced microglial activation through the TXNIP/TRX/NLRP3 Signaling pathway. DNA Cell Biol 38(8):874–879. https://doi.org/10.1089/dna.2018.4308

Francioso A, Baseggio Conrado A, Mosca L, Fontana M (2020) Chemistry and biochemistry of Sulfur natural compounds: key intermediates of metabolism and redox biology. Oxidative Med Cell Longev 2020:8294158. https://doi.org/10.1155/2020/8294158

Ganguly U, Kaur U, Chakrabarti SS, Sharma P, Agrawal BK, Saso L, Chakrabarti S (2021) Oxidative stress, neuroinflammation, and NADPH oxidase: implications in the pathogenesis and treatment of Alzheimer's disease. Oxidative Med Cell Longev 2021:7086512. https://doi.org/10.1155/2021/7086512

Gao Y, Zhuang Z, Lu Y, Tao T, Zhou Y, Liu G, Wang H, Zhang D, Wu L, Dai H, Li W, Hang C (2019) Curcumin mitigates neuro-inflammation by modulating microglia polarization through inhibiting TLR4 Axis Signaling pathway following experimental subarachnoid Hemorrhage. Front Neurosci 13:1223. https://doi.org/10.3389/fnins.2019.01223

Goyal R, Mittal P, Gautam RK, Kamal MA, Perveen A, Garg V, Alexiou A, Saboor M, Haque S, Farhana A, Papadakis M, Ashraf GM (2024) Natural products in the management of neurodegenerative diseases. Nutr Metab (Lond) 21(1):26. https://doi.org/10.1186/s12986-024-00800-4

Grabska-Kobyłecka I, Szpakowski P, Król A, Książek-Winiarek D, Kobyłecki A, Głąbiński A, Nowak D (2023) Polyphenols and their impact on the prevention of neurodegenerative diseases and development. Nutrients 15(15):3454. https://doi.org/10.3390/nu15153454

Gutiérrez-Grijalva EP, López-Martínez LX, Contreras-Angulo LA, Elizalde-Romero CA, Heredia JB (2020) Plant alkaloids: structures and bioactive properties. In: Plant-derived

bioactives: chemistry and mode of action. Springer, Singapore, pp 85–117. https://doi.org/10.1007/978-981-15-2361-8_5
Hong Byun E, Fujimura Y, Yamada K, Tachibana H (2010) TLR4 signaling inhibitory pathway induced by green tea polyphenol epigallocatechin-3-gallate through 67-kDa laminin receptor. J Immunol 185(1):33–45. https://doi.org/10.4049/jimmunol.0903742
Hu C, Yang J, Qi Z, Wu H, Wang B, Zou F, Mei H, Liu J, Wang W, Liu Q (2022) Heat shock proteins: biological functions, pathological roles, and therapeutic opportunities. MedComm (2020) 3(3):e161. https://doi.org/10.1002/mco2.161
Jia L, Liu J, Song Z, Pan X, Chen L, Cui X, Wang M (2012) Berberine suppresses amyloid-beta-induced inflammatory response in microglia by inhibiting nuclear factor-kappaB and mitogen-activated protein kinase signalling pathways. J Pharm Pharmacol 64(10):1510–1521. https://doi.org/10.1111/j.2042-7158.2012.01529.x
Jung YH, Heo J, Lee YJ, Kwon TK, Kim YH (2010) Quercetin enhances TRAIL-induced apoptosis in prostate cancer cells via increased protein stability of death receptor 5. Life Sci 86(9–10):351–357. https://doi.org/10.1016/j.lfs.2010.01.008
Kam A, Li KM, Razmovski-Naumovski V, Nammi S, Chan K, Li Y, Li GQ (2012) The protective effects of natural products on blood-brain barrier breakdown. Curr Med Chem 19(12):1830–1845. https://doi.org/10.2174/092986712800099794
Kim J (2021) Pre-clinical neuroprotective evidences and plausible mechanisms of sulforaphane in Alzheimer's disease. Int J Mol Sci 22(6):2929. https://doi.org/10.3390/ijms22062929
Kim H, Park BS, Lee KG, Choi CY, Jang SS, Kim YH, Lee SE (2005) Effects of naturally occurring compounds on fibril formation and oxidative stress of beta-amyloid. J Agric Food Chem 53(22):8537–8541. https://doi.org/10.1021/jf051985c
Kou JJ, Shi JZ, He YY, Hao JJ, Zhang HY, Luo DM, Song JK, Yan Y, Xie XM, Du GH, Pang XB (2022) Luteolin alleviates cognitive impairment in Alzheimer's disease mouse model via inhibiting endoplasmic reticulum stress-dependent neuroinflammation. Acta Pharmacol Sin 43(4):840–849. https://doi.org/10.1038/s41401-021-00702-8
Kumazoe M, Sugihara K, Tsukamoto S, Huang Y, Tsurudome Y, Suzuki T, Suemasu Y, Ueda N, Yamashita S, Kim Y, Yamada K, Tachibana H (2013) 67-kDa laminin receptor increases cGMP to induce cancer-selective apoptosis. J Clin Invest 123(2):787–799. https://doi.org/10.1172/JCI64768
Lee B, Yeom M, Shim I, Lee H, Hahm DH (2020) Protective effects of quercetin on anxiety-like symptoms and neuroinflammation induced by lipopolysaccharide in rats. Evid Based Complement Alternat Med 2020:4892415. https://doi.org/10.1155/2020/4892415
Leung JW, Cheung KK, Ngai SP, Tsang HW, Lau BW (2020) Protective effects of melatonin on neurogenesis impairment in neurological disorders and its relevant molecular mechanisms. Int J Mol Sci 21(16):5645. https://doi.org/10.3390/ijms21165645
Li Z, Geng YN, Jiang JD, Kong WJ (2014) Antioxidant and anti-inflammatory activities of berberine in the treatment of diabetes mellitus. Evid Based Complement Alternat Med 2014:289264. https://doi.org/10.1155/2014/289264
Lofrumento DD, Saponaro C, Cianciulli A, De Nuccio F, Mitolo V, Nicolardi G, Panaro MA (2011) MPTP-induced neuroinflammation increases the expression of pro-inflammatory cytokines and their receptors in mouse brain. Neuroimmunomodulation 18(2):79–88. https://doi.org/10.1159/000320027
Mack JM, Schamne MG, Sampaio TB, Pértile RA, Fernandes PA, Markus RP, Prediger RD (2016) Melatoninergic system in Parkinson's disease: from neuroprotection to the management of motor and nonmotor symptoms. Oxidative Med Cell Longev 2016:3472032. https://doi.org/10.1155/2016/3472032
Manchester LC, Qin L, Reiter RJ (2016) Melatonin: a mitochondrial targeting molecule involving mitochondrial protection and dynamics. Int J Mol Sci 17(12):2124. https://doi.org/10.3390/ijms17122124
Markus RP, Cecon E, Pires-Lapa MA (2013) Immune-pineal axis: nuclear factor κB (NF-kB) mediates the shift in the melatonin source from pinealocytes to immune competent cells. Int J Mol Sci 14(6):10979–10997. https://doi.org/10.3390/ijms140610979

Miękus N, Marszałek K, Podlacha M, Iqbal A, Puchalski C, Świergiel AH (2020) Health benefits of plant-derived Sulfur compounds, Glucosinolates, and organosulfur compounds. Molecules 25(17):3804. https://doi.org/10.3390/molecules25173804

Minnelli C, Cianfruglia L, Laudadio E, Mobbili G, Galeazzi R, Armeni T (2021) Effect of epigallocatechin-3-gallate on EGFR Signaling and migration in non-small cell lung cancer. Int J Mol Sci 22(21):11833. https://doi.org/10.3390/ijms222111833

Özduran G, Becer E, Vatansever HS, Yücecan S (2022) Neuroprotective effects of catechins in an experimental Parkinson's disease model and SK-N-AS cells: evaluation of cell viability, anti-inflammatory and anti-apoptotic effects. Neurol Res 44(6):511–523. https://doi.org/10.1080/01616412.2021.2024715

Pajares M, Rojo AI, Manda G, Boscá L, Cuadrado A (2020) Inflammation in Parkinson's disease: mechanisms and therapeutic implications. Cells 9(7):1687. https://doi.org/10.3390/cells9071687

Parsons ALM, Bucknor EMV, Castroflorio E, Soares TR, Oliver PL, Rial D (2022) The interconnected mechanisms of oxidative stress and neuroinflammation in epilepsy. Antioxidants (Basel) 11(1):157. https://doi.org/10.3390/antiox11010157

Payne A, Nahashon S, Taka E, Adinew GM, Soliman KFA (2022) Epigallocatechin-3-gallate (EGCG): new therapeutic perspectives for neuroprotection, aging, and neuroinflammation for the modern age. Biomolecules 12(3):371. https://doi.org/10.3390/biom12030371

Pezzuto JM (2019) Resveratrol: twenty years of growth. Development and Controversy Biomol Ther (Seoul) 27(1):1–14. https://doi.org/10.4062/biomolther.2018.176

Porro C, Cianciulli A, Trotta T, Lofrumento DD, Panaro MA (2019) Curcumin regulates anti-inflammatory responses by JAK/STAT/SOCS signaling pathway in BV-2 microglial cells. Biology (Basel) 8(3):51. https://doi.org/10.3390/biology8030051

Postuma RB, Lang AE, Munhoz RP, Charland K, Pelletier A, Moscovich M, Filla L, Zanatta D, Rios Romenets S, Altman R, Chuang R, Shah B (2012) Caffeine for treatment of Parkinson disease: a randomized controlled trial. Neurology 79(7):651–658. https://doi.org/10.1212/WNL.0b013e318263570d

Potes Y, Cachán-Vega C, Antuña E, García-González C, Menéndez-Coto N, Boga JA, Gutiérrez-Rodríguez J, Bermúdez M, Sierra V, Vega-Naredo I, Coto-Montes A, Caballero B (2023) Benefits of the neurogenic potential of melatonin for treating neurological and neuropsychiatric disorders. Int J Mol Sci 24(5):4803. https://doi.org/10.3390/ijms24054803

Prasanth MI, Sivamaruthi BS, Chaiyasut C, Tencomnao T (2019) A review of the role of green tea (*Camellia sinensis*) in antiphotoaging, stress resistance, neuroprotection, and autophagy. Nutrients 11(2):474. https://doi.org/10.3390/nu11020474

Rahman MH, Bajgai J, Fadriquela A, Sharma S, Trinh TT, Akter R, Jeong YJ, Goh SH, Kim CS, Lee KJ (2021) Therapeutic potential of natural products in treating neurodegenerative disorders and their future prospects and challenges. Molecules 26(17):5327. https://doi.org/10.3390/molecules26175327

Ruhee RT, Suzuki K (2020) The integrative role of sulforaphane in preventing inflammation, oxidative stress and fatigue: a review of a potential protective phytochemical. Antioxidants (Basel) 9(6):521. https://doi.org/10.3390/antiox9060521

Rui W, Li S, Xiao H, Xiao M, Shi J (2020) Baicalein attenuates neuroinflammation by inhibiting NLRP3/caspase-1/GSDMD pathway in MPTP induced mice model of Parkinson's disease. Int J Neuropsychopharmacol 23(11):762–773. https://doi.org/10.1093/ijnp/pyaa060

Santulli G, Kansakar U, Varzideh F, Mone P, Jankauskas SS, Lombardi A (2023) Functional role of taurine in aging and cardiovascular health: an updated overview. Nutrients 15(19):4236. https://doi.org/10.3390/nu15194236

Sharifi-Rad M, Lankatillake C, Dias DA, Docea AO, Mahomoodally MF, Lobine D, Chazot PL, Kurt B, Tumer TB, Moreira AC, Sharopov F, Martorell M, Martins N, Cho WC, Calina D, Sharifi-Rad J (2020a) Impact of natural compounds on neurodegenerative disorders: from preclinical to pharmacotherapeutics. J Clin Med 9(4):1061. https://doi.org/10.3390/jcm9041061

Sharifi-Rad J, Rayess YE, Rizk AA, Sadaka C, Zgheib R, Zam W, Sestito S, Rapposelli S, Neffe-Skocińska K, Zielińska D, Salehi B, Setzer WN, Dosoky NS, Taheri Y, El Beyrouthy M,

Martorell M, Ostrander EA, Suleria HAR, Cho WC, Maroyi A, Martins N (2020b) Turmeric and its major compound curcumin on health: bioactive effects and safety profiles for food, pharmaceutical, biotechnological and medicinal applications. Front Pharmacol 11:01021. https://doi.org/10.3389/fphar.2020.01021

Sharma DR, Wani WY, Sunkaria A, Kandimalla RJ, Verma D, Cameotra SS, Gill KD (2013) Quercetin protects against chronic aluminum-induced oxidative stress and ensuing biochemical, cholinergic, and neurobehavioral impairments in rats. Neurotox Res 23(4):336–357. https://doi.org/10.1007/s12640-012-9351-6

Shi Y, Holtzman DM (2018) Interplay between innate immunity and Alzheimer disease: APOE and TREM2 in the spotlight. Nat Rev Immunol 18(12):759–772. https://doi.org/10.1038/s41577-018-0051-1

Singh A, Kukreti R, Saso L, Kukreti S (2019) Oxidative stress: a key modulator in neurodegenerative diseases. Molecules 24(8):1583. https://doi.org/10.3390/molecules24081583

Sorrenti V, Contarini G, Sut S, Dall'Acqua S, Confortin F, Pagetta A, Giusti P, Zusso M (2018) Curcumin prevents acute neuroinflammation and long-term memory impairment induced by systemic lipopolysaccharide in mice. Front Pharmacol 9:183. https://doi.org/10.3389/fphar.2018.00183

Srinivasan V, Spence DW, Pandi-Perumal SR, Brown GM, Cardinali DP (2011) Melatonin in mitochondrial dysfunction and related disorders. Int J Alzheimers Dis 2011:326320. https://doi.org/10.4061/2011/326320

Takic M, Pokimica B, Petrovic-Oggiano G, Popovic T (2022) Effects of dietary α-linolenic acid treatment and the efficiency of its conversion to eicosapentaenoic and docosahexaenoic acids in obesity and related diseases. Molecules 27(14):4471. https://doi.org/10.3390/molecules27144471

Tamagno E, Guglielmotto M, Vasciaveo V, Tabaton M (2021) Oxidative stress and Beta amyloid in Alzheimer's disease. Which comes first: the chicken or the egg? Antioxidants (Basel) 10(9):1479. https://doi.org/10.3390/antiox10091479

Teleanu DM, Niculescu AG, Lungu II, Radu CI, Vladâcenco O, Roza E, Costăchescu B, Grumezescu AM, Teleanu RI (2022) An overview of oxidative stress, neuroinflammation, and neurodegenerative diseases. Int J Mol Sci 23(11):5938. https://doi.org/10.3390/ijms23115938

Tian E, Sharma G, Dai C (2023) Neuroprotective properties of berberine: molecular mechanisms and clinical implications. Antioxidants (Basel) 12(10):1883. https://doi.org/10.3390/antiox12101883

Tiong YL, Ng KY, Koh RY, Ponnudurai G, Chye SM (2019) Melatonin prevents oxidative stress-induced mitochondrial dysfunction and apoptosis in high glucose-treated Schwann cells via upregulation of Bcl2, NF-κB, mTOR, Wnt signalling pathways. Antioxidants (Basel) 8(7):198. https://doi.org/10.3390/antiox8070198

Toklu HZ, Hakan T, Celik H, Biber N, Erzik C, Ogunc AV, Akakin D, Cikler E, Cetinel S, Ersahin M, Sener G (2010) Neuroprotective effects of alpha-lipoic acid in experimental spinal cord injury in rats. J Spinal Cord Med 33(4):401–409. https://doi.org/10.1080/10790268.2010.11689719

Tordjman S, Chokron S, Delorme R, Charrier A, Bellissant E, Jaafari N, Fougerou C (2017) Melatonin: pharmacology, functions and therapeutic benefits. Curr Neuropharmacol 15(3):434–443. https://doi.org/10.2174/1570159X14666161228122115

Tripathi AK, Ray AK, Mishra SK, Bishen SM, Mishra H, Khurana A (2023) Molecular and therapeutic insights of alpha-lipoic acid as a potential molecule for disease prevention. Rev Bras 33(2):272–287. https://doi.org/10.1007/s43450-023-00370-1

Tsai CF, Chen GW, Chen YC, Shen CK, Lu DY, Yang LY, Chen JH, Yeh WL (2021) Regulatory effects of quercetin on M1/M2 macrophage polarization and oxidative/antioxidative balance. Nutrients 14(1):67. https://doi.org/10.3390/nu14010067

Unno K, Pervin M, Nakagawa A, Iguchi K, Hara A, Takagaki A, Nanjo F, Minami A, Nakamura Y (2017) Blood-brain barrier permeability of green tea catechin metabolites and their neuritogenic activity in human neuroblastoma SH-SY5Y cells. Mol Nutr Food Res 61(12). https://doi.org/10.1002/mnfr.201700294

Wang J, Song Y, Chen Z, Leng SX (2018) Connection between systemic inflammation and neuroinflammation underlies neuroprotective mechanism of several phytochemicals in neurodegenerative diseases. Oxidative Med Cell Longev 2018:1972714. https://doi.org/10.1155/2018/1972714

Winter J, Basilicata MF, Stemmler MP, Krauss S (2016) The MID1 protein is a central player during development and in disease. Front Biosci (Landmark Ed) 21(3):664–682. https://doi.org/10.2741/4413

Wollen KA (2010) Alzheimer's disease: the pros and cons of pharmaceutical, nutritional, botanical, and stimulatory therapies, with a discussion of treatment strategies from the perspective of patients and practitioners. Altern Med Rev 15(3):223–244

Xin W, Jing M, Yang J, Wang M, Hou G, Wang Q et al (2021) Baicalein exerts anti-neuroinflammatory effects by inhibiting the TLR4-ROS /NF-κB-NLRP3 inflammasome. Nat Prod Commun 16(4). https://doi.org/10.1177/1934578X211011385

Yadikar H, Torres I, Aiello G, Kurup M, Yang Z, Lin F, Kobeissy F, Yost R, Wang KK (2020) Screening of tau protein kinase inhibitors in a tauopathy-relevant cell-based model of tau hyperphosphorylation and oligomerization. PLoS One 15(7):e0224952. https://doi.org/10.1371/journal.pone.0224952

Yang Y, Tan X, Xu J, Wang T, Liang T, Xu X, Ma C, Xu Z, Wang W, Li H, Shen H, Li X, Dong W, Chen G (2020) Luteolin alleviates neuroinflammation via downregulating the TLR4/TRAF6/NF-κB pathway after intracerebral hemorrhage. Biomed Pharmacother 126:110044. https://doi.org/10.1016/j.biopha.2020.110044

Yu Y, Shen Q, Lai Y, Park SY, Ou X, Lin D, Jin M, Zhang W (2018) Anti-inflammatory effects of curcumin in microglial cells. Front Pharmacol 9:386. https://doi.org/10.3389/fphar.2018.00386

Yuan NN, Cai CZ, Wu MY, Su HX, Li M, Lu JH (2019) Neuroprotective effects of berberine in animal models of Alzheimer's disease: a systematic review of pre-clinical studies. BMC Complement Altern Med 19(1):109. https://doi.org/10.1186/s12906-019-2510-z

Zhang Y, Talalay P, Cho CG, Posner GH (1992) A major inducer of anticarcinogenic protective enzymes from broccoli: isolation and elucidation of structure. Proc Natl Acad Sci USA 89(6):2399–2403. https://doi.org/10.1073/pnas.89.6.2399

Zhang Y, Kensler TW, Cho CG, Posner GH, Talalay P (1994) Anticarcinogenic activities of sulforaphane and structurally related synthetic norbornyl isothiocyanates. Proc Natl Acad Sci USA 91(8):3147–3150. https://doi.org/10.1073/pnas.91.8.3147

Zhang M, Swarts SG, Yin L, Liu C, Tian Y, Cao Y, Swarts M, Yang S, Zhang SB, Zhang K, Ju S, Olek DJ Jr, Schwartz L, Keng PC, Howell R, Zhang L, Okunieff P (2011) Antioxidant properties of quercetin. Adv Exp Med Biol 701:283–289. https://doi.org/10.1007/978-1-4419-7756-4_38

Zhang S, Zhu Q, Chen JY, OuYang D, Lu JH (2020) The pharmacological activity of epigallocatechin-3-gallate (EGCG) on Alzheimer's disease animal model: a systematic review. Phytomedicine 79:153316. https://doi.org/10.1016/j.phymed.2020.153316

Zhang ZH, Liu JQ, Hu CD, Zhao XT, Qin FY, Zhuang Z, Zhang XS (2021) Luteolin confers cerebroprotection after subarachnoid hemorrhage by suppression of NLPR3 inflammasome activation through Nrf2-dependent pathway. Oxidative Med Cell Longev 2021:5838101. https://doi.org/10.1155/2021/5838101

Zhong LM, Zong Y, Sun L, Guo JZ, Zhang W, He Y, Song R, Wang WM, Xiao CJ, Lu D (2012) Resveratrol inhibits inflammatory responses via the mammalian target of rapamycin signaling pathway in cultured LPS-stimulated microglial cells. PLoS One 7(2):e32195. https://doi.org/10.1371/journal.pone.0032195

Natural Compounds and Mitochondrial Function

6

Mohsin Vahid Khan, Mohd. Rumman, Gulam Rabbani, Mohd Kamil Hussain, Sameer Ahmad, and Nafees Ahmad

Abstract

Approximately 1.6 billion years ago during evolution, a mortal eukaryotic host cell engulfed and permanently confined the immortal prokaryotic purple non-sulphur bacteria via endosymbiosis, and such union resulted in the creation of the energy-producing organelle known as mitochondria. It is popularly designated as the powerhouse of the cell and organelle in-charge for orchestrating several activities such as oxidation of fatty acids, energy production, cell death, cell cycle regulation and intracellular homeostasis. Mitochondria contain their own genome and synthesize some specific proteins and mitochondrial tRNA, while they entail several proteins expressed from nuclear DNA. Substrate utilization and subsequently ATP production through oxidative phosphorylation (OXPHOS) are prime functions of mitochondria and vital for performing numerous metabolic reactions within cells. Moreover, they also manage the excessive reactive oxygen species (ROS) from exerting a detrimental impact. The growing body of evidence over the past couple of decades dispelled ambiguity and demystified the link between health and mitochondrial functions. Compromised mitochondrial function, referred to as mitochondrial dysfunction due to reduced

M. V. Khan (✉) · M. Rumman
Department of Biosciences, Integral University, Lucknow, India

G. Rabbani
IT-medical Fusion Center, Gumi-si, Gyeongbuk, Republic of Korea

M. K. Hussain
Department of Chemistry, Govt. Raza PG College, Rampur, India

S. Ahmad
Department of Bioengineering, Integral University, Lucknow, India

N. Ahmad
Department of Chemistry, Integral University, Lucknow, India

S. Jahan et al. (eds.), *A Mechanistic Exploration of Natural Compounds for Neuronal Health*, https://doi.org/10.1007/978-981-95-0125-0_6

mitochondrial membrane potential and respiratory aptitude in accordance with the mitochondrial number along with elevated ROS, may lead to several deleterious impacts on the cell and ultimately to the body. Thus, designing preventive measures against mitochondrial dysfunction remains indispensable. Herein, we discuss the role and mechanism of action of different small natural compounds that act as potent rescuing or restorative agents in relation to mitochondrial dysfunction by increasing biogenesis and oxidative phosphorylation, maintaining membrane potential and reducing ROS.

Keywords

Mitochondrial dysfunction · Neurodegenerative · ROS · Natural compounds · Alzheimer's disease · Parkinson's disease

6.1 Introduction

Mitochondria are intracellular, double-membrane, semi-autonomous, ubiquitous organelles found in eukaryotic cells and conjectured to have originated from their symbiotic ancestors. The swallowing of free-living α-proteobacteria by a eukaryotic host cell is linked with the genesis of mitochondria according to the endosymbiotic theory. The theory assumes that eukaryotic host cells facilitate multiplication of bacteria by providing a suitable environment, and in turn, mitochondria enhance the supply of energy production. Mitochondria differ from other cell organelles in many aspects, and one of them is to possess their own genetic material. They contain DNA as their genetic material, which is referred to as mitochondrial DNA or mtDNA. To meet the energy requirements, mtDNA count may vary within the range of 100–10,000 in mitochondria of specific tissues. The mtDNA equips mitochondria with an autonomic translational setup with subtle deviation for encoding codons with reference to the protein synthesis of nuclear DNA. Moreover, double-stranded mtDNA shares some general features of bacterial genetic material. Like bacterial DNA, it is polycistronic (Chinnery and Hudson 2013; Frazier 2019).

Mitochondria show morphologically dynamic behaviour, i.e. consistent shape alteration like shortening and lengthening which are mediated by the fusion and fission of the organelles. The process of mitochondrial multiplication exhibits resemblance with that of bacterial division, with the cell cycle organizing mitochondrial nuclei division first, followed by matrix division, a process referred to as mitochondriokinesis (Cavalier-Smith 2006; Chan 2012; Kuroiwa 1982). Mitochondria play a central role in the cell and are referred as the powerhouse of the cell, ensuring proper energy supply via the production of the high-energy compound adenosine triphosphate (ATP) to carry out various biochemical processes (biosynthesis and degradation, etc.) to sustain the viability of cells. Although visualization of mitochondria is only possible using a compound light microscope, assessment of their detailed structure requires an electron microscope which reveals that mitochondria have a tubular network and progressively undergo division followed by fusion. The breakage of the fragile tubular network of mitochondria into fragments takes place at the time of isolation from cells. However, further sealing of these fragments of tubular network allows for the full concealment of functional potential, similar to an in vivo system, thus enabling respiration and synthesis of ATP. Additionally, mitochondria manage to perform protein import and fusion and also maintain their organizational arrangement along with membrane composition, etc. (Kuhlbrandt 2015; Schmidt 2010).

6.2 Role of Mitochindria in the Cell

The presence of mtDNA in mitochodria makes them sophisticated cell organelles within the cell. The mtDNA gene products are not competent enough to fulfill structural and functional requisition associated with mitochondria, as they contain only 37 genes. These genes could produce 13 proteins along with 2 ribosomal RNA (rRNA) and 22 transfer RNA (tRNA). Most of the proteins participate in the formation of respiratry chain subunits. For the rest of the proteins (approxiametly 99%), mitochondria are dependent on the nuclear DNA. These proteins play a vital role in the repair of mtDNA and mitochondrial membrane formation and participate in the structural organization of mitochondria (Anderson 1981; Calvo 2016).

6.2.1 Mitochindrial Dysfunction

Dysfunction is generally not merely associated with splintered or mangled mitochondria, but in a true sense, it encompasses the lack of coordination between cellular requirement with respect to the necessities and overall mitochondrial performance. On one hand, mitochondria are responsible for playing a crucial role in performing several cellular activities; on the other hand, mitochondrial dysfunction may lead to several complications (Fig. 6.1). Several cellular conditions are

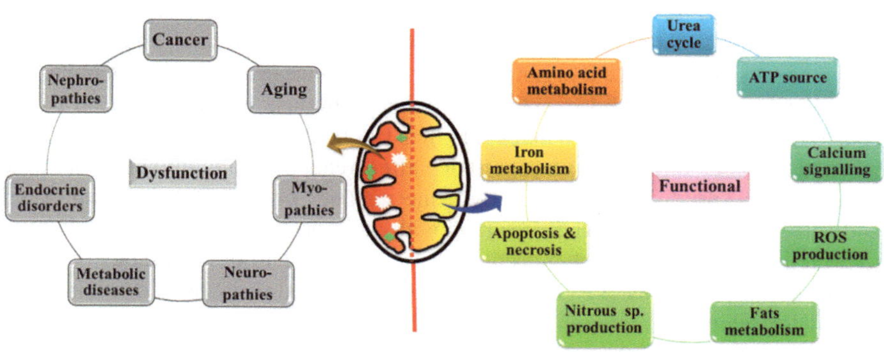

Fig. 6.1 Diagramatic representation of functions and effects associated with normal mitochondria and mitochondrial dysfunction

considered as culprits of mitochondrial dysfunction. These conditions include lessened genesis of mitochondria, hyperglycaemia, enhanced reactive oxygen species (ROS) production, superfluous intake of nutrients and elevated free fatty acid levels (Kim 2008). Accumulating evidence indicates that mainly three events are responsible for causing the mitochondrial functional impairment: diminished transport of pivotal metabolites to the mitochondria through the membrane, disruption of chemical and electrical potential management across the inner mitochondrial membrane and disturbances in the electron transport chain activity. The set two electron carriers, cytochrome c, coenzyme Q (CoQ) and respiratory chain complexes I, II, III and IV along with F_0-F_1 ATP synthase (also called as complex V), collectively constitute the oxidative phosphorylation (OXPHOX) system.

Principally, mitochondrial dysfunction is directly connected with the failure of anti-oxidant defense measures and/or disproportionate production of ROS. The production of a group of oxygen-containing reactive molecules such as O_2, OH^- and H_2O_2, at the time of exploitation of molecular oxygen by an aerobic organism, is referred to as free radicals or reactive oxygen species (ROS). Radical ROS, such as hydroxyl radicals and superoxide, and non-radical ROS, such as singlet oxygen and hydrogen peroxide, are two types of ROS species. These oxygen species are readily available for reaction with other molecules due to their high instability. In the cell, the production of free radicals as byproducts is predominantly contributed by oxidative phosphorylation corresponding to the interaction of oxygen with free electrons. The production of ROS is considered as extremely harmful for the various cellular components including mitochondria. The ROS production is a totally inevitable event for the cell, and the extent of ROS production is minimal compared to O_2 intake. The production of free radicals in the cell extensively affects the cellular structural organization, disturbs the mitochondrial membrane and even abolishes the enzyme complexes involved in several biochemical reactions (Gupta 2012; Kelso 2002).

It is imperative for cells to develop some preventive measures to avoid unsolicited damage to cellular components, which are essential for the proper functioning of associated organelles. Hence, breakdown of hydrogen peroxide in to oxygen and water is mediated by the enzyme catalase. Additionally, peroxide along with oxygen is formed by the detoxification of superoxides, which are catalyzed by several enzymes present in different organelles such as manganese superoxide dismutase in mitochondria and zinc and copper superoxide dismutase in the cytoplasm of the cells (Lebovitz 1996; McCord and Fridovich 1969). Peroxinitrite formation occurs when nitric oxide interacts with superoxide after freely diffusing into mitochondria. Consequently, production of 3-nitrotyrosine residues on subunits of complexes I and V is facilitated by peroxinitrite along with other proteins. Additionally, damage to proteins, lipids and mitochondrial DNA by nitrogen and oxygen radicals results in further intensification of ROS production, while peroxinitrite breakdown is mediated by the selenium-containing enzyme glutathione peroxidase and peroxiredoxins (Trujillo 2008). One of the major changes occuring during mitochondrial damage is the movement and concentration of Ca^{+2} into the matrix which ultimately leads to the loss of control over ion movement. This occurs due to the opening of the mitochondrial permeability transition pore (mPTP) followed by the leaking out of the cytochrome from the matrix to the cytoplasm of the cell. This situation gets compounded by the disturbed pH and membrane potential and the entry of ROS into the mitochondrial matrix. Subsequently, cells prepare for the apoptotic pathway coupled with diminished ATP production in worsening condition of damaged tissue, lipid bilayer damage and peroxidation of the cellular lipid membrane under the influence of ROS-induced enhanced stress response.

Mitochondrial dysfunction may lead to several neurological disorders (Rather et al. 2024) and non-communicable diseases (NCDs), which include many health issues affecting a large population across the globe of varied age groups (Diaz-Vegas 2020; Nicolson 2014). There is sophisticated relation between ageing and functioning of mitochondria. Inflammation is a condition of an elevated level of inflammatory mediators such as IL-6, fibrinogen and C-reactive protein (CRP) in the circulating system as a result of innate immune system activation, and inflammation is directly associated with ageing (Bratic and Larsson 2013).

6.3 Selected Complications Associated with Mitochindrial Dysfunction

A wide range of disorders and health complications are associated with mitochondrial dysfunction. These complications include non-neurodegenerative as well as neurodegenerative ailments as given in Table 6.1, and some of them are discussed below.

6.3.1 Glaucoma

Mitochondrial dysfunction induced in response to ageing is also found to be associated with glaucoma, a neurodegenerative disease characterized by death of retinal ganglionic cells (RCGs) and axon, finally causing loss of vision. The high accumulation of mitochondria around the optic nerve head has comparatively more energy demand in the form of ATP. The failure in the repair of RCGs due to the lack of proper energy availability is assumed to be linked with permanent loss of RCGs, leading towards apoptosis (Kong 2009; Zhang 2023). The in vivo study on RGCs reveals that the increased hydrostatic pressure diminishes ATP supply by dismantling the mitochondrial network and tempting mitochondrial fission. This finding substantiates that the loss of structure followed by mitochondrial function has a direct connection to the induction of glaucoma. In a similar vein, mitochondria are also associated with hereditary optic neuropathies known as Leber's hereditary optical neuropathy (LHON). This hereditary disorder is mostly reported to affect people in the 30 s to 40 s age group. In LHON, augmentation is particularly reported because of the erroneous expression of the OXPHOS complex I subunit gene due to mutation in mitochondrial DNA, ultimately leading to apoptosis of the cell (Carelli 2004; Kong 2009).

Table 6.1 Disorders or complications associated with mitochondrial dysfunction

S. no.	Diseases	Characteristic(s)	References
1.	Tuberous sclerosis complex	TSC1 and TSC2 gene alteration	Cunningham (2007)
2.	Fragile X syndrome (FXS)	FMR1 gene alteration	Wallace and Fan (2010)
3.	Angelman syndrome	Maternal chromosome 15 and UBE3A gene mutation	Su (2011)
4.	Cancer	Elevated ROS without reduced ATP, enhanced oxidative stress, mutation in mtDNA	Wallace (2005)
5.	Optic neuropathy	Point mutations mitochondrial DNA, (mtDNA); m.3460G>A, m.11778G>A, and m.14484T>C	Yu-Wai-Man (2011)
6.	Skeletal muscle hypertrophy/atrophy	Perturbed mitochondrial membrane potential, elevated ROS	Chen (2023)
7.	Atherosclerosis	Reduced OXPHOS activity, elevated ROS, altered mitophagy, mutation in mtDNA	Nakajima (2019)
8.	Aortic aneurysm	Reduced OXPHOS activity, elevated ROS	Landis (2022)
9.	Heart failure	Reduced OXPHOS activity, elevated ROS, altered mitophagy	Abudureyimu (2020)
10.	Myocardial hypertrophy	Elevated ROS, altered mitophagy, mutation in mtDNA	Yan (2022)
11.	Aortic dissection	Elevated ROS	Shi (2022)
12.	Alzheimer's disease	Aβ and tau aggregate accumulation	Antuono and Beyer (1999)
13.	Parkinson's diseases	Elevated ROS, mutation in mtDNA, nerve cell deterioration, α- synuclein amyloidosis	Esteves (2011), Tysnes and Storstein (2017)
14.	Huntington's disease	Mutation in mtDNA, neuron degeneration	Cui (2006), and Siddiqui (2012)
15.	Amyotrophic lateral sclerosis	Altered OXPHOS complex assembly, altered mitophagy	Rosen (1993), Smith (2019)
16.	Type 2 diabetes	Reduced OXPHOS activity, elevated ROS, mutation in mtDNA	Petersen (2004)

6.3.2 Cardiac Cell Death

Accumulation of $A\beta_{1-40}$ and $A\beta_{1-42}$ in coronary endothelial cells and cardiomyocytes has been reported to induce mitochondrial dysfunction and finally reduced cell viability, leading to apoptosis. The accumulated amyloid peptides in both cells create an imbalance in the capacity of calcium maintenance and disturb the mitochondrial membrane potential. An investigation of mitochondria isolated from rat heart cells confirms that calcium levels are directly proportional to mitochondrial swelling in response to $A\beta_{1-40}$ and $A\beta_{1-42}$ accumulation in both types of cells (Jang 2022).

6.3.3 Epileptic Seizures

One of the manifestations of neurodegenerative disease is seizures, characterized by impulsive behavioral variation because of momentary electrical functioning change in the brain. It has been reported that the brain shows an instantaneous need for augmented energy at the time of seizures. The onset of seizures involves almost all factors associated with mitochondrial dysfunction, specifically secondary mitochondrial dysfunction. The epileptic phenotype study suggests that epilepsy is not confined to the mitochondrial gene mutation but also with nuclear genes, which are crucial for proper mitochondrial performance (Zsurka and Kunz 2015).

6.3.4 Asthma

Conditions like bronchial hyperresponsiveness (BHR), bronchus remodelling and inflammation are linked to asthma, a relatively common condition, and chronic diseases associated with the respiratory system. Exploration of an asthmatic rat model reveals that both structural alteration and mitochondrial ATP supply are affected in asthmatic conditions. A significant reduction in ATP production has been reported by studying the murine ovalbumin asthma rat model, and lung epithelium analysis corroborates the alteration in mitochondrial structure. During ragweed pollen exposure, mast cells are induced to secrete serotonin and histamine due to the failure in the proper functioning of the electron transport chain (ETC) complex in mitochondria. Moreover, cytokine secretion along with many other mediators in asthmatic conditions affects airway smooth muscle propagation and comportment, directly concomitant with mitochondrial functioning. Similarly, bronchial smooth muscle remodelling has also been found to be connected with the dysfunction of mitochondria (Cloonan 2020; Delmotte 2017; Li 2015).

6.3.5 Skin Diseases

Ultraviolet (UV) radiation exposure to the skin shows potential to aggravate oxidative stress. UV radiation potentially induces mutation in both mitochondrial and nuclear DNA. Changes in the structure and function of skin cells have been reported in dermal fibroblast cells due to UV-mediated deletions in mitochondrial DNA, resulting in regressive mitochondrial signaling and impairment of ATP production (Feichtinger 2014; Pelle et al. 2012).

6.3.6 Skeletal Muscle Atrophy

Out of the total body mass of humans, skeletal muscles, a decidedly pliable tissue, shares about 40% of body mass and play a crucial role in movement and providing proper stance to the limbs. The onset of muscle atrophy is fatal where both mass and muscle fibre size get greatly compacted, leading to various disorders due to functional inactivation and mass reduction, followed by alteration in muscle degradation and protein synthesis coordination. A growing body of evidence reveals that mitochondrial dysfunction is connected to skeletal muscle atrophy, but the molecular mechanism behind it is still under investigation. Perturbations in mitochondrial membrane potential and enhanced ROS production have been reported to create a stressed state, resulting in reduced mitochondrial biogenesis, protein synthesis and augmented proteolysis, permeation of fat and fibrosis (Kubat 2023; Shen 2022).

6.4 Neurodegenerative Diseases Concomitant with Mitochondrial Dysfunction

In vitro studies on various model proteins have been performed, employing physical and chemical stress conditions, and the obtained data reveal that these conditions lead to protein aggregation. In vitro studies help in decoding the mechanism of amyloidosis and connecting the link between protein aggregation and both non-neurodegenerative and neurodegenerative diseases (Khan 2014, 2015, 2017, 2018a, b). The consecutive and preferred attenuation of the neuronal system in terms of physiology as well as anatomy is a representative feature of neurodegenerative disorders. Continuously growing evidence advocates that about 50% of people above the age of 85 show signs of Alzheimer's disease (AD), but the condition appears to be much more severe, with 10% of people above the age of 65 reported to develop AD. Recent studies confirm that in the pathophysiology of neurodegenerative diseases, mitochondrial dysfunction plays a principal role (Antuono and Beyer 1999; Tapias 2018). Here, we discuss some leading neurodegenerative diseases and their association with mitochondrial dysfunction.

6.4.1 Alzheimer's Disease (AD)

Alzheimer's disease (AD) is considered the most prevalent neurodegenerative disorder linked with ageing, occuring due to the accumulation of amyloidogenic protein products, which have the potential to produce ROS and severely affect mitochondria. Among the mitochondria of different organs, brain mitochondria show the highest susceptibility to dysfunction. The normal mitochondria in the brain meet all the biochemical requirements by increasing the mitochodrial number at the level of synapses and neurons. Neuronal functions remain greatly influenced by the physical and biochemical status of mitochondria originating from nerve terminals, i.e. synaptic mitrochondria, due to their extended lifespan in response to protein aggregates. The aggregation of two proteins, Aβ and tau, is considered as the main causal event leading to the onset of AD. The clinical manifestation of AD is quite slow, taking years to exhibit symptoms and be diagnosed, even though amyloidosis of tau and Aβ may have started long before. Cerebral amyloid angiopathy in leptomeningitis and blood vessels and senile plauqe formation in parenchyma by

Aβ amyloidosis, along with the development of neurofibrillar tangles due to hyperphosphorylated tau protein aggregation in neurons, are pathological hallmarks of AD (Marin and Budson 2023; Selkoe 1991).

6.4.2 Parkinson's Disease

Parkinson's disease is reported in 1 out of 100 people who are over the age of 60. The substantia nigra (SN) situated behind the crus cerebri fibres of the peduncle, located within the midbrain, is responsible for controlling motor movement. Parkinson's disease is pathologically described as a condition in which dopamine-producing nerve cells deteriorate, leading to the accumulation of aggregated α-synuclein protein-containing Lewy bodies in the substantia nigra region of the brain. Data obtained from postmortem examinations of Parkinson's patients' brains revealed diminished performance of mitochondrial complex I, which is also consistent with observations reported after isolating and purifying mitochondria from platelets (Esteves 2011; Tysnes and Storstein 2017).

Parkinson's disease is also linked with progressive mutations in mitochondrial DNA. The association of Parkinson's disease with mitochondrial dysfunction was first established when inadvertent contact with the complex I inhibitor molecule, 1-methyl-4-phenyl-1,2,3,4tetrahydropyridine (MPTP), was reported to induce Parkinson's disease (Langston 1983). Mitophagy failure in cells has been found to be concomitant with mitochondrial dysfunction as a person ages. To maintain the cellular homeostasis during the course of the autophagic pathway, i.e. self-eating of organelles, damaged mitochondria with enhanced fission rate along with altered membrane potential undergo degradation referred to as mitophagy. In order to improve sustainability of the cell by augmenting mitochondrial biogenesis, mitophagy remains essential for dysfunctional mitochondria (Hamacher-Brady and Brady 2016). The boosted ROS level besides lowered oxygen intake in pancreatic beta cells and muscle fibres in mouse models has been witnessed in dysfunctional mitochondria due to the deletion of the ATG7 autophagic protein. Collectively, lysosomal dysfunction and mitochondrial dysfunction, which lead to reduced energy supply and altered mitochondrial morphology and membrane potential due to diminished lysosomal glucocerebrosidase activity, are linked with dopamine oxidation in Parkinson's disease (Burbulla 2017; Wu 2009).

6.4.3 Huntington's Disease (HD)

The human brain is the most sophisticated organ in the body, and the neurons present there require a huge amount of energy with constant supply, which is efficiently facilitated by mitochondria. Although the weight of the central nervous system (CNS) is around 2% of the entire body, in terms of energy consumption, it requires approximately one-fifth of the total body energy. Moreover, out of the total energy supplied to the brain, almost 85% of the energy is consumed by neurons alone. The rest of the cells, such as microglia, ependymal cells, astrocytes and oligodendrocytes, are nurtured by approximately 15% of the provided energy. As a result of mutant huntingtin gene expression, an erroneous huntingtin protein is synthesized, progressing towards a neurodegenerative disease known as Huntington's disease, with dominant inherited characteristics. The erroneous huntingtin protein with extended polyglutamine unit due to the presence of multiple CAG in the huntingtin encoding gene undergoes aggregation as a result of adopting a distorted three-dimensional conformation (Howarth 2012; MacDonald et al. 1993; Sokoloff 1973).

Lymphoblasts of patients suffering from HD exhibit less Ca^{+2} requirement for depolarization, which is associated with reduced membrane potential and several neuropathological characteristics such as psychiatric ailment leading to reduced sensibility, dynamic motor impairment and altered disposition. The mitochondrial dysfunction induced in HD is the culprit of diminished bioenergetics in the body, characterized by significant weight loss even after consumption of sufficient food with balanced nutrients. It has been reported that aconitase loses its function in the basal ganglia, lactate accumulates in both the cerebral cortex and the basal ganglia, and there is a reduction in oxidative phosphorylation (OXPHOS) complexes II and III due to mitochondrial dysfunction (Choo 2004; Johri and Beal 2012). The mutation in the huntingtin gene results in the cessation of PGC-1α transcription, followed by CREB and TAF4 transcription factor-mediated activation (Cui 2006).

6.4.4 Amyotrophic Lateral Scelerosis (ALS)

The progressive deterioration of both upper and lower motor neurons, along with corticospinal tract degeneration in older adults, is a characteristic feature of amyotrophic lateral sclerosis, which is considered one of the most lethal neurodegenerative diseases. Both neurons work in a well-coordinated manner. The upper motor neuron, found in the brain, is responsible for propagating signals to the spinal cord, which are then passed on to the lower motor neuron to stimulate muscles to perform the movement. A person suffering from ALS faces severe difficulties in executing essential and basic functions such as eating, breathing, talking and walking, eventually gets paralyzed due to impaired signalling within 5 years of disease onset and ultimately dies due to atrophy, spasticity and progressive muscle weakness (Kiernan 2011; Rosen 1993). The first pathological characteristic observed in patients with ALS was with the presence of vacuolated and swollen mitochondria having anomalous cristae formation (Sasaki and Iwata 2007). Out of the total ALS cases, approximately 10% are familial and are reported to result from autosomal dominant mutations, whereas the majority (90%) appear to be sporadic. Several genes and their products have been linked to ALS, leading to alterations in mitochondrial function. An autosomal dominant mutation in the anti-oxidant enzyme gene Cu or Zn-superoxide dismutase (*SOD 1*) has been observed in almost one-fourth of ALS patients among familial cases. As a result of the mutation, the expressed protein fails to perform several enzymatic functions such as conformation alterations in the

Cu binding site caused by a couple of mutations, while in most cases, enzyme dimerization is hindered due to backbone alteration (Rosen 1993; Smith 2019).

6.5 Therapeutic Measures Against Mitochondrial Dysfunction

Designing therapeutics to address mitochondrial dysfunction is a tough challenge for scientists, pharmaceutical industries and biotechnology companies. One of the most promising approaches is the direct incorporation of fully functional mitochondria into cells experiencing mitochondrial dysfunctional. However, this approach must be experimentally validated by obtaining unambiguous positive responses extensively before generalizing for patient treatment. In vitro investigations suggest that the protein undergoing amyloidogenesis experiences a loss of functionality, which aligns with the mitochondrial dysfunction mediated by the aggregation of several proteins. Similarly, several molecules have shown protective effects against protein aggregation under harsh amyloidogenic conditions, which may provide a lead for scientists working to counter protein aggregation associated with mitochondrial dysfunction and neurodegenerative disease onset (Ajmal 2017; Chandel et al. 2017; Ishtikhar 2016; Khan 2020; Siddiqi 2018).

All treatment methods or preventive measures designed against mitochondrial dysfunction focus on achieving specific targets. These targets include proportionating the rate of mitochondrial biogenesis as per cellular energy demands, lowering the count of dysfunctional mitochondria by replacing them with normal mitochondria, exchanging inoperative mitochondrial components with functional ones, avoiding consequences after onset of dysfunction, and metabolic reconditioning. One of the approaches to combat mitochondrial dysfunction is to maintain the overall mitochondrial mass by transplanting fully functional mitochondria into specific cells or tissues, even in individuals suffering from diseases associated with mitochondrial dysfunction. This approach has produced some positive response in Parkinson's model, spinal cord injury, myocardial ischemia, etc. (Lightowlers 2020). However, this approach has some limitations; thus, we need to follow another approach in which several natural compounds (Fig. 6.2) can be used as multipotent therapeutic molecules to treat mitochondria with altered morphology and functions. Many natural compounds are promising agents which can be used to rescue mitochondrial function through different ways as depicted in Table 6.2.

6.6 Natural Molecules Affecting Mitochondrial Dyfunctions

6.6.1 Resveratrol (RES)

Resveratrol (3,5,4′-trihydroxystilbene) is a naturally occurring polyphenol with diverse therapeutic properties such as anti-inflammatory, anti-cancer, anti-obesity, anti-ageing, anti-oxidant, cardioprotective and anti-diabetic activities. Resveratrol

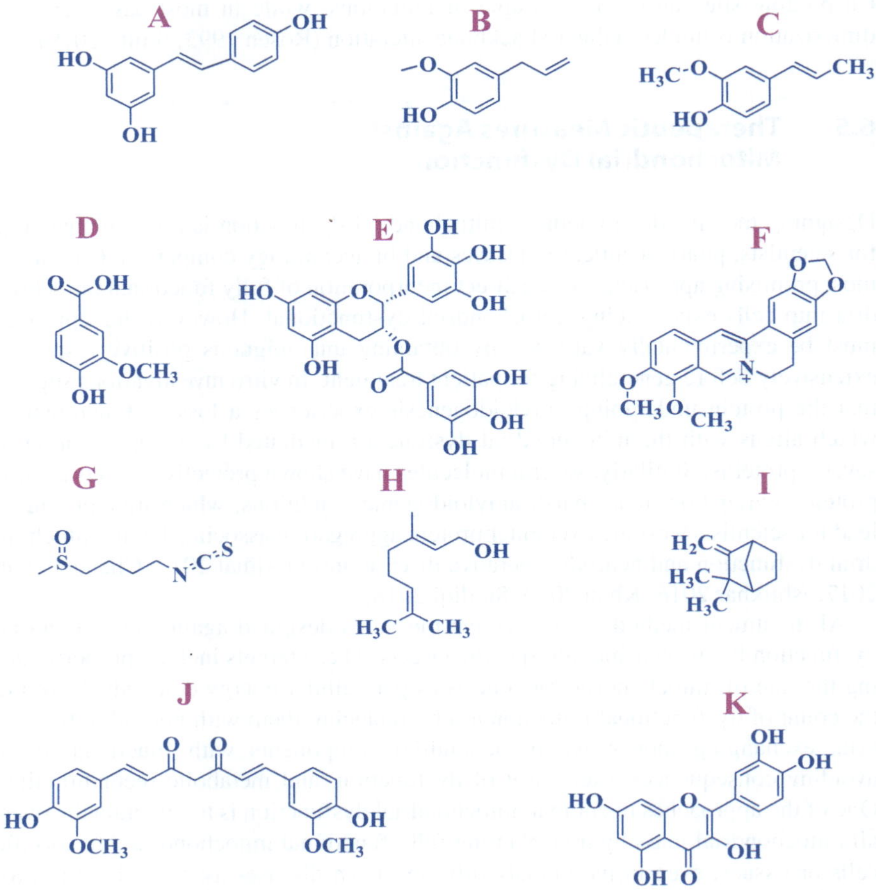

Fig. 6.2 Chemical structure of different natural compounds with protective effect against mitochondrial dysfunction **A** Resveratrol. **B** Eugenol. **C** Isoeugenol. **D** Vanillic acid. **E** EGCG. **F** Berberine. **G** Sulforaphane. **H** Geraniol. **I** Camphene. **J** Curcumin. **K** Quercetin

fulfils the enhanced energy requirement in brown adipose tissue, liver cells and skeletal muscles by accomplishing augmented mitochondrial biogenesis (de Oliveira 2016; Tao et al. 2024). The elevated ROS produced by the impaired development of the oxidative defense system and PGC-1α/Nrfs/TFAM signalling pathway result in the exposure of hyperphysiological oxygen species, leading to neonatal brain injury and apoptosis. The amplified mitochondrial biogenesis by RES treatment provides protection against brain injury induced by hyperoxia. Acting as agonist for silencing information regulator 2-related enzyme 1 (Sirt1), RES potentially augments the synthesis of Sirt1 as well as peroxisome proliferator-activated receptor gamma coactivator-1α (PGC-1α) in the brain (Yang 2023). Substantial evidence from the past few years has shown that activation of dsRNA-sensing pattern

Table 6.2 Effect of different natural molecules on mitochondria

S. no.	Compound	Chemical formula	M. wt. (g/mol)	Effect/activity	References
1.	Resveratrol	$C_{14}H_{12}O_3$	228.23	Augment mitochondrial biogenesis	de Oliveira (2016)
2.	Eugenol	$C_{10}H_{12}O_2$	164.20	Reduce ROS production	Prasad and Muralidhara (2012), Randhawa (2022)
3.	Isoeugenol	$C_{10}H_{12}O_2$	164.20	Reduce ROS production	Prasad and Muralidhara (2012)
4.	Vanillic acid	$C_8H_8O_4$	168.14	Maintain mitochondrial membrane potential, reduce ROS production	Shabani (2023)
5.	EGCG	$C_{22}H_{18}O_{11}$	458.37	Maintain mitochondrial membrane potential, reduce ROS production, impede apoptotic pathway, augment mitochondrial biogenesis	Dragicevic (2011), Shi (2018)
6.	Berberine	$C_{20}H_{18}NO_4$	336.36	Maintain mitochondrial membrane potential, reduce cytochrome c release, impede apoptotic pathway, maintain ATP production, augment mitochondrial biogenesis	Um (2023), Zhao (2019)
7.	Sulforaphane	$C_6H_{11}NOS_2$	177.29	Maintain OXPHOS, augment mitochondrial biogenesis	Bose (2020)
8.	Geraniol	$C_{10}H_{18}O$	154.25	Maintain mitochondrial membrane potential	Rekha and Inmozhi Sivakamasundari (2018)
9.	Camphene	$C_{10}H_{16}$	136.24	Reduce apoptotic factor release	Singh (2012)
10.	Curcumin	$C_{21}H_{20}O_6$	368.38	Maintain mitochondrial membrane potential, augment mitochondrial biogenesis, reduce ROS production, increase ATP production	Kuo (2012), Ray Hamidie (2015)
11.	Quercetin	$C_{15}H_{10}O_7$	302.23	Augment mitochondrial biogenesis, reduce ROS production, increase mitochondrial membrane potential	Ho (2022), Yoshida (2024)

recognition receptors mediated by double-stranded RNAs (dsRNAs) is produced by mitochondrial DNA, and in this way, mitochondria also play a crucial role in providing innate immunity. The autoimmune Sjögren's syndrome, Huntington's disease and osteoarthritis have also been found to be linked with mitochondrial dsRNAs (mt-dsRNAs), while anomalous activation of innate immunity and inflammation are triggered by various stress factors. The cessations of ATP production and presence of exogenous dsRNA are among some factors which could aggravate the immune system, but RES exhibits potential to suppress dsRNA-mediated aberrant immune activation (Yoon 2023).

6.6.2 Geraniol

Geraniol (3,7-dimethylocta-trans-2,6-dien-1-ol) is a cyclic monoterpene alcohol that possesses both anti-oxidant and antimicrobial properties and frequently used in several industries such as cosmetics, perfumery, cleaning products and flavouring agents. The essential oils produced in aromatic and spice plants are natural sources of geraniol, which exhibits neuromodulatory activity. The ROS-detoxifying property is linked with geraniol in different types of tissues in rodent models as well as in liver and brain cells. A study carried out using a *Drosophila* model substantiates that geraniol maintains the mitochondrial membrane potential, impedes the mitochondrial dysfunction and reduces the acryl amide (ACR)-induced oxidative stress. Acrylamide formation generally takes place during baking, roasting or frying foods rich in carbohydrate components. Acrylamide exhibits neurotoxic properties and possesses the ability to diminish sensory and motor functions of neurons by impairing both the peripheral and central nervous systems. Similarly, geraniol serves as a neuroprotective shield by enhancing anti-oxidant properties in mitochondria for performing normal functions during rotenone-induced oxidative stress (Prasad and Muralidhara 2014; Rekha and Inmozhi Sivakamasundari 2018; Tiwari and Kakkar 2009; Yesem and Erkan 2009).

6.6.3 Camphene

Structurally, camphene is a bicyclic monoterpenoid that is used as an expectorant and antilithic molecule and is common in the perfume and food industries in the form of an additive. The hepatotoxicity induced by nimesulide leads to the death of the liver cells as a consequence of mitochondrial dysfunction; however, the combination of camphene with geraniol has shown a protective effect on mitochondria. Nimesulide (N-(4-nitro-2-phenoxyphenyl) is a member of the non-steroidal anti-inflammatory drugs widely used for pain management but is also reported to induce hepatotoxicity. Nimesulide has been found to be responsible for causing mitochondrial dysfunction by altering the uncoupling and mitochondrial respiration via the disturbed membrane potential transition, but camphene and geraniol, when used in combination, significantly reduces the deteriorating effect of nimesulide (Aithal 2011; Singh 2012).

6.6.4 Vanillic Acid

Vanillic acid (4-hydroxy-3-methoxybenzoic acid) shows strong anti-oxidant properties and has been reported to be effective against several pathophysiological disorders such as cancer, diabetes, hepatic diseases, cardiovascular conditions, obesity and neurodegenerative ailments. Natural sources of vanillic acid are fruits, juices, intact grain and herbs (Kaur et al. 2022). The cationic lipophilic molecule methamphetamine is considered as one of the widely abused drugs in several parts of the world and is known for its associated deleterious health effects. Continuous use of methamphetamine causes problems in multiple organs, and in cases of overdose, it may lead to severe conditions such as cardiac arrest or even death. ROS production and mitochondrial dysfunction have been reported to be linked to methamphetamine use in both short and long durations, along with many other complications (Kevil 2019; Sliman 2016). The oxidative stress and mitochondrial dysfunction resulting from methamphetamine exposure, whether short or long term, cause cardiotoxicity and cardiomyopathy, against which vanillic acid shows a protective effect. Vanillic acid demonstrates exclusive activity against snake venom, along with anti-venom effects, and exhibits a wide range of therapeutic properties such as antihypertensive, antimicrobial, anti-oxidant and anti-inflammatory actions. Furthermore, vanillic acid diminishes ROS production, maintains mitochondrial membrane potential, protects against mitochondrial toxicity induced by methamphetamine and restores the cardiac mitochondrial function (Kaur et al. 2022; Shabani 2023).

6.6.5 Eugenol

Eugenol, belonging to the o-methoxy phenol class of molecules, is one of the abundant active components found in clove and, due to approval by the Food and Drug Administration, is used as a food addictive. Cloves are among the widely used spices and are also used for medicinal purposes. Besides, eugenol shows antiseptic, analgesic, anti-inflammatory, anti-oxidant, anaesthetic and hepatoprotective activities. Food rich in carbohydrates, when processed by thermal treatment, produce acrylamide, which has been reported to possess carcinogenic and neurotoxic potential (Tareke 2002). The increase in oxidative stress caused by acrylamide exposure corroborates the decrease in the reduced glutathione and increased impairment in the function of anti-oxidant enzymes and elevated ROS production. The loss of functionality of mitochondrial complexes I, II and III and succinic dehydrogenase is accompanied by acrylamide-induced mitochondrial dysfunction. The phytochemical eugenol protects mitochondria from oxidative stress, maintains the functionality of the mitochondrial complex against acrylamide-induced toxicity in a

Drosophila model and causes diminution in the periodontal ligament fibroblast (PDLF) cell apoptosis pathway related to mitochondria (Kim and Park 2015; Prasad and Muralidhara. 2012).

6.6.6 Isoeugenol

Isoeugenol, an isomer of eugenol, is a constituent of essential oils and belongs to the class of o-methoxy phenols molecule. It shares properties with eugenol and shows anti-inflammatory, anti-oxidant and anesthetic potential. Isoeugenol exhibits a protective effect against acrylamide-induced neurotoxicity (Prasad and Muralidhara. 2012).

6.6.7 Epigallocatechin-3-Gallate (EGCG)

The polyphenolic compound EGCG is the most abundant bioactive molecule present in dried tea leaves as the chief catechin. By regularly consuming green tea containing the mitochondrion-targeting medicinal compound EGCG, several beneficial effects, such as anti-oxidant properties, on human well-being can be achieved. Interestingly, EGCG among many flavonoids shows the highest potential to restore the mitochondrial function, even at very low concentrations. An analysis of a range of mitochondrial dysfunction parameters such as increased ROS production, amyloid-beta protein precursor (AβPP) and perturbed mitochondrial membrane potential in a mouse model of Alzheimer's disease reveals that EGCG reinstates almost all functional abilities impaired by amyloid-induced mitochondrial dysfunction. The extent to which EGCG treatment restores mitochondrial function may vary from approximately half to totality, which occurs without altering the levels of Aβ and AβPP in vitro and or diminishing their levels in vivo. Moreover, subarachnoid haemorrhage (SAH)-induced mitochondrial dysfunction, i.e. accumulation of blood in the subarachnoid space resulting in impairmed mitochondrial function, can be rescued by EGCG treatment. Similarly, neuronal count is maintained by EGCG-mediated reduction in cell death, which occurs due to the onset of the intrinsic apoptotic pathway associated with cytochrome-c, along with proper Drp1, Fis1, OPA1, Mfn1 and Mfn2 expression of mitochondrial fusion genes, and the reduction of oxyhemoglobin (OxyHb)-induced mitochondrial dysfunction. Collectively, the overall rescue of mitochondrial dysfunction is achieved by controlling ROS release, membrane potential depolarization and voltage-gated calcium channel-regulated Ca^{+2} influx. The fluorescence dye 5,5′,6,6′-tetrachloro-1,1,3,3′-tetraethyl-benzimidazolyl carbocyanine iodide (JC-1)-based in vitro study proves that EGCG effectively maintains the membrane potential of mitochondria and lowers the ROS production as well (Chen et al. 2018; Dragicevic 2011; Sarkar 2022).

6.6.8 Berberine

The bioactive compound berberine [Natural Yellow 18] is an organic heteropentacyclic molecule. The natural yellow attribute of berberine is associated with its needle-like morphology and yellow colour of the crystals. Berberine alkaloid has been used in traditional medicine for many centuries and is obtained from plants of the Berberidaceae family, which grow across the globe (Habtemariam 2020; Pereira 2008). Accumulating research evidence corroborates that berberine not only shows potential to selectively alter the synthesis of biomolecules such as proteins and lipids, but also modulates metabolic pathways including glycolysis and protein synthesis and hinders growth of Gram-negative as well as Gram-positive bacteria. Additionally, several pharmacological effects and activities have also been reported, such as anti-oxidant, antimicrobial, anti-tumour, antihypertensive, anti-inflammatory, anti-diarrheic and anti-arrhythmic properties (Pereira 2008; Rabbani 1987). Against neurotoxicity induced by metals and various molecular agents such as mercury, 6-hydroxydopoamine, streptozotocin, homocysteic acids and calyculin A, berberine efficiently serves as a neuroprotective molecule due to its unique ability to cross the blood-brain barrier. For appropriate plasticity and transport, processes such as maintaining redox potential, calcium release, ATP production and vesicle release. Are performed by properly functioning axonal mitochondria. However, mitochondria-directed synaptic loss and dysfunction take place because of amyloidosis of amyloid-β, characteristic feature of Alzheimer's disease. The neuroprotective property of berberine remarkably restores normal axonal mitochondrial functionality by maintaining the ATP production required for synaptic activity, restoring membrane potential, and increasing trafficking, length, density and motility. Similarly, mitochondrial dysfunction is also reversed by the natural alkaloid berberine through stimulation of mitophagy and coordination of respiratory chain components (Fang 2022; Wang 2005; Zhao 2019). Furthermore, studies conducted on PTEN-induced kinase 1 (PINK1) knockout mouse embryonic fibroblasts reveal that activation of adenosine monophosphate-activated protein kinase (AMPK) mediates the berberine-induced mitophagy, and the enhancement of mitophagy can serve as a potential neuroprotective model for treating several neurodegenerative disease such as Parkinson's and Alzheimer's.

The consequence of high glucose shock to PC12 cells results in damage and leads to mitochondrial dysfunction and elevation in oxidative stress. However, when PC12 cells were treated with berberine, it led to the reduction in cell damage and oxidative stress, upregulation of anti-oxidant enzyme gene expression, activation of the Keap1/Nrf2/ARE pathway and restoration mitochondrial function.

The pharmacological properties of berberine possibly could provide a lead for scientists in combating diabetic encephalopathy (DE), a hallmark of diabetic mellitus associated with elevated glucose levels (Um 2023; Yuan 2023). Interestingly, a high-fat diet and hyperglycaemia have been found to be linked to skeletal muscle cell mitochondrial dysfunction and may induce insulin resistance via disturbance in the fatty acid oxidation. Berberine promotes improvements in mitochondrial biogenesis, reduces the chance of diet-induced insulin resistance, activates the AMP-activated protein kinase (AMPK) and lowers gluconeogenesis (Gomes 2012).

6.6.9 Sulforaphane

The natural isothiocyanate sulforaphane is a non-toxic, sulphur-rich phytochemical found exclusively in vegetables from the cruciferous family. Many polyphenols are regularly used as Nrf-2 activators, but due to its high bioavailability and small structure, sulforaphane presents itself as a potent and effective Nrf-2 activator among them. Diverse pharmacological properties of sulforaphane have been reported based on animal model studies, such as anti-diabetic, chemopreventive, anti-cancer, anti-ageing and antimicrobial activities (Houghton 2019; Kensler 2014). After oral administration in mice, sulforaphane forms conjugates with N-acetylcysteine, glutathione and cysteine, and metabolism takes place in many organs including the colon, lung, brain, liver and kidney (Clarke 2011).

The anti-oxidant enzymes and proteins providing xenobiotic protection are expressed with the help of anti-oxidant-responsive DNA elements (AREs), also known as electrophilic response elements, which get upregulated by the Cap 'n' Collar basic leucine zipper protein (bZIP) nuclear erythroid-2-p45-related factor-2 (Nrf-2), encoded by the *NFE2l2/NRF* gene and regulate homeostasis.

The Kelch-like ECH-associating protein 1 (Keap1) is a receptor responsible for repressing the activity of Nrf-2 as part of the cullin-3 E3 ubiquitin ligase (Cul3) complex and functions during normal redox homeostasis under physiological conditions by keeping it in bound form in the cytosol, causing ubiquitination and directing it towards proteasomal degradation. The presence of sulforaphane as an ARE inducer leads to Nrf-2-cullin-3 E3 ubiquitin ligase (Cul3) complex inactivation by causing alteration in cysteine amino acid via thiol modification in Keap1. Moreover, inactivated Keap1 remains unable to hold Nrf-2 in a bound and supressed form; thus, successful translocation and accretion in the form of a heterodimer with Maf protein takes place in the nucleus. The accumulation of Nrf-2 in the presence of corresponding anti-oxidant-responsive DNA elements then mediates the transcription of genes needed for glutathione biosynthesis and cytoprotection, including anti-electrophile and anti-oxidant activities, to maintain homeostasis during oxidative stress. Furthermore, fibroblasts obtained from patients with X-associated tremor and ataxia syndrome have shown sulforaphane-mediated augmentation of mitochondrial metabolism in fragile cells (Dinkova-Kostova and Abramov 2015; Napoli 2021; Zhang 2006). In a similar vein, renal mitochondrial impairment occurs with increasing age due to diminished Nrf-2 signalling. Examination of sulforaphane-treated Fischer rats exhibits cortical Nrf-2-assisted protection from kidney damage and restoration of mitochondrial function. These activities are a result of sulforaphane-induced augmented Nrf-2 expression, followed by reduction of the Nrf-2 repressor, Keap1 protein (Mohammad 2022).

6.6.10 Curcumin

Curcumin is a polyphenol found in turmeric rhizome and *Curcuma longa*, a dietary spice commonly used in Asian countries for several hundred years. Curcumin is referred to as a nutraceutical and has been used as a traditional medicine, food additive and herbal supplement. It is well known for its anti-cancer activity associated with different organs like the ovary, brain, breast and colon (Anand 2008). Moreover, it has also shown many other pharmacological properties such as anti-mutagenic, anti-inflammatory, anti-oxidant and antimicrobial activities (Gupta 2013). The biogenesis of mitochondria gets improved in muscle cells by dietary intake of curcumin via augmentation of the cAMP level. It has been suggested that phosphodiesterase synthesis and activity reduction leads to the rise in cAMP concentration, which ultimately raises insulin secretion level and upgrades β-pancreatic cell function (Ray Hamidie 2015; Rouse 2014). The Parkinson's mouse model exhibits the preventive effect of curcumin on mitochondrial dysfunction in brain cells by reducing the production of 3-nitrotyrosine and enhancing detoxification. Similarly, the reduced oxidative and nitrosative effects of curcumin on mouse models depict the reclamation of mitochondrial function or reduction of the mitochondrial dysfunction by regulating nitrous oxide production, oxygen depletion, and increased ATPase activity as a result of diminished lipid peroxidation.

The finding of curcumin-mediated restoration of mitochondrial dysfunction in mouse liver and kidney cells may further be explored in humans, considering diabetes as well as obesity as common health concerns (Mythri 2007; Soto-Urquieta 2014). The enhanced level of free fatty acids in serum and hepatocyte steatosis and apoptosis is a characteristic of non-alcoholic fatty liver disease (NAFLD) and has been found to be linked with mitochondrial dysfunction due to oxidative stress. Elevated levels of free fatty acids hinder mitochondrial biogenesis, impair proper mitochondrial functioning and reduce ATP production in hepatocytes. Curcumin augments mitochondrial biogenesis in hepatocytes by expressing transcription factors such as nuclear respiratory factor 1 (NRF1), mitochondrial transcription factor A (Tfam) and peroxisome proliferator-activated receptor-γ coactivator 1α (PGC1α), all of which are needed for mitochondrial biogenesis.

Likewise, curcumin inhibits NF-κB, a redox-sensitive nuclear transcriptional factor, and actively regulates the coordination between gluconeogenic genes and genes for apoptosis. The inhibition of NF-κB activity by curcumin is carried out by stopping NF-κB translocation into the nucleus through downregulation of its subunit p65, present in the cytosol, which concentrate into the nucleus in the presence of high free fatty acids, resulting in lipoapoptosis and mitochondrial dysfunction. However, the NF-κB cascade inhibition leads to the recovery of mitochondrial function and cessation of gluconeogenesis and lipoapoptosis. In the presence of insulin, gluconeogenic genes, glucose 6-phosphatase (G6Pase) and phosphoenol pyruvate carboxykinase (PEPCK) remain underexpressed, while in the presence of cAMP/dexamethasone (Dex), they are expressed efficiently and induce gluconeogenesis. However, curcumin reduces the high free fatty acid-mediated expression of both G6Pase and PEPCK genes (Feldstein 2003; Kuo 2012; Rafiee 2010). Furthermore, curcumin prevents systemic skeletal infirmity and osteoporosis, which occur due to elevated oxidative stress leading to damage of bone tissues, loss of bone mass and altered bone architecture (Dai 2017).

6.6.11 Quercetin (QR)

Quercetin is a naturally occurring flavonoid found in different vegetables and fruits and is usually used as a food supplement. Substantial studies have evidenced that quercetin possesses many therapeutic capacities. It exhibits anti-Alzheimer, anti-cancer, antihypertensive, anti-atherosclerosis, anti-inflammatory and anti-oxidant properties. The QR-mediated increased expression of SIRT3 and mitochondrial function restoration has been observed alongside Ang II-induced hypertrophic conditions and the amelioration of cardiac hypertrophy in H9c2 cells and spontaneously hypertensive rats, respectively (Chen 2021). Interestingly, a mouse model study analysing the response to dietary consumption of QR showed an augmentation of mitochondrial DNA count in cells (Davis 2009). In contrast to decreasing oxidative stress markers and ROS generation during fusion, QR also manages to increase mitochondrial membrane potential, mitochondrial function markers and mitophagy. The reduction in ROS generation and oxidative stress is linked to the promoting expression of a membrane of NAD^+-dependent protein deacetylases—sirtuin1, 2 and 3 (Yoshida 2024).

Hepatotoxicity is caused by alterations in the enzymes of the respiratory chain and oxidative stress, produced by titanium dioxide (TiO_2). TiO_2 is one of the most abundant nanoparticles known for nanoparticle-induced toxicity and is present in water, the environment and food as a global pollutant. In response to TiO_2-induced hepatotoxicity, mitochondria pre-treated with QR show reduced oxidative stress, restoring the function of of respiratory chain enzymes (Waseem 2022). In combination with tanshinone, QR intervenes in oxidation and mitochondrial apoptosis through activation of the SIRT1/SIRT3-FOXO3a axis and inhibits the triptolide-induced apoptosis of KGN cells, comparable to the human granulosa cells along with regulating the expression of Bcl-2-associated X protein (Bax) and B-cell lymphoma-2 (Bcl-2).

Deacetylation of the transcriptional as well as co-transcriptional splicing regulator, FOXO3A, results in rescuing cell apoptosis (Wan 2024). Similarly, a study on methamphetamine-induced neurotoxicity in the form of depression and anxiety in a mouse model proves that upregulation of many gene expressions is associated with induced psychiatry response, accompanied by mitochondrial dysfunction and oxidative stress. The QR treatment safeguards mitochondria by exhibiting several activities, including boosting ATP production and oxygen consumption rate and diminishing mitochondria membrane potential, mitochondria-derived reactive oxygen species (mt-ROS) and total cytoplasmic ROS. Moreover, investigations on QR-supplemented astrocytes reveal that regulation of factors responsible for synaptic loss as well as neuronal apoptosis, such as inflammatory factors, optic atrophy 1 (OPA1), dynamin-related protein 1 (DRP1) and astrocyte activation, takes place

(Chen 2022). Investigation of aluminium chloride ($AlCl_3$)-induced Alzheimer's disease, characterized by cognitive function loss, depicts a dose-dependent neuroprotective QR effect. This effect is brought about by the elevation of ADAM17 gene expression in hippocampal tissues, followed by a reduction in the level of presenilin I (PSEN1), amyloid precursor protein (APP) and amyloid-converting enzyme 1 (BACE1) (Elreedy et al. 2023).

6.7 Conclusions and Future Perspective

A significant volume of cytoplasm-occupying mitochondria plays fundamental roles in the survival of cells by ensuring proper coordination of energy-producing pathways, cell signalling, etc. The translocator of inner membrane (TIM) and translocator of outer membrane (TOM) are a set of translocase variants that regulate specific entry of proteins into the mitochondria. The physical appearance such as the shape of mitochondria characterizes the cell's well-being, and incompetent or dysfunctional mitochondria caused by several factors result in the impediment of several biochemical pathways. Mitochondrial dysfunction is directly linked to the onset of a wide range of diseases including neurodegenerative diseases. Dysfunctional mitochondria-mediated apoptosis is considered a core pathological characteristic in the etiology of neurodegenerative ailments such as Alzheimer's disease, Parkinson's disease, Huntington's disease, multiple sclerosis and schizophrenia. Thus, emerging restoration and therapeutic measures targeting mitochondrial malfunction are of extreme necessity. Extensive studies demonstrate that neuroprotection involving mitochondrial dysfunction could be achieved by balancing fusion and/or fission dynamics, maintaining mitochondrial membrane potential and improving biogenesis and other mitochondrial properties. In recent years, a large number of small natural compounds have been identified as potent candidates to rescue mitochondrial from dysfunction. This chapter summarizes how various small natural molecules from different plant origins effectively manage amelioration of mitochondrial dysfunction. We hope that in the near future, analysis of purely small natural molecules or small-molecule cocktail-based treatments will open novel horizons in combating mitochondrial dysfunction, which can be incorporated into mainstream treatments.

References

Abudureyimu M (2020) Berberine promotes cardiac function by upregulating PINK1/Parkin-mediated mitophagy in heart failure. Front Physiol 11:565751. https://doi.org/10.3389/fphys.2020.565751

Aithal GP (2011) Case definition and phenotype standardization in drug-induced liver injury. Clin Pharmacol Ther 89(6):806–815. clpt201158 [pii]. https://doi.org/10.1038/clpt.2011.58

Ajmal MR (2017) Fibrillogenesis of human serum albumin in the presence of levodopa – spectroscopic, calorimetric and microscopic studies. Int J Biol Macromol 94(Pt A):301–308. S0141-8130(16)31591-4 [pii]. https://doi.org/10.1016/j.ijbiomac.2016.10.025

Anand P (2008) Curcumin and cancer: an "old-age" disease with an "age-old" solution. Cancer Lett 267(1):133–164. S0304-3835(08)00231-0 [pii]. https://doi.org/10.1016/j.canlet.2008.03.025
Anderson S (1981) Sequence and organization of the human mitochondrial genome. Nature 290(5806):457–465. https://doi.org/10.1038/290457a0
Antuono P, Beyer J (1999) The burden of dementia. A medical and research perspective. Theor Med Bioeth 20(1):3–13. https://doi.org/10.1023/a:1009915605467
Bose C (2020) Sulforaphane prevents age-associated cardiac and muscular dysfunction through Nrf2 signaling. Aging Cell 19(11):e13261. ACEL13261 [pii]. https://doi.org/10.1111/acel.13261
Bratic A, Larsson NG (2013) The role of mitochondria in aging. J Clin Invest 123(3):951–957. 64125 [pii]. https://doi.org/10.1172/JCI64125
Burbulla LF (2017) Dopamine oxidation mediates mitochondrial and lysosomal dysfunction in Parkinson's disease. Science 357(6357):1255–1261. science.aam9080 [pii]. https://doi.org/10.1126/science.aam9080
Calvo SE (2016) MitoCarta2.0: an updated inventory of mammalian mitochondrial proteins. Nucleic Acids Res 44(D1):D1251–D1257. gkv1003 [pii]. https://doi.org/10.1093/nar/gkv1003
Carelli V (2004) Mitochondrial dysfunction as a cause of optic neuropathies. Prog Retin Eye Res 23(1):53–89. S1350946203000764 [pii]. https://doi.org/10.1016/j.preteyeres.2003.10.003
Cavalier-Smith T (2006) Origin of mitochondria by intracellular enslavement of a photosynthetic purple bacterium. Proc Biol Sci 273(1596):1943–1952. R5K424547P53Q788 [pii] rspb20063531 [pii]. https://doi.org/10.1098/rspb.2006.3531
Chan DC (2012) Fusion and fission: interlinked processes critical for mitochondrial health. Annu Rev Genet 46:265–287. https://doi.org/10.1146/annurev-genet-110410-132529
Chandel TI, Khan MV, Khan RH (2017) Impact of osmolytes in conformational modulation of protein and its applications in biotechnology. In: Cellular osmolytes. Springer, pp 143–160
Chen WJ (2021) Quercetin attenuates cardiac hypertrophy by inhibiting mitochondrial dysfunction through SIRT3/PARP-1 pathway. Front Pharmacol 12:739615. 739615 [pii]. https://doi.org/10.3389/fphar.2021.739615
Chen F (2022) Quercetin ameliorates mitochondrial dysfunction and mitigates methamphetamine-induced anxiety-like behavior. bioRxiv:1–35
Chen X (2023) Mitochondrial dysfunction: roles in skeletal muscle atrophy. J Transl Med 21(1):503. 10.1186/s12967-023-04369-z [pii] 4369 [pii]. https://doi.org/10.1186/s12967-023-04369-z
Chen Y, Chen J, Sun X, Shi X, Wang L, Huang L, Zhou W (2018) Evaluation of the neuroprotective effect of EGCG: a potential mechanism of mitochondrial dysfunction and mitochondrial dynamics after subarachnoid hemorrhage. Food Funct 9(12):6349–6359
Chinnery PF, Hudson G (2013) Mitochondrial genetics. Br Med Bull 106(1):135–159. ldt017 [pii]. https://doi.org/10.1093/bmb/ldt017
Choo YS (2004) Mutant huntingtin directly increases susceptibility of mitochondria to the calcium-induced permeability transition and cytochrome c release. Hum Mol Genet 13(14):1407–1420. ddh162 [pii]. https://doi.org/10.1093/hmg/ddh162
Clarke JD (2011) Metabolism and tissue distribution of sulforaphane in Nrf2 knockout and wild-type mice. Pharm Res 28(12):3171–3179. https://doi.org/10.1007/s11095-011-0500-z
Cloonan SM (2020) Mitochondrial dysfunction in lung ageing and disease. Eur Respir Rev 29(157):200165. 29/157/200165 [pii] ERR-0165-2020 [pii]. https://doi.org/10.1183/16000617.0165-2020
Cui L (2006) Transcriptional repression of PGC-1alpha by mutant huntingtin leads to mitochondrial dysfunction and neurodegeneration. Cell 127(1):59–69. S0092-8674(06)01205-0 [pii]. https://doi.org/10.1016/j.cell.2006.09.015
Cunningham JT (2007) mTOR controls mitochondrial oxidative function through a YY1-PGC-1alpha transcriptional complex. Nature 450(7170):736–740. nature06322 [pii]. https://doi.org/10.1038/nature06322
Dai P (2017) Attenuation of oxidative stress-induced osteoblast apoptosis by curcumin is associated with preservation of mitochondrial functions and increased Akt-GSK3beta signaling. Cell Physiol Biochem 41(2):661–677. 000457945 [pii]. https://doi.org/10.1159/000457945

Davis JM (2009) Quercetin increases brain and muscle mitochondrial biogenesis and exercise tolerance. Am J Physiol Regul Integr Comp Physiol 296(4):R1071–R1077. 90925.2008 [pii]. https://doi.org/10.1152/ajpregu.90925.2008

de Oliveira MR (2016) Resveratrol and the mitochondria: from triggering the intrinsic apoptotic pathway to inducing mitochondrial biogenesis, a mechanistic view. Biochim Biophys Acta 1860(4):727–745. S0304-4165(16)30003-4 [pii]. https://doi.org/10.1016/j.bbagen.2016.01.017

Delmotte P (2017) TNFalpha decreases mitochondrial movement in human airway smooth muscle. Am J Physiol Lung Cell Mol Physiol 313(1):L166–L176. ajplung.00538.2016 [pii] L-00538-2016 [pii]. https://doi.org/10.1152/ajplung.00538.2016

Diaz-Vegas A (2020) Is mitochondrial dysfunction a common root of noncommunicable chronic diseases? Endocr Rev 41(3):5807952. [pii] bnaa005 [pii]. https://doi.org/10.1210/endrev/bnaa005

Dinkova-Kostova AT, Abramov AY (2015) The emerging role of Nrf2 in mitochondrial function. Free Radic Biol Med 88(Pt B):179–188. S0891-5849(15)00212-9 [pii]. https://doi.org/10.1016/j.freeradbiomed.2015.04.036

Dragicevic N (2011) Green tea epigallocatechin-3-gallate (EGCG) and other flavonoids reduce Alzheimer's amyloid-induced mitochondrial dysfunction. J Alzheimer's Dis 26(3):507–521. R3R830XL16527X13 [pii]. https://doi.org/10.3233/JAD-2011-101629

Elreedy H, Elfiky A, Mahmood A, Ibrahim K, Ghazy M (2023) Neuroprotective effect of quercetin through targeting key genes involved in aluminum chloride induced Alzheimer's disease in rats. Egyptian J Basic Appl Sci 10:173–184

Esteves AR (2011) Mitochondrial dysfunction: the road to alpha-synuclein oligomerization in PD. Parkinsons Dis 2011:693761. https://doi.org/10.4061/2011/693761

Fang X (2022) Research progress on the pharmacological effects of berberine targeting mitochondria. Front Endocrinol (Lausanne) 13:982145. https://doi.org/10.3389/fendo.2022.982145

Feichtinger RG (2014) Mitochondrial dysfunction: a neglected component of skin diseases. Exp Dermatol 23(9):607–614. https://doi.org/10.1111/exd.12484

Feldstein AE (2003) Diet associated hepatic steatosis sensitizes to Fas mediated liver injury in mice. J Hepatol 39(6):978–983. S0168827803004604 [pii]. https://doi.org/10.1016/s0168-8278(03)00460-4

Frazier AE (2019) Mitochondrial energy generation disorders: genes, mechanisms, and clues to pathology. J Biol Chem 294(14):5386–5395. S0021-9258(20)35499-5 [pii] R117.809194 [pii]. https://doi.org/10.1074/jbc.R117.809194

Gomes AP (2012) Berberine protects against high fat diet-induced dysfunction in muscle mitochondria by inducing SIRT1-dependent mitochondrial biogenesis. Biochim Biophys Acta 1822(2):185–195. S0925-4439(11)00231-6 [pii]. https://doi.org/10.1016/j.bbadis.2011.10.008

Gupta SC (2012) Upsides and downsides of reactive oxygen species for cancer: the roles of reactive oxygen species in tumorigenesis, prevention, and therapy. Antioxid Redox Signal 16(11):1295–1322. 10.1089/ars.2011.4414 [pii]. https://doi.org/10.1089/ars.2011.4414

Gupta SC (2013) Therapeutic roles of curcumin: lessons learned from clinical trials. AAPS J 15(1):195–218. 9432 [pii]. https://doi.org/10.1208/s12248-012-9432-8

Habtemariam S (2020) Berberine pharmacology and the gut microbiota: a hidden therapeutic link. Pharmacol Res 155:104722. S1043-6618(20)30149-3 [pii]. https://doi.org/10.1016/j.phrs.2020.104722

Hamacher-Brady A, Brady NR (2016) Mitophagy programs: mechanisms and physiological implications of mitochondrial targeting by autophagy. Cell Mol Life Sci 73(4):775–795. 10.1007/s00018-015-2087-8 [pii] 2087 [pii]. https://doi.org/10.1007/s00018-015-2087-8

Ho CL (2022) Quercetin increases mitochondrial biogenesis and reduces free radicals in neuronal SH-SY5Y cells. Nutrients 14(16) nu14163310 [pii] nutrients-14-03310 [pii]. https://doi.org/10.3390/nu14163310

Houghton CA (2019) Sulforaphane: its "coming of age" as a clinically relevant nutraceutical in the prevention and treatment of chronic disease. Oxidative Med Cell Longev 2019:2716870. https://doi.org/10.1155/2019/2716870

Howarth C (2012) Updated energy budgets for neural computation in the neocortex and cerebellum. J Cereb Blood Flow Metab 32(7):1222–1232. jcbfm201235 [pii]. https://doi.org/10.1038/jcbfm.2012.35

Ishtikhar M (2016) Anti-aggregation property of thymoquinone induced by copper-nanoparticles: a biophysical approach. Int J Biol Macromol 93(Pt A):1174–1182. S0141-8130(16)31283-1 [pii]. https://doi.org/10.1016/j.ijbiomac.2016.09.089

Jang S (2022) Beta-amyloid instigates dysfunction of mitochondria in cardiac cells. Cells 11(3) cells11030373 [pii] cells-11-00373 [pii]. https://doi.org/10.3390/cells11030373

Johri A, Beal MF (2012) Mitochondrial dysfunction in neurodegenerative diseases. J Pharmacol Exp Ther 342(3):619–630. jpet.112.192138 [pii] 3789469 [pii] 10.1124/jpet.112.192138

Kaur J, Gulati M, Singh SK, Kuppusamy G, Kapoor B, Mishra V, Gupta S, Arshad MF, O P, N.K. J, Chaitanya MVNL, Chellappan DK, Gupta G, Gupta PK, Dua K, Khursheed R, Awasthi A, Corrie L (2022) Discovering multifaceted role of vanillic acid beyond flavours: nutraceutical and therapeutic potential. Trends Food Sci Technol 122:187–200

Kelso GF (2002) Prevention of mitochondrial oxidative damage using targeted antioxidants. Ann N Y Acad Sci 959:263–274. https://doi.org/10.1111/j.1749-6632.2002.tb02098.x

Kensler TW (2014) Keap1-nrf2 signaling: a target for cancer prevention by sulforaphane. Top Curr Chem 329:163–177. https://doi.org/10.1007/128_2012_339

Kevil CG (2019) Methamphetamine use and cardiovascular disease. Arterioscler Thromb Vasc Biol 39(9):1739–1746. https://doi.org/10.1161/ATVBAHA.119.312461

Khan MV (2014) Fluoroalcohols-induced modulation and amyloid formation in conalbumin. Int J Biol Macromol 70:606–614. S0141-8130(14)00502-9 [pii]. https://doi.org/10.1016/j.ijbiomac.2014.07.027

Khan MV (2015) Non-fluorinated cosolvents: a potent amorphous aggregate inducer of metalloproteinase-conalbumin (ovotransferrin). Int J Biol Macromol 78:417–428. S0141-8130(15)00259-7 [pii]. https://doi.org/10.1016/j.ijbiomac.2015.04.021

Khan MV (2017) Polyols (glycerol and ethylene glycol) mediated amorphous aggregate inhibition and secondary structure restoration of metalloproteinase-conalbumin (ovotransferrin). Int J Biol Macromol 94(Pt A):290–300. S0141-8130(16)31914-6 [pii]. https://doi.org/10.1016/j.ijbiomac.2016.10.023

Khan MV (2018a) Protein folding, misfolding and aggregation: a tale of constructive to destructive assembly. Int J Biol Macromol 112:217–229. S0141-8130(17)34415-X [pii]. https://doi.org/10.1016/j.ijbiomac.2018.01.099

Khan MV (2018b) Cationic surfactant mediated fibrillogenesis in bovine liver catalase: a biophysical approach. J Biomol Struct Dyn 36(10):2543–2557. https://doi.org/10.1080/07391102.2017.1363085

Khan A (2020) Neuroprotection: targeting multiple pathways by naturally occurring phytochemicals. Biomedicine 8(8) biomedicines8080284 [pii] biomedicines-08-00284 [pii]. https://doi.org/10.3390/biomedicines8080284

Kiernan MC (2011) Amyotrophic lateral sclerosis. Lancet 377(9769) 942-955. S0140-6736(10)61156-7 [pii]. https://doi.org/10.1016/S0140-6736(10)61156-7

Kim JA (2008) Role of mitochondrial dysfunction in insulin resistance. Circ Res 102(4):401–414. 102/4/401 [pii]. https://doi.org/10.1161/CIRCRESAHA.107.165472

Kim YH, Park BS (2015) Antioxidant effect of eugenol in human periodontal ligament fibroblasts. Korean J Phys Anthropol 28(1):45–53

Kong GY (2009) Mitochondrial dysfunction and glaucoma. J Glaucoma 18(2):93–100. 00061198-200902000-00001 [pii]. https://doi.org/10.1097/IJG.0b013e318181284f

Kubat GB (2023) Mitochondrial dysfunction and skeletal muscle atrophy: causes, mechanisms, and treatment strategies. Mitochondrion 72:33–58. S1567-7249(23)00066-1 [pii]. https://doi.org/10.1016/j.mito.2023.07.003

Kuhlbrandt W (2015) Structure and function of mitochondrial membrane protein complexes. BMC Biol 13:89. 10.1186/s12915-015-0201-x [pii] 201 [pii]. https://doi.org/10.1186/s12915-015-0201-x

Kuo JJ (2012) Curcumin ameliorates mitochondrial dysfunction associated with inhibition of gluconeogenesis in free fatty acid-mediated hepatic lipoapoptosis. Int J Mol Med 30(3):643–649. https://doi.org/10.3892/ijmm.2012.1020

Kuroiwa T (1982) Mitochondrial nuclei. Int Rev Cytol 75:1–59. https://doi.org/10.1016/s0074-7696(08)61001-3

Landis BJ (2022) Identification of a common polymorphism in COQ8B acting as a modifier of thoracic aortic aneurysm severity. HGG Adv 3(1):100057. [pii] 10.1016/j.xhgg.2021.100057

Langston JW (1983) Chronic parkinsonism in humans due to a product of meperidine-analog synthesis. Science 219(4587):979–980. https://doi.org/10.1126/science.6823561

Lebovitz RM (1996) Neurodegeneration, myocardial injury, and perinatal death in mitochondrial superoxide dismutase-deficient mice. Proc Natl Acad Sci USA 93(18):9782–9787. https://doi.org/10.1073/pnas.93.18.9782

Li J (2015) A topical aqueous oxygen emulsion stimulates granulation tissue formation in a porcine second-degree burn wound. Burns 41(5):1049–1057. S0305-4179(14)00419-7 [pii]. https://doi.org/10.1016/j.burns.2014.11.016

Lightowlers RN (2020) Mitochondrial transplantation-a possible therapeutic for mitochondrial dysfunction?: mitochondrial transfer is a potential cure for many diseases but proof of efficacy and safety is still lacking. EMBO Rep 21(9):e50964. EMBR202050964 [pii]. https://doi.org/10.15252/embr.202050964

MacDonald ME, Ambrose MC, Duyao MP, Myers RH, Lin C, Srinidhi L, Barnes G, Taylor SA, James M, Groot N, MacFarlane H, Jenkins B, Anderson MA, Wexler NS, Gusella JF (1993) A novel gene containing a trinucleotide repeat that is expanded and unstable on Huntington's disease chromosomes. The Huntington's Disease Collaborative Research Group. Cell 72(6):971–983. 0092-8674(93)90585-E [pii]. https://doi.org/10.1016/0092-8674(93)90585-e

Marin A, Budson AE (2023) Recent advances in understanding Alzheimer's disease: diagnosis and management strategies. Fac Rev 12:24. https://doi.org/10.12703/r/12-24

McCord JM, Fridovich I (1969) Superoxide dismutase. An enzymic function for erythrocuprein (hemocuprein). J Biol Chem 244(22):6049–6055. S0021-9258(18)63504-5 [pii]

Mohammad RS (2022) Age-related mitochondrial impairment and renal injury is ameliorated by sulforaphane via activation of transcription factor NRF2. Antioxidants (Basel) 11(1):156. antiox11010156 [pii] antioxidants-11-00156 [pii]. https://doi.org/10.3390/antiox11010156

Mythri RB (2007) Mitochondrial complex I inhibition in Parkinson's disease: how can curcumin protect mitochondria? Antioxid Redox Signal 9(3):399–408. https://doi.org/10.1089/ars.2006.1479

Nakajima T (2019) Impaired mitochondrial oxidative phosphorylation capacity in epicardial adipose tissue is associated with decreased concentration of adiponectin and severity of coronary atherosclerosis. Sci Rep 9(1):3535. 10.1038/s41598-019-40419-7 [pii] 40419 [pii]. https://doi.org/10.1038/s41598-019-40419-7

Napoli E (2021) Sulforaphane improves mitochondrial metabolism in fibroblasts from patients with fragile X-associated tremor and ataxia syndrome. Neurobiol Dis 157:105427. S0969-9961(21)00176-5 [pii]. https://doi.org/10.1016/j.nbd.2021.105427

Nicolson GL (2014) Mitochondrial dysfunction and chronic disease: treatment with natural supplements. Integr Med (Encinitas) 13(4):35–43. 35–43 [pii]

Pelle E, Dong K, Layman D, Zamfir R, Yarosh DB, Pernodet N (2012) Clock gene activity in keratinocytes measured with a per1-promoter construct. J Invest Dermatol 132:60–61

Pereira CV (2008) Mechanisms of berberine (natural yellow 18)-induced mitochondrial dysfunction: interaction with the adenine nucleotide translocator. Toxicol Sci 105(2):408–417. kfn131 [pii]. https://doi.org/10.1093/toxsci/kfn131

Petersen KF (2004) Impaired mitochondrial activity in the insulin-resistant offspring of patients with type 2 diabetes. N Engl J Med 350(7):664–671. 350/7/664 [pii]. https://doi.org/10.1056/NEJMoa031314

Prasad SN, Muralidhara (2014) Neuroprotective effect of geraniol and curcumin in an acrylamide model of neurotoxicity in Drosophila melanogaster: relevance to neuropathy. J Insect Physiol 60:7–16. S0022-1910(13)00223-0 [pii]. https://doi.org/10.1016/j.jinsphys.2013.10.003

Prasad SN, Muralidhara. (2012) Evidence of acrylamide induced oxidative stress and neurotoxicity in Drosophila melanogaster – its amelioration with spice active enrichment: relevance to neuropathy. Neurotoxicology 33(5):1254–1264. S0161-813X(12)00192-1 [pii]. https://doi.org/10.1016/j.neuro.2012.07.006

Rabbani GH (1987) Randomized controlled trial of berberine sulfate therapy for diarrhea due to enterotoxigenic Escherichia coli and vibrio cholerae. J Infect Dis 155(5):979–984. https://doi.org/10.1093/infdis/155.5.979

Rafiee P (2010) Modulatory effect of curcumin on survival of irradiated human intestinal microvascular endothelial cells: role of Akt/mTOR and NF-kappaB. Am J Physiol Gastrointest Liver Physiol 298(6):G865–G877. ajpgi.00339.2009 [pii] GI-00339-2009 [pii]. https://doi.org/10.1152/ajpgi.00339.2009

Randhawa PK (2022) Eugenol attenuates ischemia-mediated oxidative stress in cardiomyocytes via acetylation of histone at H3K27. Free Radic Biol Med 194:326–336. S0891-5849(22)01033-4 [pii]. https://doi.org/10.1016/j.freeradbiomed.2022.12.007

Rather GA, Mathur V, Riyaz M, Yadav R, Nanda A, Siddiqui AJ, Rather MA, Khan A, Jahan S (2024) Mitochondrial dysfunction and its role in neurological disorders. In: Mechanism and genetic susceptibility of neurological disorders. Springer, Singapore, pp 299–325. https://doi.org/10.1007/978-981-99-9404-5_13

Ray Hamidie RD (2015) Curcumin treatment enhances the effect of exercise on mitochondrial biogenesis in skeletal muscle by increasing cAMP levels. Metabolism 64(10):1334–1347. S0026-0495(15)00190-0 [pii]. https://doi.org/10.1016/j.metabol.2015.07.010

Rekha KR, Inmozhi Sivakamasundari R (2018) Geraniol protects against the protein and oxidative stress induced by rotenone in an in vitro model of Parkinson's disease. Neurochem Res 43(10):1947–1962. 10.1007/s11064-018-2617-5 [pii]. https://doi.org/10.1007/s11064-018-2617-5

Rosen DR (1993) Mutations in Cu/Zn superoxide dismutase gene are associated with familial amyotrophic lateral sclerosis. Nature 362(6415):59–62. https://doi.org/10.1038/362059a0

Rouse M (2014) Resveratrol and curcumin enhance pancreatic beta-cell function by inhibiting phosphodiesterase activity. J Endocrinol 223(2):107–117. 223/2/107 [pii] JOE140335 [pii]. https://doi.org/10.1530/JOE-14-0335

Sarkar J (2022) Epigallocatechin-3-gallate inhibits osteoclastic differentiation by modulating mitophagy and mitochondrial functions. Cell Death Dis 13(10):908. 10.1038/s41419-022-05343-1 [pii] 5343 [pii]. https://doi.org/10.1038/s41419-022-05343-1

Sasaki S, Iwata M (2007) Mitochondrial alterations in the spinal cord of patients with sporadic amyotrophic lateral sclerosis. J Neuropathol Exp Neurol 66(1):10–16. 00005072-200701000-00002 [pii]. https://doi.org/10.1097/nen.0b013e31802c396b

Schmidt O (2010) Mitochondrial protein import: from proteomics to functional mechanisms. Nat Rev Mol Cell Biol 11(9):655–667. nrm2959 [pii]. https://doi.org/10.1038/nrm2959

Selkoe DJ (1991) The molecular pathology of Alzheimer's disease. Neuron 6(4):487–498. 0896-6273(91)90052-2 [pii]. https://doi.org/10.1016/0896-6273(91)90052-2

Shabani M (2023) Vanillic acid alleviates methamphetamine-induced mitochondrial toxicity in cardiac mitochondria via antioxidant activity and inhibition of MPT Pore opening: an in-vitro study. BMC Pharmacol Toxicol 24(1):33. 10.1186/s40360-023-00676-9 [pii] 676 [pii]. https://doi.org/10.1186/s40360-023-00676-9

Shen Y (2022) Diabetic muscular atrophy: molecular mechanisms and promising therapies. Front Endocrinol (Lausanne) 13:917113. https://doi.org/10.3389/fendo.2022.917113

Shi W (2018) The critical role of epigallocatechin gallate in regulating mitochondrial metabolism. Future Med Chem 10(7):795–809. https://doi.org/10.4155/fmc-2017-0204

Shi F (2022) Iron deficiency promotes aortic media degeneration by activating endoplasmic reticulum stress-mediated IRE1 signaling pathway. Pharmacol Res 183:106366. S1043-6618(22)00311-5 [pii]. https://doi.org/10.1016/j.phrs.2022.106366

Siddiqi MK (2018) Capreomycin inhibits the initiation of amyloid fibrillation and suppresses amyloid induced cell toxicity. Biochim Biophys Acta Proteins Proteom 1866(4):549–557. S1570-9639(18)30021-9 [pii]. https://doi.org/10.1016/j.bbapap.2018.02.005

Siddiqui A (2012) Mitochondrial DNA damage is associated with reduced mitochondrial bioenergetics in Huntington's disease. Free Radic Biol Med 53(7):1478–1488. S0891-5849(12)00343-7 [pii]. https://doi.org/10.1016/j.freeradbiomed.2012.06.008

Singh BK (2012) Natural terpenes prevent mitochondrial dysfunction, oxidative stress and release of apoptotic proteins during nimesulide-hepatotoxicity in rats. PLoS One 7(4):e34200. PONE-D-11-20752 [pii]. https://doi.org/10.1371/journal.pone.0034200

Sliman S (2016) Methamphetamine-associated congestive heart failure: increasing prevalence and relationship of clinical outcomes to continued use or abstinence. Cardiovasc Toxicol 16(4):381–389. 10.1007/s12012-015-9350-y [pii]. https://doi.org/10.1007/s12012-015-9350-y

Smith EF (2019) The role of mitochondria in amyotrophic lateral sclerosis. Neurosci Lett 710:132933. S0304-3940(17)30544-X [pii]. https://doi.org/10.1016/j.neulet.2017.06.052

Sokoloff L (1973) Metabolism of ketone bodies by the brain. Annu Rev Med 24:271–280. https://doi.org/10.1146/annurev.me.24.020173.001415

Soto-Urquieta MG (2014) Curcumin restores mitochondrial functions and decreases lipid peroxidation in liver and kidneys of diabetic db/db mice. Biol Res 47(1):74. 0717-6287-47-74 [pii] 61 [pii]. https://doi.org/10.1186/0717-6287-47-74

Su H (2011) Mitochondrial dysfunction in CA1 hippocampal neurons of the UBE3A deficient mouse model for Angelman syndrome. Neurosci Lett 487(2):129–133. S0304-3940(09)00882-9 [pii]. https://doi.org/10.1016/j.neulet.2009.06.079

Tao W, Zhang H, Jiang X, Chen N (2024) Resveratrol combats chronic diseases through enhancing mitochondrial quality. Food Sci Human Wellness 13(2):597–610

Tapias V (2018) Benfotiamine treatment activates the Nrf2/ARE pathway and is neuroprotective in a transgenic mouse model of tauopathy. Hum Mol Genet 27(16):2874–2892. 5025679 [pii] ddy201 [pii]. https://doi.org/10.1093/hmg/ddy201

Tareke E (2002) Analysis of acrylamide, a carcinogen formed in heated foodstuffs. J Agric Food Chem 50(17):4998–5006. jf020302f [pii]. https://doi.org/10.1021/jf020302f

Tiwari M, Kakkar P (2009) Plant derived antioxidants – geraniol and camphene protect rat alveolar macrophages against t-BHP induced oxidative stress. Toxicol In Vitro 23(2):295–301. S0887-2333(08)00299-3 [pii]. https://doi.org/10.1016/j.tiv.2008.12.014

Trujillo M (2008) Peroxynitrite detoxification and its biologic implications. Antioxid Redox Signal 10(9):1607–1620. https://doi.org/10.1089/ars.2008.2060

Tysnes OB, Storstein A (2017) Epidemiology of Parkinson's disease. J Neural Transm (Vienna) 124(8):901–905. 10.1007/s00702-017-1686-y [pii]. https://doi.org/10.1007/s00702-017-1686-y

Um JH (2023) Berberine induces mitophagy through adenosine monophosphate-activated protein kinase and ameliorates mitochondrial dysfunction in PINK1 knockout mouse embryonic fibroblasts. Int J Mol Sci 25(1):219. ijms25010219 [pii] ijms-25-00219 [pii]. https://doi.org/10.3390/ijms25010219

Wallace DC (2005) A mitochondrial paradigm of metabolic and degenerative diseases, aging, and cancer: a dawn for evolutionary medicine. Annu Rev Genet 39:359–407. https://doi.org/10.1146/annurev.genet.39.110304.095751

Wallace DC, Fan W (2010) Energetics, epigenetics, mitochondrial genetics. Mitochondrion 10(1):12–31. S1567-7249(09)00153-6 [pii]. https://doi.org/10.1016/j.mito.2009.09.006

Wan Q (2024) Quercetin and tanshinone prevent mitochondria from oxidation and autophagy to inhibit KGN cell apoptosis through the SIRT1/SIRT3-FOXO3a axis. Cell Mol Biol (Noisy-le-grand) 70(2):257–263. https://doi.org/10.14715/cmb/2024.70.2.36

Wang X (2005) Kinetic difference of berberine between hippocampus and plasma in rat after intravenous administration of Coptidis rhizoma extract. Life Sci 77(24):3058–3067. S0024-3205(05)00588-6 [pii]. https://doi.org/10.1016/j.lfs.2005.02.033

Waseem M (2022) Modulatory role of quercetin in mitochondrial dysfunction in titanium dioxide nanoparticle-induced hepatotoxicity. ACS Omega 7(4):3192–3202. https://doi.org/10.1021/acsomega.1c04740

Wu JJ (2009) Mitochondrial dysfunction and oxidative stress mediate the physiological impairment induced by the disruption of autophagy. Aging (Albany NY) 1(4):425–437. https://doi.org/10.18632/aging.100038

Yan M (2022) Mitochondrial damage and activation of the cytosolic DNA sensor cGAS-STING pathway lead to cardiac pyroptosis and hypertrophy in diabetic cardiomyopathy mice. Cell Death Discov 8(1):258. 10.1038/s41420-022-01046-w [pii] 1046 [pii]. https://doi.org/10.1038/s41420-022-01046-w

Yang M (2023) Protective effect of resveratrol on mitochondrial biogenesis during hyperoxia-induced brain injury in neonatal pups. BMC Neurosci 24(1):27. 10.1186/s12868-023-00797-1 [pii] 797 [pii]. https://doi.org/10.1186/s12868-023-00797-1

Yesem Y, Erkan K (2009) The carcinogenic effects of acrylamide formed during cooking of some foods. Acad J Cancer Res 2:25–32

Yoon J (2023) Resveratrol attenuates the mitochondrial RNA-mediated cellular response to immunogenic stress. Int J Mol Sci 24(8) ijms24087403 [pii] ijms-24-07403 [pii]. https://doi.org/10.3390/ijms24087403

Yoshida K (2024) Quercetin stimulates trophoblast fusion via the mitochondrial function. Sci Rep 14(1):287. 10.1038/s41598-023-50712-1 [pii] 50712 [pii]. https://doi.org/10.1038/s41598-023-50712-1

Yuan H (2023) Berberine alleviates the damage, oxidative stress and mitochondrial dysfunction of PC12 cells induced by high glucose by activating the KEAP1/Nrf2/ARE pathway. Mol Biotechnol 65(10):1632–1643. 10.1007/s12033-022-00651-5 [pii]. https://doi.org/10.1007/s12033-022-00651-5

Yu-Wai-Man P (2011) Mitochondrial optic neuropathies – disease mechanisms and therapeutic strategies. Prog Retin Eye Res 30(2):81–114. S1350-9462(10)00076-5 [pii]. https://doi.org/10.1016/j.preteyeres.2010.11.002

Zhang DD (2006) Mechanistic studies of the Nrf2-Keap1 signaling pathway. Drug Metab Rev 38(4):769–789. J6T0Q1874L787481 [pii]. https://doi.org/10.1080/03602530600971974

Zhang ZQ (2023) Mitochondrial dysfunction in glaucomatous degeneration. Int J Ophthalmol 16(5):811–823. ijo-16-05-811 [pii]. https://doi.org/10.18240/ijo.2023.05.20

Zhao C (2019) Berberine alleviates amyloid beta-induced mitochondrial dysfunction and synaptic loss. Oxidative Med Cell Longev 2019:7593608. https://doi.org/10.1155/2019/7593608

Zsurka G, Kunz WS (2015) Mitochondrial dysfunction and seizures: the neuronal energy crisis. Lancet Neurol 14(9):956–966. S1474-4422(15)00148-9 [pii]. https://doi.org/10.1016/S1474-4422(15)00148-9

Natural Compounds and Oxidative Stress Management

7

Ruchi Yadav, Namira Sana, Uzma Afaq, Chand Bibi, and Jahanarah Khatoon

Abstract

Natural compounds and their bioactive derivatives are rich sources of antioxidants, including flavonoids, anthocyanin, polyphenols, lignans, carotenoids, and vitamins. Antioxidants can manage oxidative stress-related infections and chronic inflammation by boosting the immune system. A resilient immune system will neutralize harmful free radicals in our body and reduce their levels. The trigger factor for production of oxidative stress is an imbalanced state between free radical formations to available antioxidants present in the body. Natural plant extracts exhibit multi-therapeutic effects by modulating diverse biochemical pathways. Additionally, the lower toxicity associated with natural products and their secondary metabolites reduces the risk of side effects. Antioxidants can regulate both stress-induced signaling pathways and oxidative defensive pathways. Secondary metabolites extracted from natural compounds exhibit both antioxidant and anti-inflammatory activities. Antioxidants' ability to scavenge reactive oxygen species (ROS) makes them promising candidates for treating several illnesses, such as neurodegenerative diseases, diabetes, and cancer. This book chapter highlights the therapeutic effects ingrained in natural sources of antioxidants and tries to shed light on exploring the exact mechanism behind the therapeutic effects.

Keywords

Antioxidants · Natural compounds · Oxidative stress · Inflammation · Therapeutics

R. Yadav · Uzma Afaq · C. Bibi · J. Khatoon (✉)
Department of Biosciences, Integral University, Lucknow, India
e-mail: jahanarah@iul.ac.in

N. Sana
Symbiosis Institute of Health Sciences, Pune, India

S. Jahan et al. (eds.), *A Mechanistic Exploration of Natural Compounds for Neuronal Health*, https://doi.org/10.1007/978-981-95-0125-0_7

7.1 Introduction

Natural products and their derivatives (bioactive) are the powerhouses of antioxidants, with the ability to minimize the risk of disease progression. Antioxidants are seen in chronic conditions such as heart-related disorders, diabetes, inflammatory bowel disease (IBS), neurodegenerative diseases, and cancer. Even though the mechanistic pathways of these chronic illnesses remain unexplored, scientific studies have proven their role in the advancement of these pathologies (Garcia-Sánchez et al. 2020). Scientific literature demonstrates that a wide variety of bioactive substances possess the potential to prevent these chronic disorders, yet a vast array of these natural compounds remains undiscovered and unutilized. Several studies have shown promising outcomes in validating the healing potential of these products for metabolic and cardiovascular diseases, which ultimately results in a reduction of mortality rate. The preventive role of natural products against chronic diseases has been documented; however, the bioactive potential of numerous natural products is still unexplored. Natural products protect against oxidative stress and minimizes the illness and death rates linked to long-term illness. Antioxidants aid in averting the harmful effects of free radicals by offering an electron to free radicals, without becoming reactive free radicals themselves (Jia et al. 2018).

Flavonoids are antioxidants having the potential to reduce the detrimental effect of reactive oxygen species (ROS). Oxidative stress represents a state where there exists a disparity between levels of pro-oxidants and antioxidants, with more amounts accounting for pro-oxidant activity than for antioxidant activity. ROS can negatively impact the macromolecules (DNA, proteins, and lipids), resulting in cellular dysfunction and apoptosis. The individual body can counter the action of this oxidative stress via two mechanisms: endogenous and exogenous. In the exogenous system, our body depends on external foods that are high in antioxidants to deliver antioxidants, whereas in the endogenous system, our body produces glutathione peroxidase, catalase, and superoxide dismutase (SOD) (Jena et al. 2023).

The human body is complex in that different cell compartmentalizations exist for oxidation, undergoing varying intensities of oxidative stress. Levels of imbalance are correlated with distinct structures and functions of the organs within the human body, which further makes it difficult to classify and combat cellular oxidative damage within the body, and its associated pathogeneses.

Oxidative stress is known for intensifying health problems in several disorders such as tumor, aging, allergies, and diabetes. Oxidative stress causes destruction and fragmentation of cell membranes, their ability to bind with other components, induced by irreversible DNA fragmentation and lipid peroxidation. The body's response to oxidative stress can depend on the balance between duration of oxidative stress and the resulting disparity, which may lead to over cell proliferation, cell-cycle disruption and arrest, apoptosis, and necrosis, depending on the type of ROS or RNS and their metabolites (Ahmad 2012). Reactive oxygen species (ROS) are generated during metabolic reactions (catabolic and anabolic) within the human body. ROS are associated with signaling pathways involved in the elimination of bacterial and viral pathogens, followed by the eventual elimination of these damaging foreign pathogens from the body (Hajam et al. 2022).

When ROS generation is too much, it will damage the cellular components. According to scientific studies, DNA destruction is linked with tumor formation, for example, malignant cells have higher metabolic requirements as compared to normal cells. Oxidative stress causes the destruction of cell components, their bonding with other cellular components, and death via irreversible DNA fragmentation and lipid peroxidation (Juan et al. 2021). For centuries, sources of natural products are plants, marine organisms, and microorganisms. These compounds are abundant in beneficial secondary metabolites such as polyphenols, flavonoids, alkaloids, terpenoids, and vitamins, known for their excellent free radical scavenger properties. Natural compounds can neutralize ROS, boost the body's internal antioxidant defenses, and regulate oxidative stress-associated signaling pathways (Fig. 7.1) (Chen et al. 2022). The current book chapter emphasizes the recent developments of phytochemicals as a pharmacological agent for ailments and shares information regarding the exploitation of natural products for the treatment of several disorders. Furthermore, a clear mechanistic understanding of the proper functioning and the triggering actions of natural products provides a clearer picture concerning the preventive role of natural compounds in the cure of these free radical-related diseases. The chapter will provide an updated roadmap of accurate mechanisms related to the potential application of phytochemicals in preclinical risk reduction and intervention of oxidative stress-related maladies. Therefore, exploring and studying the molecular pathways related to oxidative stress could act as a possible therapeutic target.

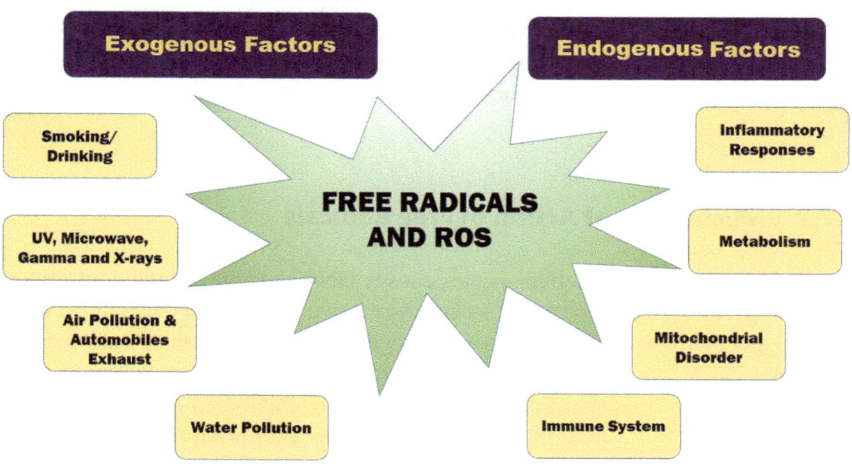

Fig. 7.1 Endogenous and exogenous triggers factors contributing to oxidative stress and subsequent cellular damage

7.2 Reactive Oxygen Species (ROS) and Free Radicals

Reactive species are characterized by the presence of a single unpaired electron. Their sources are either oxygen (ROS) or nitrogen (RNS). These reactive species are produced in our body by losing or gaining electrons. Free radicals have a highly unstable and reactive unpaired electron which acts as oxidants or reductants by either donating or accepting an electron (Di Meo and Venditti 2020). Formation of ROS in the body includes the respiratory cycle (mitochondrial), cellular metabolism, and redox reactions. Mitochondria are the main organelles for ROS production in cells. Enzymes such as Mn superoxide dismutase (MnSOD), catalase, and ascorbate oxidase maintain optimum ROS concentration necessary for normal operation of cells. ROS and oxidative stress are contributing factors in intensifying the pathogenesis of many disorders (Venditti et al. 2013). Hydroxyl radical and superoxide radical anion interact with proteins and macromolecules to produce specific aggregates, which results in pathologies such as diabetes, obesity, and neurological illness. Catalase catalyzes hydrogen peroxide into dioxygen and H_2O (Park and Han 2023).

7.2.1 Sources of Free Radicals

Reactive oxygen species (ROS) are a source of oxidative stress in an individual's body. Other organelles are also involved in this redox reaction, where the gain or loss of an electron occurs and free radicals are formed. Disorganization in electron flow during the electron transport chain will produce ROS. ROS generation is observed in both pathways as shown in Fig. 7.1. The chief source of endogenous ROS is respiration. In stress conditions, the endoplasmic reticulum releases Ca^{2+}, which will activate cytoplasmic protein, i.e., NLR-family pyrin domain-containing 3 (NLRP3). Studies have shown that vitamins A and E along with omega (ω)3 fatty acids can neutralize ROS production in an individual's body (Fig. 7.1) (Zorov et al. 2014).

7.2.2 Mechanism of Action of Free Radicals

The production of free reactive species causes DNA damage, which contributes to the etiology of various diseases. In cancer, reactive oxygen species (ROS) are the main trigger element for the onset of carcinogenesis as they cause extensive damage to cellular DNA. Mutation in DNA due to erroneous DNA proofreading can result in the conversion of a proto-oncogene into an oncogene, resulting in cancer of different types. Furthermore, initiation, promotion, and progression are the three main stages of carcinogenesis. During the initiation phase, a carcinogenic trigger (endogenous or exogenous) lays the foundation for DNA damage, which leads to an inheritable change that induces the properties of neoplastic growth in the concerned cell (Klaips et al. 2018; Scott et al. 1984). Induction of procarcinogens will impact and initiate ROS production, and the degree of DNA damage is linked to the carcinogens' potency (Waris and Ahsan 2006). ROS are considered one of the main key players for DNA damage. A higher level of DNA damage is observed in mitochondrial DNA compared to chromosomal DNA by ROS as mitochondrial DNA is located closer to the ROS generation site. Breaks in DNA strands form hydroxyl radicals that ultimately alter DNA components. The production of hydroxylated bases of DNA is a decisive process of chemical carcinogenesis. The hydroxyl radical targets purines, producing chemicals such as 8-hydroxy deoxy adenosine and 8-hydroxyoxy guanosine. Pyrimidines will form hydantoin, 5-hydroxy deoxycytidine, 5-hydroxydeoxy uridine, uracil glycol, and thymine glycol, in that sequence (Chaudhary et al. 2023).

7.3 ROS and Disease Progression

7.3.1 Liver Diseases

The liver carries out innumerable vital tasks such as nutrient uptake through the digestive system, breaking down hormones, synthesizing different substances including albumin, globulin, and clotting factors, and aiding in detoxification procedures. Alcoholics are more likely to develop liver damage since the liver produces free radicals during the metabolism of alcohol, and the breakdown of ethanol causes ROS production in liver cells (hepatocytes) (Wu and Cederbaum 2003). Alcohol molecules break down into smaller compounds during metabolism, and these chemicals then react to form ROS in the liver.

7.3.2 Lung Diseases

The atmosphere is full of oxygen, which is taken in by the lungs. Their enormous surface area and plentiful blood supply make them especially vulnerable to oxidative stress-related damage. Oxidative stress and inflammatory reactions in the lungs can be triggered by exposure to external sources of reactive oxygen species such as carbonyls/aldehydes and cigarette smoke. Furthermore, increased endogenous ROS production inside lung tissues can result from a lack of antioxidant defense in inflammatory cells. In many lung disorders, mitochondrial failure greatly affects both bioenergetic metabolism and non-energetic etiology as shown in Fig. 7.2. (Chaudhary et al. 2023). Mitochondria are essential for ROS-dependent pathways. The cytotoxic response is controlled by the mitochondrial genome.

7.3.3 Neurological Diseases

Significant loss of neurons is a characteristic feature of neurodegenerative diseases. Oxidative stress triggers neuroinflammatory responses during cerebral ischemia (CI) (Kim et al. 2014). Oxidative stress stimulates microglia and astrocyte activation, leading to a surge in levels of inflammatory mediators. These mediators are responsible for maintaining cellular integrity in the brain. Recent studies show that Parkinson's disease causes degeneration in both dopaminergic and non-dopaminergic neurons. Parkinson's disease primarily occurs due to oxidative damage and immune response activation, both of which lead to neurodegeneration. The reduction in population of dopaminergic neurons is marked by elevated T-cell infiltration, astrogliosis, and inducible nitric oxide synthase (iNOS) levels (Khan et al. 2019).

7.3.4 Cardiovascular Disorders

In our most vital organ, the heart, ROS function as a secondary messenger because they are involved in several events such as excitation, production, and differentiation. The causes of oxidative stress are cell death, hypertrophy, ischemia-reperfusion injury, and cardiac dysfunction. ROS generated by cytochrome P450 is endogenous. Additionally, ROS are accountable for the onset of certain issues linked to particular clinical contexts such as diabetic cardiomyopathy which identifies a particular kind of heart disease in individuals with diabetes who do not have other diabetes-related complications (Fig. 7.2) (Ioannidou et al. 2024).

7.3.5 Rheumatoid Arthritis

Oxidative imbalance is a key factor in the initiation of rheumatoid arthritis (RA). Furthermore, levels of mitochondrial ROS in the blood of RA patients is five times higher than that of a normal individual, indicating its pathogenic potential for RA (Gonzálezet al. 2023). RA free radicals are indirectly linked to joint damage. T-cells exposed to elevated ROS turn out to be resistant to various stimuli such as those promoting growth or death, and this may help to maintain the aberrant immune response. Free radicals, on the other hand, target proteoglycan in the joint cartilage and prevent its synthesis which causes direct degradation of the cartilage. In patients with RA, oxidative products of lipid peroxidation, oxidative damage of hyaluronic acid, and oxidation of low-density carbonyl and lipoproteins occur. Furthermore, elevated levels of 4-HNE and synovial fluid have been found in the serum of RA patients (Huang et al. 2020).

7.3.6 Renal Diseases

Renal lipids contain polyunsaturated fatty acids which increase susceptibility of the kidneys to changes caused by ROS. The renal bases of ROS are interstitial cells and stimulated macrophages (Rosales and Fabi 2023). The effects of ROS were significant in several renal diseases including glomerulosclerosis, renal failure, ischemia-reperfusion injury, and tubule-interstitial fibrosis (Huang et al. 2020). Nitric oxide, an endothelial vasodilator, reacts with free radicals to inhibit their function, causing hypertrophy of tubular cells. The products of this reaction include peroxynitrite ($ONOO^-$), produced when the highly reactive hydroxyl radical (OH) is generated (Fig. 7.2).Massive O_2 generation can result in both apoptotic cell death and proliferation in epithelial cells.

7.3.7 Cataract

A cataract is a vision-impairing partial or total opacification in or around the human lens or capsule. At present, it is considered as the leading cause of reversible blindness worldwide. Oxidation elicits the commencement of a cascade of events leading to cataracts (Zhang et al. 2021a, b). The trigger factors for induction of cataractogenesis are reduced permeability of the lens membrane, increased non-enzymatic glycosylation and peroxidation of lipids, and reduced antioxidant defense mechanism. These findings demonstrate that oxidative stress plays a crucial role in cataract development, and antioxidant molecules can help prevent or lessen its effect (Santos et al. 2022).

7.4 Antioxidants

Antioxidants are produced as a result of metabolic activities taking place within the body. They are the body's natural defense mechanism against injury caused by reactive oxygen species. Depending on the sources, the body has two kinds of antioxidants: endogenous and exogenous. Sources of natural antioxidants include food items such as fresh fruits, vegetables, plant materials, and several medicinal plants. Furthermore, these antioxidants help prevent tumor, aging, inflammation, etc. These days, even synthetic antioxidants also possess these above-mentioned properties. Artificial dietary antioxidants are more economical than natural ones and are gaining popularity in the market. The high-tech pharmaceutical companies are in constant search for new antioxidants (Kopustinskiene et al. 2020). The commonly extracted antioxidants from plant-based compounds include secondary metabolites such as polyphenols, carotenoids, and vitamins. Phenolic acids, flavonoids, anthocyanins, and stilbenes are among the different classes of polyphenols.

As they gain popularity as great health supplements, gradual increases in their demands are seen in the market. Recent research mainly focus on efficient extraction and assessment methods, including pulsed electric field extraction, ultrasound-assisted and microwave-assisted extraction, and high-voltage electrical discharge extraction (Zhang et al. 2021a, b). Currently, continuous investigation on clean technology is being conducted that will convert waste from fruits and vegetables into beneficial products. Recent studies have shown that approximately 25–30% of waste material is generated in juices, starch, or sugar preparation. Higher levels of phenols and ascorbic acids have been found in these discarded scalps compared to their pulp, and they are in unripe form rather than ripened form. Fruit peels contain 2–27 times greater amounts of antioxidants than their pulp, for example, banana pulp possesses 25% higher phenolic components than banana peels. During food and therapeutic preparations, many bioactive substances with strong antioxidant properties are extracted from food wastes (Kim 2021). Hence, these antioxidants obtained from food waste are used as raw materials for the manufacture of functional foods and sometimes as food additives (Aranda-Rivera et al. 2022).

Mango peel is loaded with great amounts of antioxidants like carotenoids, vitamin C, and fibers, having great antioxidative potential against diseases (Alzheimer's disease, tumors, and Parkinson's disease). The antioxidants present in wine waste such as skin, stem, and seeds are shown to have a therapeutic role in degenerative diseases. Another example is from the coffee industry, where its wastes contain around 6% polyphenols and about 4% tannins (Rasmus and Kozłowska 2023).

7.5 Active Compounds Present in Antioxidants Derived from Natural Sources

7.5.1 Polyphenols

Polyphenols are compounds loaded with a lot of antioxidants. The antioxidant properties of these polyphenols prevent lipid oxidation. Chemically, most polyphenols are conjugates of mono- and polysaccharides linked through one or more groups of phenol rings. Sources of the above-mentioned natural antioxidants are grapes and green and red teas (Table 7.1). The biological potential of polyphenols present in teas is far greater than that in fruits (Antonious 2018).

7.5.2 Flavonoids

Flavonoids are polyphenols, found abundantly in human food such as red berries, potatoes, tomatoes, and wheat. Flavonoids are classified into six subclasses: flavones, flavonols, flavanones, isoflavones, anthocyanidins, and flavanols (catechins). Anthocyanin is a potent antioxidant copiously found in berries and red wine. Polyphenols exhibit their antioxidant activity by halting the oxygenation of low-density lipoproteins (LDL) (Manchali et al. 2012). Quite few polyphenols hindered the oxidation of some enzymes, hence preserving their activity. Carotenoids are also important antioxidants, found in potatoes, carrots, papayas, apricots, and other vegetables (Tables 7.1 and 7.2). Apart from their antioxidant properties, flavonoids have

Table 7.1 Antioxidant source, chemical constituent, and mode of action

Antioxidant	Source	Type	Mechanism of action	References
Polyphenols	Fruits, vegetables, tea, coffee	Phenolic compound	Scavenging free radicals, metal chelation	Shahidi et al. (2015)
Flavonoids	Fruits, vegetables, beverages	Phenolic compound	Scavenging free radicals, modulating enzymes	Pietta (2000)
Vitamin C	Citrus fruits, strawberries	Vitamin	Scavenging free radicals, enhancing peppers, enzymes	Mandl et al. (2009)
Vitamin E	Nuts, seeds, vegetable oils	Vitamin	Scavenging free radicals, protecting lipids	Traber et al. (2011)
Carotenoids	Carrots, tomatoes, leafy vegetables	Terpenoids	Scavenging free radicals, protecting membranes	Maiani et al. (2009)
Glutathione	Fruits, vegetables, meats	Peptide	Enhancing endogenous antioxidant	Wu et al. (2004)
Coenzyme Q10	Meat, fish, whole grains	Quinone	Scavenging free radicals, energy	Littarru et al. 2007)

anti-inflammatory and anti-proliferative activities, too. Flavonoids regulate the adipocyte count and size, cytokine assembly, gene expression, and Toll-like receptor (TLR) blockage. Moreover, flavonoids also regulate transcription factors, thus modulating the immune response and regulating the energy balance by preventing inflammation of the hypothalamus (Murillo and Mehta 2001). Flavonoids have shown therapeutic effects such as quenching ROS, triggering apoptosis initiation, and exhibiting anti-proliferative activity. Flavonoids possess strong phenolic hydroxyl groups, which are responsible for their antioxidant activity and reducing ROS production in an individual's body (Kopustinskiene et al. 2020).

7.5.3 Vitamins C and E

The next powerful antioxidant is vitamin C, chemically called ascorbic acid, a potent water-soluble antioxidant. Fresh fruits and vegetables such as Indian gooseberry, citrus fruits, green and red peppers, kiwi fruit, strawberries, green leafy vegetables such as broccoli, and fortified cereals are its sources. Vitamin E is a fat-soluble, nonpolar vitamin, belonging to the tocopherol family of antioxidants and found in lipid-rich fruits and vegetables, including olives, sunflowers, and nuts. The bioavailability of vitamin E is higher than that of vitamin C (Table 7.1) (Santos et al. 2022).

7.5.4 Isoflavones

Isoflavones are phytochemicals present in vegetables, pulses, and cereals. Soybeans are the main source of isoflavones. Genistein, daidzein, and glycitein are examples of isoflavones. They may have anti-adipogenic and anti-lipogenic effects and play a

Table 7.2 Natural antioxidants and their bioactive compounds employed as preventive and pharmacological agents for oxidative stress-related disorders

S. no.	Nature of compound	Compound	Experiment approaches	Results	References
1.	Flavonoids	Catechin	Type 2 diabetic erythrocytes	Decreased MDA. Increased GSH	Shahidi (2003)
		Hesperetin	Acetaminophen-induced hepatotoxicity	Increased levels of glutathione. Increased SOD and CAT activities. Reduced MDA levels	Pietta (2000)
		Quercetin	Subacute chlorpyrifos-induced oxidative stress	Decreased malondialdehyde levels. Enhanced SOD, CAT, and GPx	Mandl et al. (2009)
		Rutin	Alloxan-induced diabetic nephropathy	Increased SOD and catalase. Reduced lipid peroxidation. Downregulated endoplasmic reticulum stress markers GRP78 and CHOP	
2.	Terpenoids	Carvacrol	Aluminum-induced oxidative stress	Increased GSH, CAT, and SOD concentration. Decreased MDA	Littarru et al. (2007)
		Camphor	Atrazine-induced toxicity	Increased SOD activity and reduces MDA levels	
		Citral	Streptozotocin/ high-fat diet-induced diabetic dyslipidemic rats	Reduction in MDA levels. Decreased protein carbonyl content. Higher activity of SOD	Wu et al. (2004)
		Geraniol	Isoproterenol-induced cardiotoxicity	Increased GSH levels. Elevated GPx, CAT, and SOD activities. Upregulation of Nrf2. Upregulation of HO-1 expression	
3.	Phenolic acid	Gallic acid	Cerebral ischemia/ reperfusion induced by middle cerebral artery occlusion	Lowers the MDA levels	Maiani et al. (2009)
		Rosmarinic acid	Aluminum-induced oxidative stress	Increased GSH concentration. Decreased MDA concentration. Increased CAT and SOD activities	
		Caffeic acid	Medium-term rat hepatocarcinogenesis model	Decreased lipid peroxidation	
		Vanillic acid	Ischemia-reperfusion-induced oxidative stress in isolated rat heart	Decreased MDA. Elevated SOD, CAT, and GPx activities	

role against inflammation, obesity, and oxidative stress. They interact with estrogen receptors and PPARγ, which regulates fatty acid metabolism and insulin sensitivity (Zhang et al. 2021a, b). Genistein acts as an antioxidant and is responsible for chelating metals, elevating reduced/oxidized glutathione ratio (GSH/GSSG), and improving mitochondrial membrane potential. Genistein also normalizes the activity of antioxidant enzymes and lowers ROS formation, pro-inflammatory cytokines, endothelial NOS (eNOS), and inflammation by stimulating JNK and blocking NF-κB, TNF-α, and IL-6 secretion. Genistein in postmenopausal women reduces body mass index (BMI), body fat mass, and waist size and increases blood HDL (Kim 2021).

7.5.5 Carotenoids

Carotenoids are antioxidants extracted from plants, animals, and microorganisms, having antioxidant and anti-inflammatory potential. They are responsible for providing immunity against disorders such as metabolic syndrome and tumors. The pro-vitamin A compounds such as β-cryptoxanthin and α-carotene are examples of carotenoids. Studies reveal that obese people show lower levels of serum carotenoid and that intake of carrots, pumpkin, spinach, and mandarins, is associated with reduced levels of inflammation, oxidative stress, cardiovascular disorders, and related mortality (Table 7.1) (Aranda-Rivera et al. 2022). Astaxanthin, a well-known carotenoid extracted from microscopic fungi, has a defensive action against oxidative stress. It prevents lipid peroxidation and enhances antioxidant defense mechanisms. Furthermore, lycopene consumption lowers the oxidative stress and blood pressure in hypertensive individuals. The levels of α-carotene, β-carotene, lutein, and cryptoxanthin are inversely related to hypertension occurrence (Rasmus and Kozłowska 2023).

7.5.6 Capsaicinoids and Capsinoids

Capsaicinoids and casinoids are chemically alkaloids, discovered in peppers, with activity against tumors, inflammation, and obesity. Their intake increases energy consumption, lipid oxidation, appetite loss, and weight loss. Obesity-induced inflammation is regulated by capsaicin by reducing expression levels of cytokines (Antonious 2018).

7.5.7 Isothiocyanates and Catechins

Broccoli, cauliflower, Brussels sprouts, and cabbage—all cruciferous vegetables with chemopreventive potential—are sources of isothiocyanates (ITCs). Sulforaphane and phenethyl are the main chemicals responsible for their properties (Manchali et al. 2012). Catechins are under the class of flavonols, extracted from tea, cocoa, and red wine. The presence of catechins in the body induces elevated levels of SOD and catalase, which lowers the expression levels of NF-κB, TNF-α, Nox, and iNOS, causing reduced production of oxidative stress and inflammation in the body. Catechins in the human body reduces blood pressure, LDL cholesterol, and heart-related risk factors (Mokra et al. 2022).

7.5.8 Resveratrol

Resveratrol is a polyphenol and phytoalexin produced by plants against an injury or attack by pathogens. Chemically 3,4′,5-trihydroxystilbene, it has antibacterial, antioxidant, anti-inflammatory, and anticancer activities. Resveratrol is abundantly found in peanuts, apples, and seeds of red grapes. Studies have shown its preventive role in fatty liver, diabetes, atherosclerosis, aging, and neural and cardiovascular system disorders. Resveratrol is a great antioxidant and reduces the occurrence of obesity, metabolic disorders, type 2 diabetes, CVD, cancer, and aging (Ma et al. 2018). However, research focusing on its underlying mechanisms of action is still lacking. Resveratrol triphosphate, when used as a supplement in the body, is associated with a reduction in oxidative stress, redox-sensitive genes in blood cells, lipogenesis, and inflammatory mediators. It is evident from recent studies that resveratrol regulates PPARγ expression and physiological activities such as lipolysis, apoptosis, and mitochondria genes (Varoni et al. 2016). When resveratrol is added as a food supplement it is linked to reduced insulin sensitivity in type 2 diabetes. Resveratrol also augments the levels of adiponectin and improves heart health, reducing the blocking of atherothrombotic signals in blood mononuclear cells. In patients with type 2 diabetes, hypertension, and coronary artery disease, resveratrol regulates the levels of cytokines and inflammation in peripheral blood mononuclear cells (Raj et al. 2021).

7.5.9 Quercetin

Quercetin is a flavonol found in green tea, onions, scallions, and broccoli. Quercetin is chemically 3,5,7,3′,4′-pentahydroxyflavone and has anti-inflammatory and antimutagenic activities. In *Gingo biloba* extract, quercetin is a major phytochemical. Quercetin regulates levels of inflammation by modulating TNF-α, nitric oxide, and IL-6 levels (Pinheiro et al. 2021). Quercetin lowers the levels of metalloproteinase 1 and oxidative stress production. Furthermore, regulating the levels of LDL stabilizes plaque (Table 7.2).

Quercetin induces higher expression of GLUT4 glucose transporters which results in lowering insulin resistance. Quercetin supplementation reduces dyslipidemia, hypertension, and hyperinsulinemia (Table 7.2) (Ansari et al. 2022). Future investigation is required to discover the bioavailability and bioactive effects of quercetin in reducing obesity.

7.5.10 Curcumin

Haldi, scientifically known as *Curcuma longa*, is a rhizome. Curcumin is a polyphenol found in *Curcuma longa* and has several biochemical properties against tumors, body fat, diabetes, and inflammation. Curcumin enhances the levels of a potent antioxidant known as glutathione. It is also correlated with the reduced activity of transcription factors NF-kB, ERK1/2, and p38 MAPK, hence modulating the gene activity of inflammatory cytokines and growth factors. Curcumin induces minor oxidative and metabolic stress. It halts macrophage intrusion and NF-κB activation in adipose tissue, and it increases high-fat diet-induced insulin sensitivity and reduces obesity in the liver by blocking lipogenesis. Studies have shown that curcumin also possesses anti-cancer potential because of its estrogen-like activity (Zhang et al. 2021a, b).

7.5.11 Ferulic Acid and Phloretin

Ferulic acid is a phenolic acid found in apples, oranges, chocolate, whole wheat, sage, and oregano. Ferulic acid regulates the secretion of insulin and the production of lipogenic enzymes and antioxidants (Zhang et al. 2021a, b). Phloretin is an antioxidant obtained from the leaves of apple trees and Manchurian apricots and has a role in lowering obesity and hepatic lipid accumulation and in maintaining metabolic and glucose homeostasis. Phloretin is associated with weight loss in obese animals by blocking weight gain (Nakhate et al. 2022).

7.6 Molecular Mechanisms of Antioxidants

The main source of ROS production in an individual is the ongoing chemical processes that include both catabolic and anabolic reactions. When nucleobases and lipids interact with radicals, free radicals are generated as a result of these chemical reactions. Lipids are associated with the generation of free radicals in cells through lipid peroxidation. Free radicals are formed by scavenging hydrogen in the presence of oxygen, and then these radicals make bonds (Nakai and Tsuruta 2021). Myeloperoxidase can produce hypochlorous acid by utilizing hydrogen peroxide as a substrate. Free radicals have unpaired electrons so they become highly reactive and can readily oxidize the protein components of cells. When nitric acid and superoxide react in a cell, they form the peroxynitrite radical, which is a highly reactive species. Furthermore, protonation of nitric acid will produce peroxynitrous acid, and this peroxynitrous acid will bring hemolytic cleavage, leading to production of nitrogen dioxide and hydroxyl radicals (Martemucci et al. 2022).

7.6.1 Preventive Antioxidants

Several approaches are employed to counter the damage caused by free reactive species which are continuously generated in our bodies in the form of oxidative stress. The human body is blessed with antioxidant enzymes which are produced internally, called endogenous antioxidant enzymes, which act as a defensive mechanism against reactive free species.

7.6.2 Free Radical Scavengers

The metabolic reactions occurring in the body during physiological processes will utilize oxygen and generate reactive oxygen species. Mitochondria are powerhouse of cells because they deal with ATP production, which involves several reactions where oxygen is metabolized and water is produced as a product. As we know during ATP generation in mitochondria via the electron transfer chain (ETC cycle), incomplete reduction of oxygen will take place, resulting in biosynthesis of reactive oxygen species. These ROS are highly damaging because of the presence of unpaired electrons (Juan et al. 2021). The superoxide dismutase (SOD) metalloenzyme converts superoxide radicals into hydrogen peroxide, hence reducing their toxicity. Green tea is another potent antioxidant, which will help counter the detrimental effect of superoxide ions. Usually, the Cu^+/H_2O_2 or Fe^{2+} Fenton reaction system produces hydroxyl radicals when H_2O_2 and $FeSO_4$ are cultivated in an aqueous solution. It is a well-known fact that for macromolecules (DNA, lipids, and proteins), these radicals are harmful. The polyphenols and polyenes present in fruits and vegetables exhibit preventive activity against the damaging effects of free radicals (Rai et al. 2019). Lipid peroxidation is another metabolic reaction, related to the biosynthesis of free radicals. Lipids present in the human body need to be studied precisely to understand the mechanisms and calculate the damage caused by lipid peroxidation. Polyunsaturated fatty acids (PUFAs) have bis-allylic methylene groups, which are involved in the peroxidation of lipid groups. The bond dissociation energy of hydrogen-carbon bonds in methylene units is deficient, which helps in the easy abstraction of hydrogen atoms in radical reactions (Ayala et al. 2014). Antioxidants such as vitamin E, t-butyl hydroxyanisole (BHA), and t-butylhydroxytoluene (BHT) block lipid peroxidation in rat liver microsomes (Valgimigli 2023). Furthermore, rosmarinic acid, a naturally occurring compound, can permeate cell membranes and reduce lipid peroxidation (Valgimigli 2023).

7.7 Prevention of DNA Damage

Most DNA destruction in humans is triggered by oxidative stress, which is facilitated by several variables thought to increase ROS levels and free radicals. Smoking, eating junk food or fried meals, stress, and lifestyle changes are major causes of overproduction of free radicals in the body. To meet the body's natural demands, one can receive antioxidants through nutrition, which includes both internal and exterior forms (Chaudhary et al. 2023). Hence, antioxidants neutralize these free radicals, helping to prevent their harmful effects on the body. Methanolic extracts of *Pseudomugii furcatus*, *Centella asiatica*, *Adhatoda vasica*, and *Tamarind indica* and the ethanolic fraction of germinated peanut also have protective action against DNA damage (Limmongkon et al. 2019).

7.8 Prevention of Protein Modification

Hydroxyl radicals (•OH), peroxyl, alkoxyl, hydroperoxyl, and superoxide anions ($O^{2•-}$), as well as nonradical species, including singlet oxygen, hypochlorous acid (HOCl), ozone (O^3), peroxynitrite ($OONO^-$), and hydrogen peroxide (H_2O_2), have a major role in protein oxidation. These reactive species will cause protein-protein cross-linking, as a result of the oxidation of various amino acids inside the protein, which can lead to the cascade of events such as protein denaturation, loss of protein function, degradation of transport proteins and receptors, and decreased enzymatic activity. Because of their extreme reactivity, hydroxyl radicals can interact with both organic and inorganic molecules, resulting in serious cellular damage (Chaudhary et al. 2023). Several studies on cell lines have proven that biliverdin and bilirubin have higher antioxidant potential than alpha-tocopherol in deterring oxidation of protein in the rat brain caused by peroxynitrite and peroxyl radicals (Miazek et al. 2022).

7.9 Therapeutic Limitations and Challenges of Natural Antioxidants

Natural antioxidants are essential for maintaining health and avoiding diseases caused by oxidative stress. They are extremely useful in fostering general well-being due to their variety of sources and modes of action. Natural antioxidants and their bioactive compounds employ numerous mechanisms of action and have shown preventive roles in several stress-induced diseases without causing side effects. Disproportionate ingestion of antioxidant supplements results in side effects, as high quantities of antioxidants can act as pro-oxidants. There is a crucial distinction between obtaining antioxidants from food and taking them as isolated supplements. Many compounds show positive effects in in vitro settings, but they may not have the same impact when consumed by humans due to poor bioavailability. For instance, the levels of polyphenols are often so low in the bloodstream that no substantial effects are observed. Further investigation into natural antioxidants will clarify their potential and open the door to new treatment approaches. Their diverse sources and mechanisms of action make them invaluable in promoting overall well-being. Continued research into natural antioxidants will further elucidate their potential and lay the foundation for novel therapeutic strategies.

7.10 Conclusion and Future Perspectives

Oxidative stress negatively impacts human health because of free radical overproduction. Research has proven free radical involvement in the development and progression of several pathologies from cancer to cardiovascular disease. Oxidative stress is combated by antioxidants, diminishing its negative effects on human health. The biomedical research community paid close attention to these compounds because they demonstrated a higher level of efficacy in treating and preventing these diseases. Research has shown that synthetic antioxidants are harmful to an organism's health. Thus, in the past few years, there has been greater effort to find a natural non-toxic compound with higher antioxidant activity. Based on previous research and studies, it is possible to conclude that oxidative stress can be used as a therapeutic tool—provided that the mechanisms controlling it within a living organism are understood. Soon newer strategies that combine the current technology and cooperative research with well-established orthodox health practices can be used to improve health status, particularly for those without access to more expensive therapeutic drugs.

References

Ahmad S (2012) Oxidative stress and antioxidant defenses in biology. Springer

Ansari P, Choudhury ST, Seidel V, Rahman AB, Aziz MA, Richi AE, Rahman A, Jafrin UH, Hannan JMA, Abdel-Wahab YHA (2022) Therapeutic potential of quercetin in the management of Type-2 diabetes mellitus. Life (Basel, Switzerland) 12(8):1146. https://doi.org/10.3390/life12081146

Antonious GF (2018) Capsaicinoids and vitamins in hot pepper and their role in disease therapy. In: Capsaicin and its human therapeutic development. IntechOpen

Aranda-Rivera AK, Cruz-Gregorio A, Arancibia-Hernández YL, Hernández-Cruz EY, Pedraza-Chaverri J (2022) RONS and oxidative stress: An overview of basic concepts. Oxygen. 2(4):437–478. https://doi.org/10.3390/oxygen2040030

Ayala A, Muñoz MF, Argüelles S (2014) Lipid peroxidation: production, metabolism, and signaling mechanisms of malondialdehyde and 4-hydroxy-2-nonenal. Oxidative Med Cell Longev 2014:360438. https://doi.org/10.1155/2014/360438

Chaudhary P, Janmeda P, Docea AO, Yeskaliyeva B, Abdull Razis AF, Modu B, Calina D, Sharifi-Rad J (2023) Oxidative stress, free radicals and antioxidants: potential crosstalk in the pathophysiology of human diseases. Front Chem 11:1158198. https://doi.org/10.3389/fchem.2023.1158198

Chen B, Zhao J, Zhang R, Zhang L, Zhang Q, Yang H, An J (2022) Neuroprotective effects of natural compounds on neurotoxin-induced oxidative stress and cell apoptosis. Nutr Neurosci 25(5):1078–1099

Di Meo S, Venditti P (2020) Evolution of the knowledge of free radicals and other oxidants. Oxidative Med Cell Longev 2020(1):9829176
García-Sánchez A, Miranda-Díaz AG, Cardona-Muñoz EG (2020) The role of oxidative stress in physiopathology and pharmacological treatment with pro- and antioxidant properties in chronic diseases. Oxidative Med Cell Longev 2020:2082145. https://doi.org/10.1155/2020/2082145
González P, Lozano P, Ros G, Solano F (2023) Hyperglycemia and oxidative stress: an integral, updated and critical overview of their metabolic interconnections. Int J Mol Sci 24(11):9352. https://doi.org/10.3390/ijms24119352
Hajam YA, Rani R, Ganie SY, Sheikh TA, Javaid D, Qadri SS, Reshi MS (2022) Oxidative stress in human pathology and aging: molecular mechanisms and perspectives. Cells 11(3):552
Huang M, Zheng L, Xu H, Tang D, Lin L, Zhang J et al (2020) Oxidative stress contributes to vascular calcification in patients with chronic kidney disease. J Mol Cell Cardiol 138:256–268
Ioannidou S, Tsiakalidou A, Kazeli K, Ginoudis A, Fouza A, Daoudaki M, Lymperaki E (2024) Evaluation of biochemical and oxidative stress markers in the early stages of rheumatoid arthritis in a comparative study of two different therapeutic approaches. Oxygen 4(3):253–265. https://doi.org/10.3390/oxygen4030014
Jena AB, Samal RR, Bhol NK, Duttaroy AK (2023) Cellular red-ox system in health and disease: the latest update. Biomed Pharmacother 162:114606
Jia Z, Anandh Babu PV, Chen W, Sun X (2018) Natural products targeting on oxidative stress and inflammation: mechanisms, therapies, and safety assessment. Oxidative Med Cell Longev 2018:6576093. https://doi.org/10.1155/2018/6576093
Juan CA, Pérez de la Lastra JM, Plou FJ, Pérez-Lebeña E (2021) The chemistry of reactive oxygen species (ROS) revisited: outlining their role in biological macromolecules (DNA, lipids and proteins) and induced pathologies. Int J Mol Sci 22(9):4642. https://doi.org/10.3390/ijms22094642
Khan AU, Akram M, Daniyal M, Zainab R (2019) Awareness and current knowledge of Parkinson's disease: a neurodegenerative disorder. Int J Neurosci 129(1):55–93
Kim I-S (2021) Current perspectives on the beneficial effects of soybean Isoflavones and their metabolites for humans. Antioxidants 10(7):1064. https://doi.org/10.3390/antiox10071064
Kim JY, Kawabori M, Yenari MA (2014) Innate inflammatory responses in stroke: mechanisms and potential therapeutic targets. Curr Med Chem 21(18):2076–2097
Klaips CL, Jayaraj GG, Hartl FU (2018) Pathways of cellular proteostasis in aging and disease. J Cell Biol 217(1):51–63
Kopustinskiene DM, Jakstas V, Savickas A, Bernatoniene J (2020) Flavonoids as anticancer agents. Nutrients 12(2):457. https://doi.org/10.3390/nu12020457
Limmongkon A, Pankam J, Somboon T, Wongshaya P, Nopprang P (2019) Evaluation of the DNA damage protective activity of the germinated peanut (Arachis hypogaea) in relation to antioxidant and anti-inflammatory activity. LWT 101:259–268
Littarru GP, Tiano L (2007) Bioenergetic and antioxidant properties of coenzyme Q10: recent developments. Molecularbiotechnology, 37(1):31–37. https://doi.org/10.1007/s12033-007-0052-y
Ma DS, Tan LTH, Chan KG, Yap WH, Pusparajah P, Chuah LH et al (2018) Resveratrol—potential antibacterial agent against foodborne pathogens. Front Pharmacol 9:102
Maiani et al (2009) Carotenoids: actual knowledge on food sources, intakes, stability and bioavailability and their protective role in humans. Molecular nutrition and food research, 53 Suppl 2:S194–S218. https://doi.org/10.1002/mnfr.200800053
Manchali S, Murthy KNC, Patil BS (2012) Crucial facts about health benefits of popular cruciferous vegetables. J Funct Foods 4(1):94–106
Mandl J, Szarka A, Banhegyi G (2009) Vitamin C: update on physiology and pharmacology. British j of pharmacology, 157(7):1097–1110
Marić M, Grassino AN, Zhu Z, Barba FJ, Brnčić M, Brnčić SR (2018) An overview of the traditional and innovative approaches for pectin extraction from plant food wastes and by-products: ultrasound-, microwaves-, and enzyme-assisted extraction. Trends Food Sci Technol 76:28–37

Martemucci G, Costagliola C, Mariano M, D'andrea L, Napolitano P, D'Alessandro AG (2022) Free radical properties, source and targets, antioxidant consumption and health. Oxygen 2(2):48–78. https://doi.org/10.3390/oxygen2020006

Miazek K, Beton K, Śliwińska A, Brożek-Płuska B (2022) The effect of β-carotene, tocopherols and ascorbic acid as anti-oxidant molecules on human and animal in vitro/in vivo studies: a review of research design and analytical techniques used. Biomolecules 12(8):1087. https://doi.org/10.3390/biom12081087

Mokra D, Adamcakova J, Mokry J (2022) Green tea polyphenol (−)-Epigallocatechin-3-Gallate (EGCG): a time for a new player in the treatment of respiratory diseases? Antioxidants 11(8):1566. https://doi.org/10.3390/antiox11081566

Murillo G, Mehta RG (2001) Cruciferous vegetables and cancer prevention. Nutr Cancer 41(1–2):17–28

Nakai K, Tsuruta D (2021) What are reactive oxygen species, free radicals, and oxidative stress in skin diseases? Int J Mol Sci 22(19):10799. https://doi.org/10.3390/ijms221910799

Nakhate KT, Badwaik H, Choudhary R, Sakure K, Agrawal YO, Sharma C, Ojha S, Goyal SN (2022) Therapeutic potential and pharmaceutical development of a multitargeted flavonoid phloretin. Nutrients 14(17):3638. https://doi.org/10.3390/nu14173638

Park JE, Han JS (2023) Scopoletin protects INS-1 pancreatic β cells from glucotoxicity by reducing oxidative stress and apoptosis. Toxicol In Vitro 93:105665

Pietta PG (2000) Flavonoids as antioxidants. J of natural products, 63(7):1035–1042

Pinheiro RGR, Pinheiro M, Neves AR (2021) Nanotechnology innovations to enhance the therapeutic efficacy of quercetin. Nanomaterials (Basel, Switzerland) 11(10):2658. https://doi.org/10.3390/nano11102658

Rai RK, Chalana A, Karri R, Das R, Kumar B, Roy G (2019) Role of hydrogen bonding by thiones in protecting biomolecules from copper (I)-mediated oxidative damage. Inorg Chem 58(10):6628–6638

Raj P, Thandapilly SJ, Wigle J, Zieroth S, Netticadan T (2021) A comprehensive analysis of the efficacy of resveratrol in atherosclerotic cardiovascular disease, myocardial infarction and heart failure. Molecules 26(21):6600. https://doi.org/10.3390/molecules26216600

Rasmus P, Kozłowska E (2023) Antioxidant and anti-inflammatory effects of carotenoids in mood disorders: an overview. Antioxidants (Basel, Switzerland) 12(3):676. https://doi.org/10.3390/antiox12030676

Rosales TKO, Fabi JP (2023) Valorization of polyphenolic compounds from food industry by-products for application in polysaccharide-based nanoparticles. Front Nutr 10:1144677

Santos KL, Bragança VA, Pacheco LV, Ota SS, Aguiar CP, Borges RS (2022) Essential features for antioxidant capacity of ascorbic acid (vitamin C). J Mol Model 28:1–8

Scott, R. E., Wille Jr, J. J., & Wier, M. L. (1984). Mechanisms for the initiation and promotion of carcinogenesis: a review and a new concept. In Mayo Clinic proceedings (Vol. 59, No. 2, pp. 107–117). Elsevier

Shahidi F, Ambigaipalan P (2015) Phenolics and polyphenolics in foods, beverages and spices: Antioxidant activity and health effects–A review. J of functional foods, 18:820–97

Traber MG, Stevens JF (2011) Vitamins C and E: beneficial effects from a mechanistic perspective. Free radical biology and medicine, 51(5):1000–1013

Valgimigli L (2023) Lipid peroxidation and antioxidant protection. Biomolecules 13(9):1291. https://doi.org/10.3390/biom13091291

Varoni EM, Lo Faro AF, Sharifi-Rad J, Iriti M (2016) Anticancer molecular mechanisms of resveratrol. Front Nutr 3:8

Venditti P, Di Stefano L, Di Meo S (2013) Mitochondrial metabolism of reactive oxygen species. Mitochondrion 13(2):71–82

Waris G, Ahsan H (2006) Reactive oxygen species: role in the development of cancer and various chronic conditions. J Carcinog 5:14. https://doi.org/10.1186/1477-3163-5-14

Wu D, Cederbaum AI (2003) Alcohol, oxidative stress, and free radical damage. Alcohol Res Health 27(4):277–284

Wu G, Fang YZ, Yang S, Lupton JR, Turner ND (2004) Glutathione metabolism and its implications for health. The J of nutrition, 134(3):489–492. https://doi.org/10.1093/jn/134.3.489
Zhang L, Han Z, Granato D (2021a) Polyphenols in foods: classification, methods of identification, and nutritional aspects in human health. Adv Food Nutr Res 98:1–33
Zhang Z, Shi J, Nice EC, Huang C, Shi Z (2021b) The multifaceted role of flavonoids in cancer therapy: leveraging autophagy with a double-edged sword. Antioxidants 10(7):1138. https://doi.org/10.3390/antiox10071138
Zorov DB, Juhaszova M, Sollott SJ (2014) Mitochondrial reactive oxygen species (ROS) and ROS-induced ROS release. Physiol Rev 94(3):909–950

Natural Compounds as Epigenetic Modulators to Treat Neurodegenerative Disorders

8

Mohammad Rumman and Mohsin Vahid Khan

Abstract

Neurodegenerative disorders largely affect individuals at an advanced age; however, early-onset neurodegenerative diseases have also been reported. Alzheimer's disease (AD) is the most common neurodegenerative disorder that affects millions of people worldwide, along with Huntington's disease, Parkinson's disease (PD), multiple system atrophy, and progressive supranuclear palsy. The etiology and pathological mechanisms of neurodegenerative disorders are diverse; however, dysregulated epigenetic modifications have frequently been observed in these disorders. In AD, DNA methylation is extensively suppressed in the neuronal cells of the affected regions. Similarly, abnormal DNA methylation pattern has been associated with dopaminergic neuron degeneration in PD. Besides, alterations in histone modification patterns have also been reported in the brains of animal models as well individuals with neurodegenerative disorders. Thus, targeting the epigenetic regulators is one of the strategies for managing and treating neurodegenerative diseases. Plant-derived natural compounds hold an immense potential in the field of drug discovery, as these compounds are abundantly available and exhibit minimal to no side effects. The anticancer, anti-diabetic, anti-inflammatory, cardio-protective, hepato-protective, and anti-obesity qualities of these substances have all been thoroughly investigated. Mechanistically, these natural compounds can modulate diverse cellular processes, including transcription, oxidative stress, translation, cellular trafficking, and post-translational modification. A large group of herbal natural compounds are shown to regulate the activity of several epigenetic regulators. By modulating the activity of specific epigenetic regulators, these natural compounds can revert the epigenetic signatures, thereby preventing neurodegenerative disease progres-

M. Rumman (✉) · M. V. Khan
Department of Biosciences, Integral University, Lucknow, Uttar Pradesh, India
e-mail: rumman@iul.ac.in

S. Jahan et al. (eds.), *A Mechanistic Exploration of Natural Compounds for Neuronal Health*, https://doi.org/10.1007/978-981-95-0125-0_8

sion. This chapter discusses the possible role of natural compounds as epigenetic modulators in neurodegenerative conditions.

Keywords
Epigenetics · Neurodegeneration · Phytochemicals · Histones · DNA methylation

8.1 Introduction

The goal of the quickly expanding field of epigenetics research is to comprehend how structural variations in nucleotide bases and related proteins, independent of DNA sequence, control gene expression. Traditionally, it was believed that the genetic code that makes us who we are is fixed and unchangeable. However, epigenetics has shown that our genes are not always set in stone and can be affected by external factors, most importantly the environment (Weinhold 2006; Allis and Jenuwein 2016). Whereas genetic changes are irreversible, studies have shown that epigenetic changes are generally reversible. Gene expression can be greatly impacted by epigenetics, which may contribute to the onset of illnesses or have an effect on our physical and mental health (Moosavi and Motevalizadeh Ardekani 2016; Fuggle et al. 2022). Chemical changes to the nitrogenous bases found in DNA, known as histones, are one of the main processes behind epigenetic regulation of gene expression. These modifications can either activate or suppress gene expression, ultimately determining which proteins are produced in our cells. Epigenetic regulation is essential for the proper functioning of our cells, and alternations in epigenetics mechanisms have been observed in several pathological conditions, including cardiovascular diseases, cancer, neurodegenerative diseases, and inflammation, suggesting its potential role in the onset and progression of diseases (Farsetti et al. 2023; Mazzone et al. 2019; Moosavi and Motevalizadeh Ardekani 2016). Environmental influences on our genes have also been clarified by epigenetics. Research indicates that environmental factors, including stress, diet, and exposure to toxins, can result in epigenetic modifications that can be inherited by subsequent generations. This process is known as trans-generational epigenetic inheritance and has been observed in animals and humans. For instance, research has demonstrated that feeding pregnant mice a high-fat diet can alter the offspring's epigenetic makeup, raising the risk of obesity and other metabolic diseases (Harmancıoğlu and Kabaran 2023).

8.2 Epigenetic Modifications

DNA methylation, the most prevalent epigenetic alteration, is the addition of a methyl group to particular DNA bases (Fig. 8.1), typically leading to the silencing of the corresponding gene (Kiselev et al. 2021). DNA methyl transferases (DNMTs) catalyze the addition of methyl groups to specific cytosine residues within the cytosine-guanine sequence in the DNA, commonly known as the CpG island. These

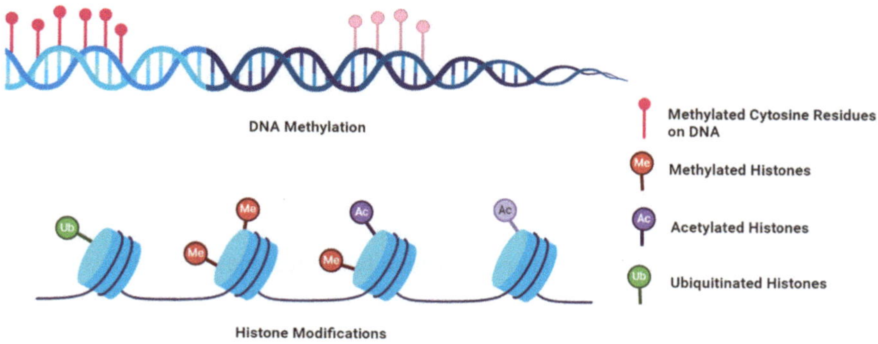

Fig. 8.1 Various types of epigenetic modifications on DNA bases and histone residues

CpG islands are commonly dispersed throughout the promoter regions of genes. Methylated cytosines inhibit the interaction between the transcription factors and DNA while recruiting transcriptional suppressors to inhibit transcription (Medvedeva et al. 2014; Kiselev et al. 2021). DNA methylation can be reversed through a process called demethylation and typically involves the TET family of enzymes. DNA methylation generally causes gene expression to be "OFF," whereas demethylation causes it to be "ON." Numerous pathological conditions have been linked to abnormal DNA methylation, for example, it has been discovered that the promoters of proto-oncogenes are typically hypomethylated and those of tumor suppressor genes are typically hypermethylated in a variety of cancer types, both of which contribute to carcinogenesis (C. Chen et al. 2022).

Another epigenetic mechanism is histone modification, which involves acetylation, methylation, and phosphorylation of the histone proteins that package and organize DNA (Fig. 8.1) (Y. Zhang et al. 2021). Gene expression may also be impacted by these processes, and anomalies in histone modifications have been connected to a number of illnesses, such as mental illnesses, cardiovascular conditions, and cancer (Santana et al. 2023; Alaskhar Alhamwe et al. 2018). DNA is tightly bound to histones to form structures known as nucleosomes. Histones and DNA interact to control how transcription regulatory proteins approach the promoter and regulator regions. When histones are bound tightly to DNA, proteins that initiate transcription cannot access the DNA, turning the gene expression to the "OFF" state. When histones are loosely packed, more DNA is exposed and can be accessed by proteins that initiate the gene transcription, turning the gene expression "ON." Specific amino acid restudies of histone proteins can be chemically modified to alter the nucleosome structure (Marmorstein and Zhou 2014). For example, the addition of acetyl groups to the positively charged lysine 27 and lysine 9 residues of histone H3 (H3K27ac and H3K9ac) weakens the histone-DNA interactions, thereby loosening the nucleosome and facilitating the recruitment of transcriptional machinery. Similarly, methylation of various lysine and arginine residues is also an epigenetic mechanism involved in modulating transcriptional activity. Only mono- or di-methylation is possible for arginine, but lysine can be mono-, di-, or tri-methylated.

These histone methylations have varied effects on nucleosome structure and gene expression. When methyl groups are added to the lysine 4 residue on histone H3 (H3K4me), transcription is activated; when lysine 27 residue on histone H3 is trimethylated (H3K27me3), transcription is suppressed (Marmorstein and Zhou 2014; Smith and Denu 2009). Phosphorylation of histones is another epigenetic modification; however, the significance of this modification is not well understood. Histone H2A(X) phosphorylation is crucial for DNA damage response and repair (Rossetto et al. 2012). Recently, ubiquitination, a novel histone modification has been reported. The addition of ubiquitin monomers to histone lysine residues is known as ubiquitylation, and it is facilitated by ubiquitin ligase E3, ubiquitin-activating enzyme E1, and ubiquitin-conjugating enzyme E2. Ubiquitylation of histone H3 lysine 119 residue inhibits the transcription process, while ubiquitylation of histone H2 lysine 123 residue enhances transcriptional activity (J. Cao and Yan 2012).

Another important mechanism involved in regulating epigenetics is through non-coding RNAs (ncRNAs), particularly microRNAs (miRNAs). Within the human genome, approximately 2% encodes proteins, while the remaining 98% is converted into regulatory non-coding RNAs, such as small non-coding RNAs (20 nts) which include miRNAs, siRNAs, piRNAs, and other types of long non-coding RNAs (Avtandilov et al. 1983). The ncRNAs that have been researched the most are microRNAs, which target the three-prime end of mRNAs, causing their suppression or degradation. MicroRNAs induce mRNA degradation by binding imperfectly to the 3′-untranslated region and inhibiting translation through the formation of an RNA-induced silencing complex. Some microRNAs can enhance gene expression by promoting the translation of mRNA and initiating the activation of the target gene by attaching to its promoter.

8.3 Epigenetics and Neurological Complications

Epigenetic changes are important players in regulating the expression of several genes. Cellular epigenetic mechanisms ensure that only proteins that are required are produced while those which are not required are not produced with cells. It has been observed in several pathological conditions that epigenetic regulatory mechanisms are perturbed which leads to disease development and progression (Fig. 8.2a) (Tzika et al. 2018; M. P. Fu et al. 2023; Zoghbi and Beaudet 2016; Santana et al. 2023; Alaskhar Alhamwe et al. 2018). The next sections describe the observed epigenetic changes in neurological disorders and the role and underlying mechanism of natural compounds in modulating epigenetic regulators in a wide array of neurodegenerative conditions (Fig. 8.2b).

There are several studies on the methylation patterns of DNA in the brain tissues of patients with Alzheimer's disease (AD). Increased levels of DNA methylation have been observed in the middle frontal and temporal gyrus of patients with AD. Furthermore, hypermethylation of ankyrin 1 (*ANK1*), ATP binding cassette subfamily A member 7 (*ABCA7*), and bridging integrator 1 (*BIN1*) has also been reported in the affected regions of AD brains (P. L. De Jager et al. 2014). Compared

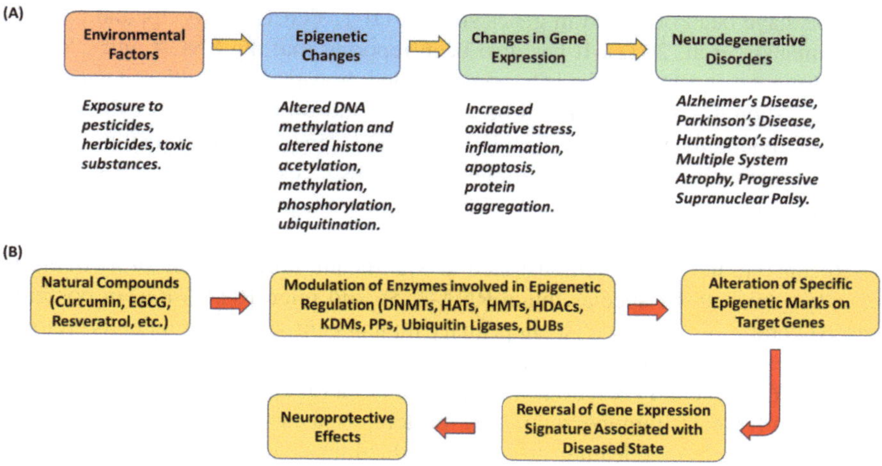

Fig. 8.2 (**a**) Environmental factors can alter the epigenetic signature of cells, leading to changes in gene expression which ultimately causes disease development/progression. (**b**) Natural compounds can modulate the activity of epigenetic regulatory enzymes which in turn can fine tune the expression of genes, leading to neuroprotective effects. EGCG, epigallocatechin-3-gallate; DNMTs, DNA methyltransferases; HATs, histone acetyl transferases; HMTs, histone methyl transferases; HDACs, histone deacetylases; KDMs, lysine demethylases; PPs, protein phosphatases; DUBs, deubiquitinating enzymes

to healthy controls, significant differences in the DNA methylation patterns of amyloid beta precursor protein (*APP*), microtubule-associated protein tau (*MAPT*), and glycogen synthase kinase-3 beta (*GSK3B*) have been observed in the inferior temporal lobe, superior parietal lobe, and cerebellum of AD individuals (Iwata et al. 2014). Modifications in the histone proteins are shown to be correlated with cognitive functions, synaptic plasticity, neurogenesis, as well as cell survival. It has been reported that H3K27 acetylation levels were increased in the entorhinal cortex of AD patient brains (Marzi et al. 2018). In the temporal lobe of patients with AD, H3K27 and H3K9 acetylation levels were increased while the levels of H3K122 acetylation were reduced, which affected the expression of genes that regulate normal physiological roles of neurons. Similarly, in a transgenic mice model of AD, histone H4 acetylation levels were significantly reduced in the hippocampus and frontal cortex, which correlates with compromised spatial learning and memory impairment in these animals (T. Cao et al. 2018).

Saliva, whole blood, leukocytes, and post-mortem brain tissues from Parkinson's disease (PD) patients have all been shown to have aberrant methylation patterns (Chuang et al. 2017; Masliah et al. 2013; Kaut et al. 2022). In the blood of PD patients, seven differentially expressed genes linked to 62 differentially methylated CpG sites have been found (Henderson et al. 2021). These differentially methylated genes were found to be associated with motor and cognitive functions and were involved in regulating Wnt signaling, apoptosis, mitochondrial functions, and synaptic plasticity. In the parietal cortex of PD patient brains, changed methylated patterns have been observed in genes such as DJ-1, VGLUT2, IA-2β, and NURR1 (Kochmanski et al. 2022). It has also been reported that the SNCA and CYP2E1 gene promoter in PD samples is hypomethylated; however, the evidence is not convincing. Though less research has been done on the subject, there is evidence that suggests dysregulated histone acetylation in PD brains. Increased histone acetylation has been observed in PD samples with H3K27 exhibiting the most notable enhancement, which is consistent with other neurodegenerative diseases, such as late-onset AD (Marzi et al. 2018; Toker et al. 2021).

Recent studies have demonstrated a significant upregulation of two brain-abundant microRNA species, microRNA-125b and microRNA-146a, in postmortem AD brain tissues as well as in primary human neuronal and glial cells cultured under stressful conditions in the presence of proinflammatory cytokines and Aβ42 peptide (Yashooa and Nabi 2022). Complement factor H, the major inflammation-related and innate-immune regulatory protein, is a target of these miRNAs in NF-κB-activated cells, leading to notable reductions in the expression of complement factor H. Evidence suggests that human neuronal-glial cells significantly upregulate microRNA-125b and microRNA-146a in response to IL-1β and Aβ42-peptide-induced stress (Alexandrov et al. 2014; Banzhaf-Strathmann et al. 2014). These studies confirm that epigenetics play an important role in the development and progression of neurodegenerative disorders.

8.4 Natural Compounds as Epigenetic Regulators in Neurodegenerative Disorders

Nature's vast chemical diversity provides a rich source of potential therapeutic agents, including those that can modulate epigenetic processes. It has been discovered that certain natural substances obtained from fungi, plants, and marine life have epigenetic activity, meaning they can change the expression of microRNA, histone modifications, and DNA methylation (Fig. 8.3). These compounds offer several advantages over synthetic drugs, including natural origin, diverse chemical structures, and traditional use. They are often associated with fewer side effects and improved tolerability compared to synthetic compounds and offer a wide range of potential targets and mechanisms of action, increasing the likelihood of identifying effective therapies. Many natural compounds have a history of traditional medicinal use, providing a foundation for their safety and efficacy. The next section describes selected natural compounds with epigenetic potential in various disease models.

8.4.1 Curcumin

Curcumin can alter the activity of some enzymes that control epigenetic modifications, including histone acetyl transferases (HATs), histone deacetylase 1 (HDAC1), HDAC3, and HDAC8, according to in vitro research (S. Fu and Kurzrock 2010; Ming et al. 2022; Hassan et al. 2019). A growing body of evidence suggests that

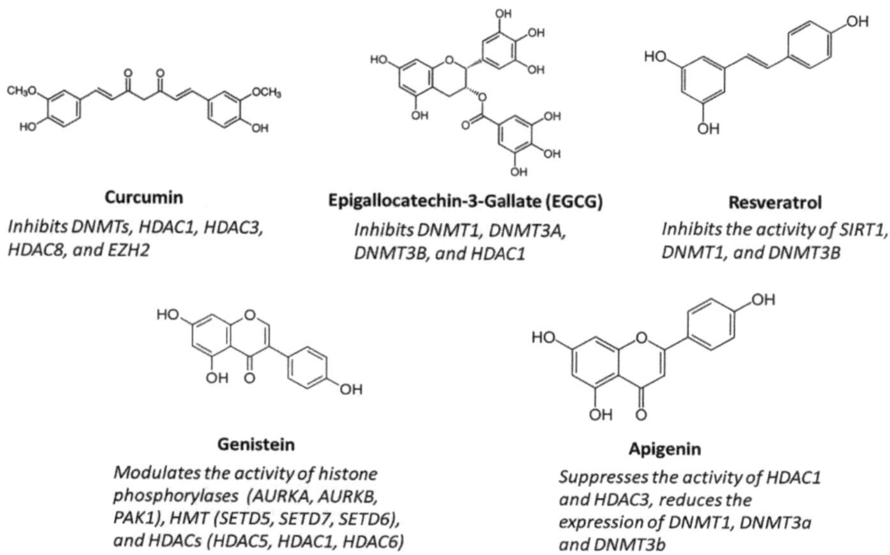

Fig. 8.3 Chemical structure of some natural compounds that can be used as epigenetic modulators

curcumin can inhibit DNMT, thereby inhibiting DNA hypermethylation in leukemia MV4–11 cells and suppressing tumor cell proliferation (Ming et al. 2022). By inhibiting DNMT1, curcumin suppresses methylation in *DLC1* promoter which consequently suppresses the expression of DLC1 in breast cancer cells and inhibits its growth. Interestingly, curcumin has also been shown to increase the methylation in the promoter region of the oncogenic genes, including mTOR (mammalian target of rapamycin) gene in RPMI-8226 and NCI-H929 myeloma cells which resulted in suppression of cancer cell proliferation (J. Chen et al. 2019). Previous studies have shown that curcumin can inhibit the expression of HDAC1, HDAC3, and HDAC8, which consequently leads to increased histone H4 acetylation (Sanaei et al. 2020; Guo et al. 2021). Moreover, curcumin can also bind to p300 histone acetyl transferase to suppress its activity, ultimately leading to altered acetylation of H3 and H4.

By modulating histone acetylation, curcumin can enhance the expression of tumor suppressor genes in cancer cells. Curcumin has the ability to suppress the growth and migration of term triple-negative breast cancer cells by blocking the activity of the EZH2 chromatin remodeler, which results in increased trimethylation at histone H3 lysine 27 of the DLC1 promoter (Zhou et al. 2020).

These studies confirm the epigenetic regulatory role of curcumin in mammalian cells which could be further exploited in treating neurodegenerative disorders. In fact, several preclinical studies have shown that curcumin epigenetically provides neuroprotection ex vivo and in vitro, as well as in preclinical models. Curcumin reduces the production of amyloid aggregates by lowering the methylation levels of histone H3 at lysine 27 (H3K27me3) near the BDNF promoter region in an AD mouse model (J. Li et al. 2021). Moreover, a previous study in another AD model showed that curcumin can modulate the activity of H3K27 demethylase Jumonji domain-containing protein 3 (JMJD3) to exert the neuroprotective effect via the BDNF pathway (Wang et al. 2022). BDNF is essential for several neurological functions, including brain development, synaptic plasticity, and neuron survival, and it has been shown that exogenous BDNF supplementation could enhance mitochondrial function. Curcumin treatment decreases H3K27me3 levels at the BDNF promoter region to enhance BDNF expression and prevent Aβ aggregates from building up. Curcumin can control the activities of HAT and HDAC enzymes to regulate the expression of genes. Previous studies have shown that the expression of genes related to memory and learning can be regulated by histone acetylation (Peixoto and Abel 2013). In the case of familial AD, curcumin can selectively suppress H3 acetylation in the promoter region of AD-related genes PS1 and beta-site APP cleaving enzyme (BACE1), thereby reducing their gene expression (Lu et al. 2014). It is well known that specific brain regions and cell types are affected differently by changes in DNA methylation patterns associated with AD. Curcumin can inhibit the activity of DNMT1, which lowers the levels of DNA methylation levels globally. Moreover, curcumin analogues can increase the activity of neprilysin (NEP), an Aβ-degrading enzyme, by inhibiting the activity of DNMT1 (P.-T. Chen et al. 2016). Dysregulation of Wnt signaling pathways has been linked to AD pathology, including tau phosphorylation, increased Aβ aggregation, and inflammation. In animal models of neurodegenerative disorders, curcumin has been shown to modulate the Wnt pathway to exert neuroprotective effects (Vallée 2022). Curcuminoid compounds like dimethoxy-curcumin and bisdemethoxycurcumin can demethylate the WIF-1 promoter region by directly suppressing the activity of DNMT1, thus blocking the canonical Wnt pathway (Liu et al. 2011). Given its potent ability to regulate epigenetic machinery, curcumin-based formulations can be tested clinically for the development of drugs to treat neurodegenerative diseases. However, further studies in advanced animal models are required to delineate precise molecular mechanisms since unwanted epigenetic changes in the genome might be deleterious.

8.4.2 Resveratrol

Resveratrol is a polyphenol naturally found in grapes, red wine, and other plant sources. It has gained significant attention for its potential health benefits, including neuroprotective properties. Resveratrol exerts neuroprotective effects by acting as an antioxidant, anti-inflammatory, mito-protective, anti-apoptotic, and neurotrophic agent (Capiralla et al. 2012; Kumar et al. 2006; Moyano-Mendez et al. 2014). Resveratrol has been found to reduce amyloid-beta plaque formation, improve cognitive function, and protect against neuronal loss in animal models of AD (Rahman et al. 2020). Resveratrol may slow the progression of PD by protecting against oxidative damage and neuroinflammation.

Resveratrol can induce epigenetic modifications by directly regulating the activities of DNMTs, HDACs, and HATs, which ultimately affect the expression of several important genes expressed in various tissues, including the brain (Izquierdo et al. 2021; Kurzava Kendall et al. 2024; Kumar et al. 2006). In breast cancer cells, resveratrol can suppress the activity and reduce mRNA levels of *DNMT1*, *DNMT3A*, and *DNMT3B*, thereby significantly reducing genome-wide DNA methylation. Resveratrol can reduce DNMT1 activity specifically at the BRCA1 promoter which ultimately reduces the DNA methylation and increases the expression of this tumor suppressor to induce cell death (Papoutsis et al. 2010).

One of the most studied gene that is epigenetically regulated by resveratrol is SIRT1. Resveratrol can activate SIRT1 in animal models of neurodegeneration, thereby suppressing NF-κB signaling, the major pathway regulating neuroinflammation (Ma et al. 2019). Moreover, activation of SIRT1 prevents β-amyloid-induced microglial death. Further studies are required to validate the neuroprotective ability of this epigenetic modulator before clinical assessments in patients. Interestingly, epigenetic inheritance has been reported in resveratrol-supplemented mice. Offspring of such mice exhibited significant increase in the global methylation of DNA along with improved cognitive functions and differential expression of DNMT3a/b and Tet2, supporting the notion that resveratrol can modulate epigenetic changes to exert neuroprotective effects (Izquierdo et al. 2019; Isac et al. 2017). Another research highlighted that offspring of rats supplemented with resveratrol during perinatal asphyxia demonstrated improved memory functions due to reduced expression of inflammatory markers, such as TNF-α, S-100 calcium-binding protein B (S-100B), and interleukin-1β (IL-1β), the expression of which is controlled by particular epigenetic mechanisms (Isac et al. 2017). These studies confirm that the epigenetic modulation induced by resveratrol is heritable, which is beneficial in some hereditary neurodegenerative diseases. Previous studies have shown that ε-viniferin, a substance molecularly similar to resveratrol, increases the expression of SIRT3 and deacetylation of forkhead box O3 (FOXO3) gene (S. Zhang et al. 2020), suggesting that structurally similar compounds can also be tested and developed as potential drugs.

8.4.3 Apigenin

Flavonoids such as apigenin (4′,7-trihydroxyflavone) are widely distributed in a number of herb parts, including leaves, stem, fruits, roots, and vegetables. A few examples of these include thyme, chamomile, onion, parsley, and oranges. It has been demonstrated that apigenin alters the histone signatures in a variety of illnesses, including neurodegenerative conditions (Salehi et al. 2019). By lowering HDAC activity and reviving miRNA expression, it achieves this. Apigenin improves general cognitive functions by reducing neuroinflammation (Ahmedy et al. 2022). In vitro studies have shown that apigenin can suppress the expression of DNMT1, DNMT3a, DNMT3b, and HDACs1–8 in the epidermal cancer cell line JB6 P+ (Paredes-Gonzalez et al. 2014). In animal models of postoperative cognitive dysfunction (POCD) and PD, apigenin administration can reduce oxidative stress-induced neuronal apoptosis and neuroinflammation (L. Chen et al. 2017; Anusha et al. 2017). NF-kB gene expression has been shown to be upregulated, and inflammation in the substantia nigra pars compacta (SNpc) was inhibited by this flavonoid. Additionally, in PD models, it inhibits the production of cytokines that cause inflammation, including IL-β, TNF-α, IL-6, caspase-1, and the enzyme inducible nitric oxide synthase (iNOS-1). It also suppresses the expression of IL-4, IL-2, and IL-10 in models of POCD (Patel and Singh 2022; L. Chen et al. 2017; Anusha et al. 2017). Moreover, apigenin treatment lowers α-synuclein levels while upregulating dopamine D2 receptor and tyrosine hydrolase (TH) protein expression, suggesting that apigenin possesses neuroprotective capabilities and can be used as an alternative treatment for neurodegenerative diseases such asPD (L. Chen et al. 2017; Anusha et al. 2017). The acetylation of histones and changes in the expression of genes linked to normal cognitive functions are common outcomes of neurodegenerative disorders. BDNF is one of the histone acetylation target genes that helps restore normal cognitive functions. In AD, recombinant BDNF can increase hippocampal synaptic density and cognitive function. Differential histone acetylation impacts BDNF expression levels (Ji et al. 2014). According to reports, apigenin can enhance the acetylation of histone 3 lysine 9 and histone 4 lysine 12 at the promoter region of the brain-derived neurotrophic factor gene, leading to increased gene transcription (L. Chen et al. 2017). This, in turn, can lead to the activation of downstream signaling pathways, such as extracellular signal-regulated kinase (p-ERK), cAMP response element-binding protein (p-CREB), and calcium/calmodulin-stimulated protein kinase II (p-CAMK II) (L. Chen et al. 2017). The acetylation of histones is essential for healthy memory functions in addition to cognitive functions. Aging-related changes in histone acetylation and chromatin plasticity are frequently linked to memory decline. Apigenin stimulates H3K9 and H4K12 acetylation by reducing HDAC activity and HDAC1 and HDAC3 protein expression in various tissues. Apigenin has been shown to be essential for controlling histone acetylation in the hippocampal regions. This is demonstrated by apigenin's ability to restore the levels of acetylation of both histone 3 and histone 4 via the downregulation of HDAC2 and upregulation of CREB-binding protein (CBP) (Peleg et al. 2010).

There is a correlation between miRNAs and the control of neurogenesis and memory. Since AD brains have been demonstrated to have downregulated miR-15a, increased expression of miR-132, miR-15a, and miR-219 can improve general cognitive functions and spatial/navigational memory in a mice model of AD (Yang et al. 2020; Taha et al. 2023). Rho-associated protein kinase-1 (ROCK-1) controls microglial activation and has downregulatory effects on ERK1/2, CREB, and BDNF. This is important for the translation of pro-inflammatory cytokines in microglia and their release (Koch et al. 2018; Hensel et al. 2015). The microRNA miR-15a reduces the expression of ROCK-1 gene in hippocampal neuronal cells, which lowers aberrantly increased tau phosphorylation and neuronal cell death, according to a study on AD (Koch et al. 2018). Apigenin was found to have a neuroprotective effect because it dramatically raised the expression of the miR-15a gene, lowered the expression of ROCK-1, reversed the activation of microglia, and reduced ionization. Notably, apigenin may 1 day be used as a treatment option for neurodegenerative illnesses like PD and AD through epigenetic therapy.

8.4.4 Genistein

Products associated with soybeans often contain genistein (4′,5′,7-trihydroxyisoflavone), a flavonoid belonging to the isoflavone class. Many studies have demonstrated its potent anticancer properties; it functions as an inhibitor of the tyrosine kinase activity of the epidermal growth factor receptor (Subedi et al. 2017). Apart from its anticancer effects, genistein has also been shown to have anti-disease effects against various disorders of the central nervous system (CNS), including dementia, AD, PD, depression, and epilepsy (Viña et al. 2022; M. Chang et al. 2021; Arbabi et al. 2016). In vitro and in vivo, genistein mimics neuroprotective effects through its anti-inflammatory activity, showing promising blood-brain barrier (BBB) permeability (Fuloria et al. 2022).

Genistein can specifically inhibit the activity of DNMT1 in the promoter regions of several tumor suppressor genes such as adenomatous polyposis coli (APC), ataxia telangiectasia mutated (ATM), phosphatase and tensin homolog (PTEN), and mammary serpin peptidase inhibitor (SERPINB5), thereby lowering global DNA methylation levels in mammalian cells (Xie et al. 2014). In mice, dietary genistein can decrease DNA methylation at the BRCA1 gene's promoter region in a dose-dependent manner, increasing BRCA1 expression (Donovan et al. 2019). Furthermore, genistein has the ability to control the expression and activity of histone phosphorylases (AURKA, AURKB, PAK1), HMT (SETD5, SETD7, SETD6), and HDACs (HDAC5, HDAC1, HDAC6), which in turn controls the epigenetic configuration of cells (Sundaram et al. 2019).

Numerous studies on neural cells have demonstrated the neuroprotective properties of estrogens. However, there are possible risks associated with estrogen therapy, including an enhanced risk for certain tumors and damage to patients who are prone to forming blood clots (Lobo 1995). Thus, further studies to discover phytoestrogens that mimic estrogen but lack any harmful properties is required. Given its molecular resemblance to estrogen, genistein is acknowledged as a strong phytoestrogen that binds to estrogen receptors α and β (ERα and Erβ) as an agonist (Morito et al. 2001). Nigrostriatal dopaminergic neurons die as a result of oxidative stress and neuroinflammation. It has been discovered that genistein modifies inflammatory molecules to ameliorate the circumstances related to these processes. According to recent research, genistein protects neurons in a PD model involving human neuronal SH-SY5Y cells by acting on ERs and Nrf2 channels (Wu et al. 2018). Heme oxygenase 1 (HMOX1) is upregulated when Nrf2 is activated by genistein, providing neuronal protection in addition to glutathione peroxidase. Furthermore, it has been shown that genistein can mitigate or even reverse the harm and death of neuronal cells caused by mitochondrial oxidative stress (Y. Li et al. 2022a). Moreover, genistein reduces high levels of malondialdehyde (MDA) but does not change the levels of nitrite or the functioning of superoxide dismutase (SOD), suggesting that it can reverse memory impairments brought on by Aβ by reducing oxidative stress (Sutrisno et al. 2022). Similar effects of genistein on cerebral infarction, neuronal damage, and apoptosis are reported in a study by Li et al. in neonatal hypoxic-ischemic brain damage mice (Y. Li et al. 2022b). By activating the Nrf2 and HMOX-1 pathways and attenuating the nuclear-factor kappa B inflammatory pathway, this phytoestrogen reduces oxidative stress and neuroinflammation and completely eliminates inflammation-inducing cytokines such as IL-6, IL-1β, and TNF-α (Y. Li et al. 2022b).

Few studies have examined how genistein affects memory and the expression of genes in the brain through epigenetic regulation. When it comes to interaction with the G protein-coupled receptor 30 (GPR30) and improving the working of spatial memory, genistein and 17β-estradiol have similar effects (Hammond et al. 2009). Additionally, it enhances the expression of the genes BDNF, miR-132, and insulin-like growth factor 1 (IGF-1) within the hippocampal region of the brain of rats that have had their ovaries removed (Habibi et al. 2017). Research indicates that angiogenesis, synaptic formation, and neuronal cell growth are related to miR-132 expression (Scott et al. 2012). Through the reduced expression of p250GAP, a protein that regulates Rac1-PAK-mediated dendrogenesis and spinogenesis and enhancement of actin polymerization in the perirhinal cortex's dendritic region, miR-132 contributes to memory and learning cognitive function (Impey et al. 2010; Scott et al. 2012). These results suggest that genistein can cause epigenetic changes by regulating miRNA expression and thus slow down the advancement of neurodegenerative illnesses.

8.4.5 Epigallocatechin-3-Gallate (EGCG)

Epigallocatechin-3-gallate (EGCG), one the main tea polyphenols, has been shown to inhibit DNMT (DNMT1, DNMT3A, and DNMT3B)-mediated methylation of DNA in a dose-dependent manner, resulting in DNA demethylation and enhanced gene expression (Fang et al. 2007). EGCG influences DNA methylation either directly or indirectly. It reduces the activity of DNMT by non-covalently binding with its active site. EGCG treatment enhances the acetylation (H3K9/14ac, H3ac) as well as methylation of histones (H3K4me3, H3K9me3) in human endothelial cells (Ciesielski et al. 2020). Furthermore, by blocking HMTs, EGCG aids in the post-translational modifications of histones (Link et al. 2010). As shown in neuronal cells, EGCG functions as an inhibitor of HDAC and suppresses the transcription of APP (Hu et al. 2015). Additionally, EGCG increased NEP expression in M146L cells while lowering Aβ levels in AD mouse models by inhibiting the activity of HDAC1 (X. Chang et al. 2015). Along with being an inhibitor, EGCG may also lessen HDAC1's competition with the APP intracellular domain (AICD) for interacting with the promoter sequence of NEP, increasing the induction of the neprilysin enzyme and lowering levels of APP and sAPPα (Nalivaeva and Turner 2019; Grimm et al. 2013).

8.4.6 Nosustrophine

Nosustrophine, a bioproduct produced from porcine, contains glutamic and aspartic acids; various vitamins including E, B3, B2, B3, and D3; minerals including iron, calcium, zinc, and magnesium; and other compounds including dopamine, noradrenaline/norepinephrine, dopamine, BDNF, serotonin, neuropeptide Y, and neurotrophic tyrosine kinase receptor type 3 (NTRK3/TrkC) (Martínez-Iglesias et al. 2022). Studies showed that nosustrophine targets the epigenetic machinery and controls the expression of genes related to AD by modifying DNA methylation, SIRT activity, and HDAC expression. The enzyme adenosyl-homocysteinase (AHCY), which regulates the methylation of DNA, and nicotinamide phosphoribosyltransferase (NAMPT), a compound that produces the NAD+ precursor and increases the activity of SIRT1, were found to be present in nosustrophine extracts by mass spectrometric analysis. Moreover, SIRT1 substrate histone H3 is deacetylated by nosustrophine. Similar to how it drives resveratrol's neuroprotective qualities, SIRT1 activation seems to be a major mechanism that enhances survival via reduction of Bax-dependent cell death and the upregulation of transcription factors that promote apoptosis. Thus, nosustrophine is a new bio-nutraceutical with epigenetic modulatory characteristics (epinutraceutical), and the latest studies suggest that it can exert epigenetic effects in animal models of neurodegenerative disorders (Martínez-Iglesias et al. 2022). With respect to AD, it is possible that nosustrophine could be used as a primary or early secondary interventional treatment; however, further studies are required for a more conclusive evidence.

8.4.7 Vitamins

Increased B vitamin uptake is shown to have neuroprotective effects in several degenerative models. Some of the underlying mechanisms of vitamins might be through modulation of epigenetic cellular machinery. It has been speculated that dietary deficiencies in vitamins can cause hyper-homocysteinemia, which in turn can cause a drop in the concentration of S-adenosylmethionine, a methyl donor required during DNA methylation. This, in turn, leads to demethylation of genes implicated in AD pathology, resulting in their overexpression and disease progression (Nalivaeva and Turner 2019).

Cognitive function, neurological disorders, and AD can all be linked to vitamin B12 deficiency (Sanchez-Mut and Gräff 2015). Supplementing with nutrients like B12, B6, and folate lowers homocysteine concentrations and prevents cognitive decline in people with mild cognitive impairment, specifically in those with elevated homocysteine levels, according to a randomized controlled trial conducted in humans (Mohajeri et al. 2015). In people with normal folate and homocysteine levels, an immediate correlation among AD and low B12 and folate levels have been reported (C. A. de Jager et al. 2012). Dietary B12 deficiency can increase the expression of presenilin (PSEN1) and β-site APP cleaving enzyme (β-secretase) (BACE), which causes their gene promoters to become hypomethylated, resulting in Aβ deposition. BACE and PSEN1 are both associated with AD. Interestingly, it has been reported that the activity of DNMT1, DNMT3a, and DNMT3b is influenced by dietary B12 (Marques and Outeiro 2013).

According to certain population-based studies, dementia may be associated with low folate levels and hyperhomocysteinemia (Araújo et al. 2015). Folate deficiency was linked to the formation of NFT in human neuroblastoma cells through the dysregulation of protein phosphatase 2A (W. Li et al. 2015). In addition to a correlation between serum folate levels, methylation potential, and plasma S-adenosylmethionine concentrations, a recent study on patients with type 2 diabetes found a link between reduced folate and S-adenosylmethionine levels and impaired cognitive functions (Zheng et al. 2014). A study in postmenopausal women examining the relationship between red blood cell folate levels, global DNA methylation, and various biomarkers of the single carbon cycle metabolism, as well as between DNA methylation in leukocytes and required folate fortification produced some intriguing results. Instead of showing a linear relationship, the correlation between DNA methylation and folate levels displayed a reverse U-shaped curve. This is due to the fact that excessive consumption of folate leads to the accumulation of unmetabolized folate, which in turn leads to reduced methylation of DNA and negatively affects folate metabolism (Bae et al. 2014).

The antioxidant qualities of vitamins A, E, and C are widely known. These qualities are crucial for epigenetics because oxidative lesions impede DNA's capacity to act as the best DNMT substrate, which leads to global reduction in DNA methylation (Bollati et al. 2014). Histone acetylation mechanisms can also be altered by oxidation. In HDACs, increased levels of nitration of tyrosine residues and other redox modifications can tip the balance of acetylation of histone and other proteins toward enhanced acetylation even in highly oxidative conditions, which results in decreased deacetylase activity (Vadnal et al. 2012). Therefore, due to their antioxidant qualities, vitamins A, E, and C may lessen such aberrant histone acetylation modifications.

Sustaining appropriate levels of vitamin C in plasma may offer protection against AD (Mohajeri et al. 2015). Vitamin C is a cofactor for enzymes involved in DNA demethylation, such as TET2 (Minor et al. 2013). Demethylation of specific gene regions can lead to transcriptional activation and the induction of neuroprotective genes. Vitamin C can also affect histone modifications, such as histone acetylation and methylation, and can regulate the expression of non-coding RNAs, including miRNAs. Under normal pH, intracellular vitamin C exists as a negatively charged ascorbate, a cofactor for enzymes methylcytosine dioxygenases that mediate demethylation of DNA at specific sites. The enzymes that regulate the demethylation of histones, known as Jumonji C domain-containing histone demethylases, also require the ascorbate anion as a cofactor (Camarena and Wang 2016).

Significant research in humans suggests that vitamin D intake may lower the risk of developing AD (Mohajeri et al. 2015) and that vitamin D deficiency may be linked to the pathology of AD (Littlejohns et al. 2014; Sadowska-Bartosz and Bartosz 2014). Numerous mechanisms have been proposed to account for vitamin D neuroprotective benefits. Several mechanisms have been thoroughly studied, including antioxidative action, calcium homeostasis, neurotrophic regulation, Aβ metabolism, and anti-inflammatory action (Banerjee et al. 2015). Vitamin D epigenetic effects on AD have not yet been thoroughly investigated. Nonetheless, there are significant findings from research on other illnesses, such as cancer. It is interesting to note that since DNA methylation can silence important vitamin D synthesis genes, vitamin D influences epigenetics as well as epigenetics influences vitamin D. Vitamin D acts through the vitamin D receptor protein (VDR), and studies have shown that the expression of VDR can be epigenetically modified by histone acetylation. Usually, retinoid X receptors and VDR form heterodimers. The VDR-retinoid X receptor dimer may alter DNA methylation and influence gene transcription through its interactions with HATs. Consequently, it is thought that VDR ligands even control DNA methylation and alter gene expression (Fetahu et al. 2014; Karlic and Varga 2011). It is imperative that these effects are validated in AD models as well.

8.5 Conclusion

The potential of natural substances as epigenetic modulators offers a promising path toward the creation of innovative treatments for neurodegenerative illnesses. Their ability to influence gene expression, modulate cellular pathways, and regulate neuronal function offers a unique opportunity to address the complex pathogenesis of these debilitating disorders. The potential of natural compounds must be turned into effective patient treatments through more research aimed at comprehending their mechanisms of action, enhancing their bioavailability, and carrying out clinical trials. By exploring nature's vast chemical diversity, we can unlock new possibilities for combating neurodegenerative diseases and improving the lives of millions affected by these devastating conditions.

Acknowledgements The authors would like to thank the Department of Biosciences and Integral Center of Excellence for Interdisciplinary Research (ICEIR), Integral University, Lucknow, U.P., India.

References

Ahmedy OA, Abdelghany TM, El-Shamarka MEA, Khattab MA, El-Tanbouly DM (2022) Apigenin attenuates LPS-induced neurotoxicity and cognitive impairment in mice via promoting mitochondrial fusion/mitophagy: role of SIRT3/PINK1/parkin pathway. Psychopharmacology 239(12):3903–3917. https://doi.org/10.1007/s00213-022-06262-x

Alaskhar Alhamwe B, Khalaila R, Wolf J, von Bülow V, Harb H, Alhamdan F, Hii CS et al (2018) Histone modifications and their role in epigenetics of atopy and allergic diseases. Allergy, Asthma Clin Immunol 14(1):39. https://doi.org/10.1186/s13223-018-0259-4

Alexandrov PN, Dua P, Lukiw WJ (2014) Up-regulation of MiRNA-146a in progressive, age-related inflammatory neurodegenerative disorders of the human CNS. Front Neurol 5:181. https://doi.org/10.3389/fneur.2014.00181

Allis CD, Jenuwein T (2016) The molecular hallmarks of epigenetic control. Nat Rev Genet 17(8):487–500. https://doi.org/10.1038/nrg.2016.59

Anusha C, Sumathi T, Joseph LD (2017) Protective role of apigenin on rotenone induced rat model of Parkinson's disease: suppression of neuroinflammation and oxidative stress mediated apoptosis. Chem Biol Interact 269(May):67–79. https://doi.org/10.1016/j.cbi.2017.03.016

Araújo JR, Martel F, Borges N, Araújo JM, Keating E (2015) Folates and aging: role in mild cognitive impairment, dementia and depression. Ageing Res Rev 22(July):9–19. https://doi.org/10.1016/j.arr.2015.04.005

Arbabi E, Hamidi G, Talaei SA, Salami M (2016) Estrogen agonist genistein differentially influences the cognitive and motor disorders in an ovariectomized animal model of Parkinsonism. Iran J Basic Med Sci 19(12):1285–1290. https://doi.org/10.22038/ijbms.2016.7911

Avtandilov GG, Kuligin AG, Reĭtblat MA, Telianer IL (1983) Combined histochemical and histostereometric evaluation of the morphogenesis of human aortic atherosclerotic plaques. ArkhivPatologii 45(7):64–66

Bae S, Ulrich CM, Bailey LB, Malysheva O, Brown EC, Maneval DR, Neuhouser ML et al (2014) Impact of folic acid fortification on global DNA methylation and one-carbon biomarkers in the women's health initiative observational study cohort. Epigenetics 9(3):396–403. https://doi.org/10.4161/epi.27323

Banerjee A, Khemka VK, Ganguly A, Roy D, Ganguly U, Chakrabarti S (2015) Vitamin D and Alzheimer's disease: neurocognition to therapeutics. Int J Alzheimers Dis 2015:192747. https://doi.org/10.1155/2015/192747

Banzhaf-Strathmann J, Benito E, May S, Arzberger T, Tahirovic S, Kretzschmar H, Fischer A, Edbauer D (2014) Micro RNA -125b induces tau hyperphosphorylation and cognitive deficits in Alzheimer's disease. EMBO J 33(15):1667–1680. https://doi.org/10.15252/embj.201387576

Bollati V, Favero C, Albetti B, Tarantini L, Moroni A, Byun H-M, Motta V et al (2014) Nutrients intake is associated with DNA methylation of candidate inflammatory genes in a population of obese subjects. Nutrients 6(10):4625–4639. https://doi.org/10.3390/nu6104625

Camarena V, Wang G (2016) The epigenetic role of vitamin C in health and disease. Cell Mol Life Sci CMLS 73(8):1645–1658. https://doi.org/10.1007/s00018-016-2145-x

Cao J, Yan Q (2012) Histone ubiquitination and deubiquitination in transcription, DNA damage response, and cancer. Front Oncol 2:26. https://doi.org/10.3389/fonc.2012.00026

Cao T, Zhou X, Zheng X, Cui Y, Tsien JZ, Li C, Wang H (2018) Histone deacetylase inhibitor alleviates the neurodegenerative phenotypes and histone dysregulation in presenilins-deficient mice. Front Aging Neurosci 10:137. https://doi.org/10.3389/fnagi.2018.00137

Capiralla H, Vingtdeux V, Zhao H, Sankowski R, Al-Abed Y, Davies P, Marambaud P (2012) Resveratrol mitigates lipopolysaccharide- and aβ-mediated microglial inflammation by inhibiting the TLR4/NF-KB/STAT signaling cascade. J Neurochem 120(3):461–472. https://doi.org/10.1111/j.1471-4159.2011.07594.x

Chang M, Zhang L, Dai H, Sun L (2021) Genistein acts as antidepressant agent against chronic mild stress-induced depression model of rats through augmentation of brain-derived neurotrophic factor. Brain Behav 11(8):e2300. https://doi.org/10.1002/brb3.2300

Chang X, Rong C, Chen Y, Yang C, Qian H, Mo Y, Zhang C et al (2015) (-)-Epigallocatechin-3-gallate attenuates cognitive deterioration in Alzheimer's disease model mice by upregulating neprilysin expression. Exp Cell Res 334(1):136–145. https://doi.org/10.1016/j.yexcr.2015.04.004

Chen C, Wang Z, Ding Y, Wang L, Wang S, Wang H, Qin Y (2022) DNA methylation: from cancer biology to clinical perspectives. Front Biosci (Landmark Ed) 27(12):326. https://doi.org/10.31083/j.fbl2712326

Chen J, Ying Y, Zhu H, Zhu T, Chunsheng Q, Jiang J, Fang B (2019) Curcumin-induced promoter hypermethylation of the mammalian target of rapamycin gene in multiple myeloma cells. Oncol Lett 17(1):1108–1114. https://doi.org/10.3892/ol.2018.9662

Chen L, Xie W, Xie W, Zhuang W, Jiang C, Liu N (2017) Apigenin attenuates isoflurane-induced cognitive dysfunction via epigenetic regulation and neuroinflammation in aged rats. Arch Gerontol Geriatr 73(November):29–36. https://doi.org/10.1016/j.archger.2017.07.004

Chen P-T, Chen Z-T, Hou W-C, Lung-Chih Y, Chen RP-Y (2016) Polyhydroxycurcuminoids but not curcumin upregulate neprilysin and can be applied to the prevention of Alzheimer's disease. Sci Rep 6(July):29760. https://doi.org/10.1038/srep29760

Chuang Y-H, Paul KC, Bronstein JM, Bordelon Y, Horvath S, Ritz B (2017) Parkinson's disease is associated with DNA methylation levels in human blood and saliva. Genome Med 9(1):76. https://doi.org/10.1186/s13073-017-0466-5

Ciesielski O, Biesiekierska M, Balcerczyk A (2020) Epigallocatechin-3-gallate (EGCG) alters histone acetylation and methylation and impacts chromatin architecture profile in human endothelial cells. Molecules (Basel, Switzerland) 25(10). https://doi.org/10.3390/molecules25102326

Donovan MG, Selmin OI, Doetschman TC, Romagnolo DF (2019) Epigenetic activation of BRCA1 by Genistein in vivo and triple negative breast cancer cells linked to antagonism toward aryl hydrocarbon receptor. Nutrients 11(11):2559. https://doi.org/10.3390/nu11112559

Fang M, Chen D, Yang CS (2007) Dietary polyphenols may affect DNA methylation. J Nutr 137(1 Suppl):223S–228S. https://doi.org/10.1093/jn/137.1.223S

Farsetti A, Illi B, Gaetano C (2023) How epigenetics impacts on human diseases. Eur J Intern Med 114(August):15–22. https://doi.org/10.1016/j.ejim.2023.05.036

Fetahu IS, Höbaus J, EnikőKállay. (2014) Vitamin D and the epigenome. Front Physiol 5:164. https://doi.org/10.3389/fphys.2014.00164

Fu MP, Merrill SM, Sharma M, Gibson WT, Turvey SE, Kobor MS (2023) Rare diseases of epigenetic origin: challenges and opportunities. Front Genet 14(February). https://doi.org/10.3389/fgene.2023.1113086

Fu S, Kurzrock R (2010) Development of curcumin as an epigenetic agent. Cancer 116(20):4670–4676. https://doi.org/10.1002/cncr.25414

Fuggle NR, Laskou F, Harvey NC, Dennison EM (2022) A review of epigenetics and its association with ageing of muscle and bone. Maturitas 165(November):12–17. https://doi.org/10.1016/j.maturitas.2022.06.014

Fuloria S, Yusri MAA, Sekar M, Gan SH, Rani NNIM, TengLum P, Ravi S et al (2022) Genistein: a potential natural lead molecule for new drug design and development for treating memory impairment. Molecules (Basel, Switzerland) 27(1). https://doi.org/10.3390/molecules27010265

Grimm MOW, Mett J, Stahlmann CP, Haupenthal VJ, Zimmer VC, Hartmann T (2013) Neprilysin and Aβ clearance: impact of the APP intracellular domain in NEP regulation and implications in Alzheimer's disease. Front Aging Neurosci 5(December):98. https://doi.org/10.3389/fnagi.2013.00098

Guo L, Li H, Fan T, Ma Y, Wang L (2021) Synergistic efficacy of curcumin and anti-programmed cell death-1 in hepatocellular carcinoma. Life Sci 279(August):119359. https://doi.org/10.1016/j.lfs.2021.119359

Habibi P, Babri S, Ahmadiasl N, Yousefi H (2017) Effects of genistein and swimming exercise on spatial memory and expression of MicroRNA 132, BDNF, and IGF-1 genes in the hippocampus of ovariectomized rats. Iran J Basic Med Sci 20(8):856–862. https://doi.org/10.22038/IJBMS.2017.9106

Hammond R, Mauk R, Ninaci D, Nelson D, Gibbs RB (2009) Chronic treatment with estrogen receptor agonists restores acquisition of a spatial learning task in young ovariectomized rats. Horm Behav 56(3):309–314. https://doi.org/10.1016/j.yhbeh.2009.06.008

Harmancıoğlu B, Kabaran S (2023) Maternal high fat diets: impacts on offspring obesity and epigenetic hypothalamic programming. Front Genet 14:1158089. https://doi.org/10.3389/fgene.2023.1158089

Hassan F-U, Rehman MS-U, Khan MS, Ali MA, AroosaJaved AN, Yang C (2019) Curcumin as an alternative epigenetic modulator: mechanism of action and potential effects. Front Genet 10:514. https://doi.org/10.3389/fgene.2019.00514

Henderson AR, Wang Q, Meechoovet B, Siniard AL, Naymik M, De Both M, Huentelman MJ, Caselli RJ, Driver-Dunckley E, Dunckley T (2021) DNA methylation and expression profiles of whole blood in Parkinson's disease. Front Genet 12:640266. https://doi.org/10.3389/fgene.2021.640266

Hensel N, Rademacher S, Claus P (2015) Chatting with the neighbors: crosstalk between rho-kinase (ROCK) and other signaling pathways for treatment of neurological disorders. Front Neurosci 9:198. https://doi.org/10.3389/fnins.2015.00198

Hu Q, Chang X, Yan R, Rong C, Yang C, Cheng S, Gu X et al (2015) (-)-Epigallocatechin-3-gallate induces cancer cell apoptosis via acetylation of amyloid precursor protein. Med Oncol (Northwood, London) 32(1):390. https://doi.org/10.1007/s12032-014-0390-0

Impey S, Davare M, Lesiak A, Fortin D, Ando H, Varlamova O, Obrietan K, Soderling TR, Goodman RH, Wayman GA (2010) An activity-induced microRNA controls dendritic spine formation by regulating Rac1-PAK signaling. Mol Cell Neurosci 43(1):146–156. https://doi.org/10.1016/j.mcn.2009.10.005

Isac S, Panaitescu AM, Spataru A, Iesanu M, Totan A, Udriste A, Cucu N, Peltecu G, Zagrean L, Zagrean A-M (2017) Trans-resveratrol enriched maternal diet protects the immature hippocampus from perinatal asphyxia in rats. Neurosci Lett 653(July):308–313. https://doi.org/10.1016/j.neulet.2017.06.003

Iwata A, Nagata K, Hatsuta H, Takuma H, Bundo M, Iwamoto K, Tamaoka A, Murayama S, Saido T, Tsuji S (2014) Altered CpG methylation in sporadic Alzheimer's disease is associated with APP and MAPT dysregulation. Hum Mol Genet 23(3):648–656. https://doi.org/10.1093/hmg/ddt451

Izquierdo V, Palomera-Ávalos V, López-Ruiz S, Canudas A-M, Pallàs M, Griñán-Ferré C (2019) Maternal resveratrol supplementation prevents cognitive decline in senescent mice offspring. Int J Mol Sci 20(5). https://doi.org/10.3390/ijms20051134

Izquierdo V, Palomera-Ávalos V, Pallàs M, Griñán-Ferré C (2021) Resveratrol supplementation attenuates cognitive and molecular alterations under maternal high-fat diet intake: epigenetic inheritance over generations. Int J Mol Sci 22(3):1453. https://doi.org/10.3390/ijms22031453

Jager CA, Oulhaj A, Jacoby R, Refsum H, Smith AD (2012) Cognitive and clinical outcomes of homocysteine-lowering B-vitamin treatment in mild cognitive impairment: a randomized controlled trial. Int J Geriatr Psychiatry 27(6):592–600. https://doi.org/10.1002/gps.2758

de Jager PL, Srivastava G, Lunnon K, Burgess J, Schalkwyk LC, Lei Y, Eaton ML et al (2014) Alzheimer's disease: early alterations in brain DNA methylation at ANK1, BIN1, RHBDF2 and other loci. Nat Neurosci 17(9):1156–1163. https://doi.org/10.1038/nn.3786

Ji MH, Dong L, Jia M, Liu WX, Zhang MQ, Ju LS, Yang JJ, Xie Z, Yang JJ (2014) Epigenetic enhancement of brain-derived neurotrophic factor signaling pathway improves cognitive impairments induced by isoflurane exposure in aged rats. Mol Neurobiol 50(3):937–944. https://doi.org/10.1007/s12035-014-8659-z

Karlic H, Varga F (2011) Impact of vitamin D metabolism on clinical epigenetics. Clin Epigenetics 2(1):55–61. https://doi.org/10.1007/s13148-011-0021-y

Kaut O, Schmitt I, Stahl F, Fröhlich H, Hoffmann P, Gonzalez FJ, Wüllner U (2022) Epigenome-wide analysis of DNA methylation in Parkinson's disease cortex. Life (Basel, Switzerland) 12(4). https://doi.org/10.3390/life12040502

Kiselev IS, Kulakova OG, Boyko AN, O OFavorova. (2021) DNA methylation as an epigenetic mechanism in the development of multiple sclerosis. Acta Nat 13(2):45–57. https://doi.org/10.32607/actanaturae.11043

Koch JC, Tatenhorst L, Roser A-E, Saal K-A, Tönges L, Lingor P (2018) ROCK inhibition in models of neurodegeneration and its potential for clinical translation. Pharmacol Ther 189(September):1–21. https://doi.org/10.1016/j.pharmthera.2018.03.008

Kochmanski J, Kuhn NC, Bernstein AI (2022) Parkinson's disease-associated, sex-specific changes in DNA methylation at PARK7 (DJ-1), SLC17A6 (VGLUT2), PTPRN2 (IA-2β), and NR4A2 (NURR1) in cortical neurons. NPJ Parkinson's Dis 8(1):120. https://doi.org/10.1038/s41531-022-00355-2

Kumar P, Padi SSV, Naidu PS, Kumar A (2006) Effect of resveratrol on 3-nitropropionic acid-induced biochemical and behavioural changes: possible neuroprotective mechanisms. Behav Pharmacol 17(5–6):485–492. https://doi.org/10.1097/00008877-200609000-00014

Kurzava Kendall L, Ma Y, Yang T, Lubecka K, Stefanska B (2024) Epigenetic effects of resveratrol on oncogenic signaling in breast cancer. Nutrients 16(5):699. https://doi.org/10.3390/nu16050699

Li J, Wang S, Zhang S, Cheng D, Yang X, Wang Y, Yin H et al (2021) Curcumin slows the progression of Alzheimer's disease by modulating mitochondrial stress responses via JMJD3-H3K27me3-BDNF Axis. Am J Transl Res 13(12):13380–13393

Li W, Jiang M, Xiao Y, Zhang X, Cui S, Huang G (2015) Folic acid inhibits tau phosphorylation through regulation of PP2A methylation in SH-SY5Y cells. J Nutr Health Aging 19(2):123–129. https://doi.org/10.1007/s12603-014-0514-4

Li Y, Zhang J-J, Chen R-J, Ling Chen S, Chen X-FY, Min J-W (2022a) Genistein mitigates oxidative stress and inflammation by regulating Nrf2/HO-1 and NF-KB signaling pathways in hypoxic-ischemic brain damage in neonatal mice. Ann Transl Med 10(2):32. 10.21037/atm-21-4958

Li Y, Zhang J-J, Chen R-J, Chen L, Chen S, Yang X-F, Min J-W (2022b) Genistein mitigates oxidative stress and inflammation by regulating Nrf2/HO-1 and NF-KB signaling pathways in hypoxic-ischemic brain damage in neonatal mice. Ann Transl Med 10(2):32. https://doi.org/10.21037/atm-21-4958

Link A, Balaguer F, Goel A (2010) Cancer chemoprevention by dietary polyphenols: promising role for epigenetics. Biochem Pharmacol 80(12):1771–1792. https://doi.org/10.1016/j.bcp.2010.06.036

Littlejohns TJ, Henley WE, Lang IA, Annweiler C, Beauchet O, Chaves PHM, Fried L et al (2014) Vitamin D and the risk of dementia and Alzheimer disease. Neurology 83(10):920–928. https://doi.org/10.1212/WNL.0000000000000755

Liu Y-L, Yang H-P, Gong L, Tang C-L, Wang H-J (2011) Hypomethylation effects of curcumin, demethoxycurcumin and bisdemethoxycurcumin on WIF-1 promoter in non-small cell Lung cancer cell lines. Mol Med Rep 4(4):675–679. https://doi.org/10.3892/mmr.2011.473

Lobo RA (1995) Benefits and risks of estrogen replacement therapy. Am J Obstet Gynecol 173(3 Pt 2):982–989. https://doi.org/10.1016/0002-9378(95)90247-3

Lu X, Deng Y, Daohai Y, Cao H, Wang L, Liu L, Caijia Y, Zhang Y, Guo X, Yu G (2014) Histone acetyltransferase P300 mediates histone acetylation of PS1 and BACE1 in a cellular model of Alzheimer's disease. PLoS One 9(7):e103067. https://doi.org/10.1371/journal.pone.0103067

Ma XR, Sun ZK, Han X, Li S, Jiang X, Chen S, Zhang J, Hong L (2019) Neuroprotective effect of resveratrol via activation of Sirt1 signaling in a rat model of combined diabetes and Alzheimer's disease. Front Neurosci 13:1400. https://doi.org/10.3389/fnins.2019.01400

Marmorstein R, Zhou M-M (2014) Writers and readers of histone acetylation: structure, mechanism, and inhibition. Cold Spring Harb Perspect Biol 6(7):a018762–a018762. https://doi.org/10.1101/cshperspect.a018762

Marques S, Outeiro TF (2013) Epigenetics in Parkinson's and Alzheimer's diseases. Subcell Biochem 61:507–525. https://doi.org/10.1007/978-94-007-4525-4_22

Martínez-Iglesias O, Naidoo V, Carrera I, Corzo L, Cacabelos R (2022) Nosustrophine: an epinutraceutical bioproduct with effects on DNA methylation, histone acetylation and sirtuin expression in Alzheimer's disease. Pharmaceutics 14(11). https://doi.org/10.3390/pharmaceutics14112447

Marzi SJ, Leung SK, Ribarska T, Hannon E, Smith AR, Pishva E, Poschmann J et al (2018) A histone acetylome-wide association study of Alzheimer's disease identifies disease-associated H3K27ac differences in the entorhinal cortex. Nat Neurosci 21(11):1618–1627. https://doi.org/10.1038/s41593-018-0253-7

Masliah E, Dumaop W, Galasko D, Desplats P (2013) Distinctive patterns of DNA methylation associated with Parkinson disease: identification of concordant epigenetic changes in brain and peripheral blood leukocytes. Epigenetics 8(10):1030–1038. https://doi.org/10.4161/epi.25865

Mazzone R, Zwergel C, Artico M, Taurone S, Ralli M, Greco A, Mai A (2019) The emerging role of epigenetics in human autoimmune disorders. Clin Epigenetics 11(1):34. https://doi.org/10.1186/s13148-019-0632-2

Medvedeva YA, Khamis AM, Kulakovskiy IV, Ba-Alawi W, Bhuyan MSI, Kawaji H, Lassmann T, Harbers M, Forrest ARR, Bajic VB (2014) Effects of cytosine methylation on transcription factor binding sites. BMC Genomics 15(1):119. https://doi.org/10.1186/1471-2164-15-119
Ming T, Tao Q, Tang S, Zhao H, Yang H, Liu M, Ren S, Haibo X (2022) Curcumin: an epigenetic regulator and its application in cancer. Biomed Pharmacother 156(December):113956. https://doi.org/10.1016/j.biopha.2022.113956
Minor EA, Court BL, Young JI, Wang G (2013) Ascorbate induces ten-eleven translocation (Tet) methylcytosine dioxygenase-mediated generation of 5-hydroxymethylcytosine. J Biol Chem 288(19):13669–13674. https://doi.org/10.1074/jbc.C113.464800
Mohajeri MH, Troesch B, Weber P (2015) Inadequate supply of vitamins and DHA in the elderly: implications for brain aging and alzheimer-type dementia. Nutrition (Burbank, Los Angeles County, Calif.) 31(2):261–275. https://doi.org/10.1016/j.nut.2014.06.016
Moosavi A, Motevalizadeh Ardekani A (2016) Role of epigenetics in biology and human diseases. Iran Biomed J 20(5):246–258. https://doi.org/10.22045/ibj.2016.01
Morito K, Hirose T, Kinjo J, Hirakawa T, Okawa M, Nohara T, Ogawa S, Inoue S, Muramatsu M, Masamune Y (2001) Interaction of phytoestrogens with estrogen receptors alpha and beta. Biol Pharm Bull 24(4):351–356. https://doi.org/10.1248/bpb.24.351
Moyano-Mendez JR, Fabbrocini G, De Stefano D, Mazzella C, Mayol L, Scognamiglio I, Carnuccio R, Ayala F, La Rotonda MI, De Rosa G (2014) Enhanced antioxidant effect of trans-resveratrol: potential of binary systems with polyethylene glycol and cyclodextrin. Drug Dev Ind Pharm 40(10):1300–1307. https://doi.org/10.3109/03639045.2013.817416
Nalivaeva NN, Turner AJ (2019) Targeting amyloid clearance in Alzheimer's disease as a therapeutic strategy. Br J Pharmacol 176(18):3447–3463. https://doi.org/10.1111/bph.14593
Papoutsis AJ, Lamore SD, Wondrak GT, Selmin OI, Romagnolo DF (2010) Resveratrol prevents epigenetic silencing of BRCA-1 by the aromatic hydrocarbon receptor in human breast cancer cells. J Nutr 140(9):1607–1614. https://doi.org/10.3945/jn.110.123422
Paredes-Gonzalez X, Fuentes F, Zheng-Yuan S, Kong A-NT (2014) Apigenin reactivates Nrf2 anti-oxidative stress signaling in mouse skin epidermal JB6 P + cells through epigenetics modifications. AAPS J 16(4):727–735. https://doi.org/10.1208/s12248-014-9613-8
Patel M, Singh S (2022) Apigenin attenuates functional and structural alterations via targeting NF-KB/Nrf2 signaling pathway in LPS-induced parkinsonism in experimental rats: apigenin attenuates LPS-induced parkinsonism in experimental rats. Neurotox Res 40(4):941–960. https://doi.org/10.1007/s12640-022-00521-7
Peixoto L, Abel T (2013) The role of histone acetylation in memory formation and cognitive impairments. Neuropsychopharmacology 38(1):62–76. https://doi.org/10.1038/npp.2012.86
Peleg S, Sananbenesi F, Zovoilis A, Burkhardt S, Bahari-Javan S, Agis-Balboa RC, Cota P et al (2010) Altered histone acetylation is associated with age-dependent memory impairment in mice. Science (New York, N.Y.) 328(5979):753–756. https://doi.org/10.1126/science.1186088
Rahman MH, Akter R, Bhattacharya T, Abdel-Daim MM, Alkahtani S, Arafah MW, Al-Johani NS et al (2020) Resveratrol and neuroprotection: impact and its therapeutic potential in Alzheimer's disease. Front Pharmacol 11:619024. https://doi.org/10.3389/fphar.2020.619024
Rossetto D, Avvakumov N, Côté J (2012) Histone phosphorylation: a chromatin modification involved in diverse nuclear events. Epigenetics 7(10):1098–1108. https://doi.org/10.4161/epi.21975
Sadowska-Bartosz I, Bartosz G (2014) Effect of antioxidants supplementation on aging and longevity. Biomed Res Int 2014:404680. https://doi.org/10.1155/2014/404680
Salehi B, Venditti A, Sharifi-Rad M, Kręgiel D, Sharifi-Rad J, Durazzo A, Lucarini M et al (2019) The therapeutic potential of Apigenin. Int J Mol Sci 20(6). https://doi.org/10.3390/ijms20061305
Sanaei M, Kavoosi F, Arabloo M (2020) Effect of Curcumin in Comparison with Trichostatin A on the Reactivation of Estrogen Receptor Alpha Gene Expression, Cell Growth Inhibition and Apoptosis Induction in Hepatocellular Carcinoma Hepa 1–6 Cell LLine. Asian Pacific Journal of Cancer Prevention APJCP 21(4):1045–1050. https://doi.org/10.31557/APJCP.2020.21.4.1045

Sanchez-Mut JV, Gräff J (2015) Epigenetic alterations in Alzheimer's disease. Front Behav Neurosci 9:347. https://doi.org/10.3389/fnbeh.2015.00347

Santana DA, de Arruda M, Smith C, Chen ES (2023) Histone modifications in Alzheimer's disease. Genes 14(2). https://doi.org/10.3390/genes14020347

Scott HL, Tamagnini F, Narduzzo KE, Howarth JL, Lee Y-B, Wong L-F, Brown MW, Warburton EC, Bashir ZI, Uney JB (2012) MicroRNA-132 regulates recognition memory and synaptic plasticity in the perirhinal cortex. Eur J Neurosci 36(7):2941–2948. https://doi.org/10.1111/j.1460-9568.2012.08220.x

Smith BC, Denu JM (2009) Chemical mechanisms of histone lysine and arginine modifications. Biochimicaet Biophysica Acta 1789(1):45–57. https://doi.org/10.1016/j.bbagrm.2008.06.005

Subedi L, Ji E, Shin D, Jin J, Yeo JH, Kim SY (2017) Equol, a dietary Daidzein gut metabolite attenuates microglial activation and potentiates neuroprotection in vitro. Nutrients 9(3). https://doi.org/10.3390/nu9030207

Sundaram MK, Unni S, Somvanshi P, Bhardwaj T, Mandal RK, Hussain A, Haque S (2019) Genistein modulates signaling pathways and targets several epigenetic markers in HeLa cells. Genes 10(12). https://doi.org/10.3390/genes10120955

Sutrisno S, Miryani I, Dwijayasa PM, Rini Suprobo N, Wayan Arsana Wiyasa I (2022) Genistein administration increases the level of superoxide dismutase and glutathione peroxidase in the endometriosis mice model: an experimental study. Int J Reprod Biomed 20(10):873–882. https://doi.org/10.18502/ijrm.v20i10.12271

Taha M, Eldemerdash OM, Elshaffei IM, Yousef EM, Soliman AS, Senousy MA (2023) Apigenin attenuates hippocampal microglial activation and restores cognitive function in methotrexate-treated rats: targeting the MiR-15a/ROCK-1/ERK1/2 pathway. Mol Neurobiol 60(7):3770–3787. https://doi.org/10.1007/s12035-023-03299-7

Toker L, Tran GT, Sundaresan J, Tysnes O-B, Alves G, Haugarvoll K, Nido GS, Dölle C, Tzoulis C (2021) Genome-wide histone acetylation analysis reveals altered transcriptional regulation in the Parkinson's disease brain. Mol Neurodegener 16(1):31. https://doi.org/10.1186/s13024-021-00450-7

Tzika E, Dreker T, Imhof A (2018) Epigenetics and metabolism in health and disease. Front Genet 9(September). https://doi.org/10.3389/fgene.2018.00361

Vadnal J, Houston S, Bhatta S, Freeman E, McDonough J (2012) Transcriptional signatures mediated by acetylation overlap with early-stage Alzheimer's disease. Exp Brain Res 221(3):287–297. https://doi.org/10.1007/s00221-012-3172-y

Vallée A (2022) Curcumin and Wnt/B-catenin signaling in exudative age-related macular degeneration (review). Int J Mol Med 49(6). https://doi.org/10.3892/ijmm.2022.5135

Viña J, Escudero J, Baquero M, Cebrián M, Carbonell-Asíns JA, Muñoz JE, Satorres E et al (2022) Genistein effect on cognition in prodromal Alzheimer's disease patients. The GENIAL clinical trial. Alzheimer's Res Ther 14(1):164. https://doi.org/10.1186/s13195-022-01097-2

Wang Y, Zhang H, Hua L, Wang Z, Geng S, Zhang H, Zeng Z, Zhao J, Wang X, Wang Y (2022) Curcumin prevents Alzheimer's disease progression by upregulating JMJD3. Am J Transl Res 14(8):5280–5294

Weinhold B (2006) Epigenetics: the science of change. Environ Health Perspect 114(3):A160–A167. https://doi.org/10.1289/ehp.114-a160

Wu H-C, Qun-Liang H, Zhang S-J, Wang Y-M, Jin Z-K, Lv L-F, Zhang S, Liu Z-L, Hong-Lian W, Cheng O-M (2018) Neuroprotective effects of genistein on SH-SY5Y cells overexpressing A53T mutant α-synuclein. Neural Regen Res 13(8):1375–1383. https://doi.org/10.4103/1673-5374.235250

Xie Q, Bai Q, Zou L-Y, Zhang Q-Y, Zhou Y, Chang H, Yi L, Zhu J-D, Mi M-T (2014) Genistein inhibits DNA methylation and increases expression of tumor suppressor genes in human breast cancer cells. Genes Chromosom Cancer 53(5):422–431. https://doi.org/10.1002/gcc.22154

Yang S, Wang G, Ma ZF, Qin L-Q, Zhai Y-J, Yu Z-L, Xue M, Zhang Y-H, Wan Z (2020) Dietary advanced glycation end products-induced cognitive impairment in aged ICR mice: protective role of quercetin. Mol Nutr Food Res 64(3):e1901019. https://doi.org/10.1002/mnfr.201901019

Yashooa RK, Nabi AQ (2022) The MiR-146a-5p and MiR-125b-5p levels as biomarkers for early prediction of Alzheimer's disease. Hum Gene Ther 34(December):201129. https://doi.org/10.1016/j.humgen.2022.201129

Zhang S, Ma Y, Feng J (2020) Neuroprotective mechanisms of ε-viniferin in a rotenone-induced cell model of Parkinson's disease: significance of SIRT3-mediated FOXO3 deacetylation. Neural Regen Res 15(11):2143–2153. https://doi.org/10.4103/1673-5374.282264

Zhang Y, Sun Z, Jia J, Du T, Zhang N, Tang Y, Fang Y, Fang D (2021) Overview of histone modification. Adv Exp Med Biol 1283:1–16. https://doi.org/10.1007/978-981-15-8104-5_1

Zheng M, Zhang M, Yang J, Zhao S, Qin S, Chen H, Gao Y, Huang G (2014) Relationship between blood levels of methyl donor and folate and mild cognitive impairment in Chinese patients with type 2 diabetes: a case-control study. J Clin Biochem Nutr 54(2):122–128. https://doi.org/10.3164/jcbn.13-89

Zhou X, Jiao D, Dou M, Zhang W, Lv L, Chen J, Li L, Wang L, Han X (2020) Curcumin inhibits the growth of triple-negative breast cancer cells by silencing EZH2 and restoring DLC1 expression. J Cell Mol Med 24(18):10648–10662. https://doi.org/10.1111/jcmm.15683

Zoghbi HY, Beaudet AL (2016) Epigenetics and human disease. Cold Spring Harb Perspect Biol 8(2):a019497. https://doi.org/10.1101/cshperspect.a019497

Targeting Endoplasmic Reticulum to Treat Neurodegenerative Disorders

9

Chetan Singh Rajpurohit and Vipin K Yadav

Abstract

The chapter thoroughly explores the comprehensive relationship between endoplasmic reticulum stress (ERS)-induced unfolded protein response (UPR) in neurological diseases, alongside an investigation into the modulatory effects of natural compounds on ERS. ERS-induced UPR is pivotal in neurodegenerative diseases like Alzheimer's and Parkinson's, where aberrant protein accumulation challenges cellular homeostasis. Under pathological conditions, UPR is activated to restore endoplasmic reticulum (ER) function. However, sustained or dysregulated UPR exacerbates neuronal dysfunction, contributing to neurodegeneration. This intricate process involves chaperone activation and a reduction in protein synthesis to alleviate ERS. Natural compounds, such as phytochemicals and herbal extracts, have emerged as promising modulators of ERS. These compounds exhibit diverse mechanisms, including antioxidant properties and the induction of ER chaperones, to mitigate ERS and support cellular resilience. By attenuating neuroinflammation and promoting overall neuronal health, natural compounds present potential therapeutic avenues. The exploration of ERS-induced UPR and the influence of natural compounds represent a compelling area for innovative interventions targeting protein homeostasis in the ER, offering hope for neuroprotective strategies in the complex landscape of neurological disorders. This chapter serves as an essential resource for researchers, clinicians, and enthusiasts seeking a deeper understanding of the mechanistic intricacies

C. S. Rajpurohit (✉)
Huffington Center on Aging, Baylor College of Medicine, Houston, TX, USA
e-mail: chetan.rajpurohit@bcm.edu

V. K. Yadav
Department of Tumor Microenvironment and Metastasis, Moffitt Cancer Center and Research Institute, Tampa, FL, USA

S. Jahan et al. (eds.), *A Mechanistic Exploration of Natural Compounds for Neuronal Health*, https://doi.org/10.1007/978-981-95-0125-0_9

underlying ERS in neurodegenerative disorders and the therapeutic potential of natural compounds in restoring neuronal health.

Keywords

ER stress · Unfolded protein responses · Neurodegeneration · Natural compounds · Protein aggregation · Herbal extracts · Neuroinflammation · Personalized medicine

9.1 Introduction

The landscape of therapeutic strategies for neurodegenerative diseases continues to evolve as we better understand their molecular and cellular mechanisms. The endoplasmic reticulum stress (ERS) and the unfolded protein response (UPR) are the ones that surface first as significant contributors in various neurodegenerative dispositions. They act as the cell protective systems that the neuron uses to counteract misfolded protein aggregates and to keep its protein machinery functioning. However, when the pathways of these systems are mis-regulating, they could even lead to the manifestation of problems in the neurons and eventually cell death in cases like Alzheimer's, Parkinson's, and Amyotrophic Lateral Sclerosis (ALS). Therefore, a proper understanding of both ERS and UPR is a key to the discovery of new therapeutic targets.

This chapter mainly focuses on ERS and UPR, the pivotal processes of neurodegenerative diseases, explaining their involvement in the pathways of neurodegeneration and potential for treatment. We consider the duality of health-producing and mortality-causing pathways that are inherent in relationships. We also give a detailed account of UPR actions with other cell stress responses which include inflammation and oxidative stress. Furthermore, the chapter highlights remarkable prospects that natural compounds, such as phytochemicals and natural extracts, can offer in terms of reprogramming ERS and thus promoting neuroprotection.

9.2 The Role of the Endoplasmic Reticulum in Cellular Function

The endoplasmic reticulum (ER) is a central organelle of the cell that carries out various functions that are important for regular cellular activities. One of the primary actions of the ER is protein folding, and it is also the site for lipid synthesis and calcium storage, which all are necessary for the diverse functions of the cell (Wu et al. 2023; Yuan et al. 2024). The process of protein synthesis is when the riboprotein complex moves into the ER lumen—undergoing their folding and post-translational modifications like glycosylation. The large network of molecular chaperones and protein folding enzymes that are found in the ER is responsible for the correct folding, and lastly, the final 3D structure of the proteins is achieved. When proteins are properly folded in the ER, they then undergo one more quality assurance action in the Golgi apparatus and finally move to their functional sites.

However, the ER's inability to fold proteins appropriately due to genetic mutations, environmental stressors, or high protein levels leads to the increase of misfolded or unfolded proteins (Hetz and Mollereau 2014). This condition triggers a cellular response known as endoplasmic reticulum stress (ERS). The response of cells to ERS is to employ the unfolded protein response (UPR) that is a signaling pathway which aims to restore ER homeostasis by boosting the ER's protein-folding capacity, eliminating misfolded proteins, and decreasing the total amount of protein that is synthesized. Apart from protein folding, the ER is also a primary lipid biosynthesis organelle, where it manufactures phospholipids, cholesterol, and ceramides that are the main components of cellular membranes. Other than that, ER is involved in lipid metabolism, by constructing lipid droplets that are to be used in neutral lipids for energy production and for the synthesis of membranes (Yoon et al. 2021). The overproduction of lipids, associated with the ER, poses serious health problems, often resulting in metabolic disorders and contributing to the development of diseases like non-alcoholic fatty liver and atherosclerosis.

The ER also acts as an important intracellular calcium reservoir by modulating the cytoplasmic calcium levels. Calcium ions are necessary for several processes that take place in cells such as contraction in muscles, signal transmission, and enzymatic catalysis. The calcium-binding proteins and transporters build the backbone of the network through which the ER controls calcium homeostasis. Alterations in ER calcium levels may produce a communicative and functional disturbance of cells which in turn may result in diseases such as cardiac hypertrophy and neurodegeneration (Hamilton and Terentyev 2018). The research presented by Hetz and Mollereau (2014) was dedicated to ERS and UPR in neurodegenerative disorders, bringing to light how chronic ERS that is triggered by misfolded protein aggregates affects neurons, resulting in cellular death. The study of Petersen and Shulman (2018) also brought to the fore the crucial role of ER lipid synthesis in the maintenance of metabolic health, which correlated the imbalance in lipid metabolism with insulin resistance and type 2 diabetes. In addition, Hamilton and Terentyev (2018, 2022) dissected the contribution of ER calcium stores to cardiac performance, concluding that disarrangement in ER calcium handling leads to cardiac hypertrophy and heart failure. Furthermore, Cubillos-Ruiz et al. (2017) proposed that ERS assumes both a positive and a negative role in cancer in that chronic ERS can either cause cell death or stimulate tumor growth and survival by activating adaptive UPR pathways. These works illustrate the varied roles of the ER in cellular processes and homeostasis status, and the necessity to target ERS and UPR pathways with therapeutic strategies to heal a variety of diseases.

9.2.1 Endoplasmic Reticulum Stress and Unfolded Protein Response

The endoplasmic reticulum (ER) is unusually active in folding and processing secretory proteins; however, it can experience stress (ERS) if the process is disrupted in severe conditions. The primary reason for ERS is the cytosolic misfolded protein aggregates that are produced in excess quantity, and these can be triggered by chaperone or lysosome dysfunction. This physiological condition caused by mistakes that were made in the secretory pathway is opposed by the unfolded protein response (UPR), a protective mechanism of the cell, built from integrated multilayered signaling to restore the function of the ER (Fig. 9.1), at the expense of the whole protein synthesis reduction, increased protein folding capacity, and degradation of misfolded proteins (Negroni et al. 2014; Cao 2015).

The disbalanced protein homeostasis under ER stress may arise from multiple causes including genetic mutations (Lukasiak and Zajac 2021), environmental stressors (Adav and Sze 2020; Dawood and Altobje 2020), and metabolic disturbances (Zatterale et al. 2019; Yong et al. 2021). Moreover, viruses can also exploit the host's ER for protein synthesis, causing ERS (Rios-Ocampo et al. 2020). Activation of the UPR path is turned on by the detection of ERS, which is a process involving numerous signaling pathways that are aimed at restoring ER homeostasis. This UPR process is started by the accumulation of misfolded proteins, which are then recognized by three signaling pathways mediated by the transmembrane proteins of the ER: (1) IRE1 pathway, (2) PERK pathway, and (3) ATF6 pathway (Fig. 9.2).

9.2.1.1 IRE1 Pathway

IRE1 is formed through the accumulation of misfolded proteins within the ER, which, after activation, splits the mRNA encoding the protein that binds the X-box. The spliced XBP1 mRNA is a transcription factor of a high order, the XBP1s. XBP1s boost transcription of the genes involved in protein folding, such as chaperones, and ER-associated degradation (ERAD) components (Siwecka et al. 2021).

9.2.1.2 PERK Pathway

ERS activates PERK which then oligomerizes prior to the autophosphorylation that occurs. The activated PERK then phosphorylates the eukaryotic initiation factor 2 alpha (eIF2α), which in turn reduces the total amount of new proteins available to the ER. ATF4 genes may be also upregulated, and thus, antioxidant responses, the amino acid metabolism, and apoptosis are triggered (Liu et al. 2015).

9.2.1.3 ATF6 Pathway

The ATF6 story is more or less the same as the previous two; upon its activation, it is translocated from the ER to the Golgi apparatus, from which it is cleaved by site-1 and site-2 proteases, resulting in the release of its cytoplasmic domain. The cleaved cytoplasmic domain of ATF6 is the transcription factor, and it increases the synthesis of genes that code for ER chaperones and components of the ERAD pathway,

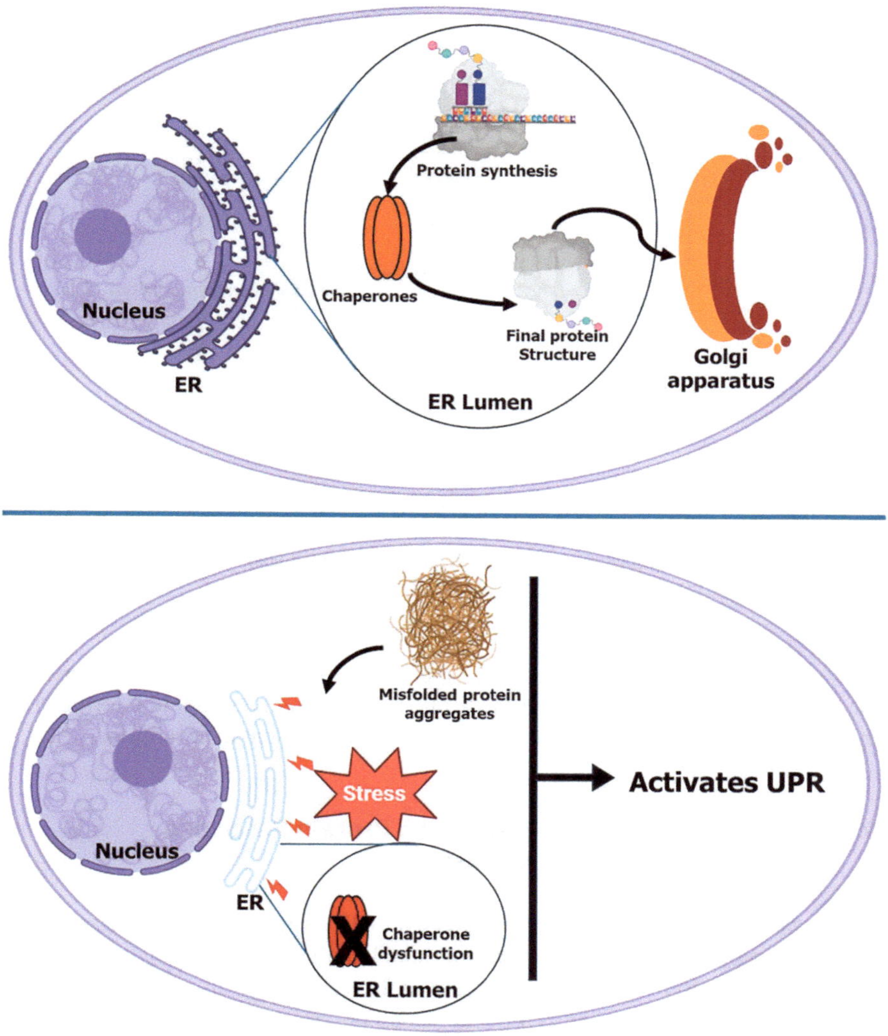

Fig. 9.1 Depicting ER homeostasis under normal and stressful conditions

thus enhancing the cell's capacity to refold misfolded proteins and degrade irreparably damaged ones (Wang et al. 2009; Stauffer et al. 2020).

Nevertheless, if there are any cases of strong ERS, then the compensatory mechanisms not only become active but also hyperactive, which is what causes apoptosis and neuronal cell death; the same was also shown in 2014 by Hetz and Mollereau (2014). The UPR's role in managing ERS is enlightening and contributes to understanding not only neurodegenerative diseases and diabetes but also cancer. The low molecular weight compounds that can affect the signaling of UPR are being assessed for their efficiency in relieving ERS and in improving cellular function in

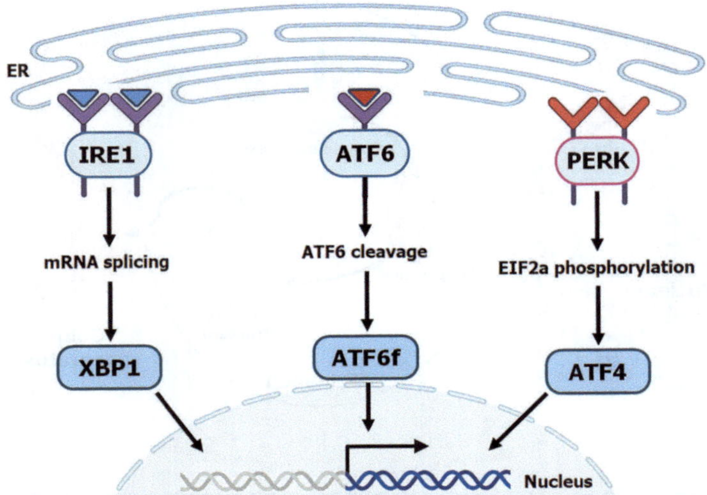

Fig. 9.2 The three major pathways of ERS-induced UPR, which leads to the transcriptional activation of genes involved in maintaining ER homeostasis

disorganization. Such studies reinforce the idea that it is essential to comprehend the key factors responsible for the cause of ERS while at the same time formulating the therapeutical policy to alter the UPR for clinical ailment intervention.

9.2.2 The Dual Role of the Unfolded Protein Response in Neurodegenerative Diseases

While the main objective of UPR is to recover ER function, its initiation may lead to a dual result depending on the intensity and length of the ERS. Generally, adaptive UPR activation resulting from acute and mild ERS is usually the case, but maladaptive UPR that causes cell death is also possible. For example, at the early stages of the disease, the unfolded response can, for a short time, deal with the huge number of misfolded proteins. This helps to maintain the neuronal function by providing a shortage of misfolded proteins temporarily. Likewise, the long-term or harsh forms of ER stress can reverse the effects of UPR, which causes misfolded proteins to get rid of the damaged ones (Corazzari et al. 2017). The high metabolic demand and the extensive protein synthesis of neurons make them particularly sensitive to ERS. The continual activation of the pathways of ERS has finally overtaken the capacity of the adaptive UPR which results in UPR dysregulation, a major cause in the progression of various neurodegenerative diseases, including Alzheimer's disease, Parkinson's disease, and Amyotrophic Lateral Sclerosis (ALS).

9.3 Mechanisms of ERS-Induced UPR in Neurodegenerative Diseases

The neurodegenerative diseases caused by ERS-induced UPR mechanisms depend on the continuous stimulation of ERS pathways, leading to neuronal dysfunction. Knowing about these mechanisms opens up the possible targets for medication to UPR, thus rerouting the progression of diseases. This section gives particulars of the role of UPR in the misfolding of proteins, the formation of amyloid plaque, and the consequent detrimental effects on the neurons in Alzheimer's disease; the accumulation of aberrantly folded alpha-synuclein proteins, in turn, causes neurodegeneration and affects ERS in Parkinson's disease; and finally, mutations of SOD1 that lead to motor neuron death due to specific interactions with every IgG was explored in detail in these studies. We will also discuss the UPR-related functions that deal with other stressed cellular states, such as oxidative stress and the inflammatory response, and their shared influences on the health of neurons.

9.3.1 ERS in Alzheimer's Disease

Alzheimer's disease (AD) is a disorder characterized by the accumulation of amyloid-beta (Aβ) plaques and neurofibrillary tangles made of hyperphosphorylated tau protein, which can largely disturb the ER. The research by Ajoolabady et al. has revealed a higher rate of ERS along with an increase in the UPR activation within the regions of the brains of individuals who died from Alzheimer's, a sign of a prolonged ERS disturbance (Ajoolabady et al. 2022). The ERS-invoked UPR is generally a stress protector, programmed to protect neurons, but in the long-term phase, chronic UPR activation ends up being no good to neurons and might even lead them to die off. For instance, Ekundayo et al. in their research showed the way chronic ERS causes the activation of C/EBP homologous protein (CHOP), a pro-apoptotic factor, that makes the brain cells die (Ekundayo et al. 2024). Additionally, the irreversible UPR signal disturbs calcium balance and oxidative stress, which also complicates the neurodegenerative condition.

The research has recently highlighted the feasibility of UPR modulation as the new therapeutic strategy by targeting the specific UPR components to dampen ERS and to deter neurodegeneration in the model of AD. An example was the work of Radford et al., which revealed that the inhibition of the PERK pathway could not only reduce neurodegeneration but also protect against disease. Besides, using PERK inhibitors in their study led to lower levels of eIF2α dephosphorylation. This also reduced protein synthesis inhibition, and neuronal survival improved in mice with AD. An equal example is the selective modulation of the IRE1 pathway, which has also shown a potential (Radford et al. 2015). As stated in the study (Ye et al. 2022), inhibiting endoribonuclease activity of IRE1 diminished XBP1 splicing, which subsequently lowered the expression levels of pro-apoptotic genes and ultimately restored neuronal function in AD models (Ye et al. 2022). This evidence emphasizes the possibility of UPR regulation being a treatment avenue in

Alzheimer's disease and demonstrates that the fine-tuning of UPR to its optimal level is essential to increase its protective role alongside the decrease of its contribution to neuronal injury.

9.3.2 ERS in Parkinson's Disease

With the gradual termination of dopamine-producing neurons in the substantia nigra and the gathering of α -synuclein (SNCA), which is toxic to neurons, Parkinson's disease (PD) deteriorates further and further. The work by Credle et al. suggested that SNCA aggregates can disrupt the normal function of the ER membrane, causing the development of ERS and the impairment of UPR signaling (Credle et al. 2015). The SNCA gene mutations are the primary drivers for UPR activation, persistently increasing ERS. When the amount of misfolded SNCA simultaneously rises in the ER, it triggers an adaptive response to the protein load overall, but chronic activation of such a message can cause cellular dysfunction and death. For example, Kim et al. found out that the mutations in SNCA lead to amplified ERS and the following activation of the PERK pathway, which in turn phosphorylates eIF2α and results in decreased protein synthesis and increased neural vulnerability (Kim et al. 2022).

The recent research efforts have concentrated on the call of inhibiting UPR pathways as the therapeutic approach in the models of PD. Specifically, targeting the PERK and ATF6 pathways has given good results. For example, a study carried out by Mercado et al. found that using pharmacological inhibition against the PERK pathway in PD models resulted in reduced phosphorylation of eIF2α; thus, neuronal death was lower while motor function had a better outcome (Mercado et al. 2018). Another research by Credle et al. showed that the modulatory effect on ATF6 activity might enhance the cell's ability to cope with ERS, therefore minimizing α-synuclein toxicity and subsequently aiding neuronal survival (Credle et al. 2015). Thus, this evidence points out that targeting the defects via the tuning of UPR can significantly alleviate chronic ERS, thus increasing the chances of survival for dopaminergic neurons and giving hope for enhanced health in Parkinson's disease.

9.3.3 ERS in Amyotrophic Lateral Sclerosis

Amyotrophic Lateral Sclerosis (ALS) is a condition that involves the progressive loss of motor neurons, muscle weakness, and paralysis. The mutations in genes such as SOD1, TARDBP (TDP-43), and FUS are involved in the exercise of ERS in ALS. These gene mutations cause the accumulation of misfolded proteins in the ER, which then triggers the UPR. The UPR is protective at first, but activation over a long period of time can lead to the death of motor neurons from the induction of involved apoptotic pathways. An example is the study by Rozpedek et al. who found that SOD1 mutant proteins cause the ERS and activate the PERK pathway that results in an increase in levels of CHOP, which is a pro-apoptotic factor and thus contributes to the death of motor neurons (Rozpedek et al. 2016). Also, in a study by Rajpurohit et al., they noted that the ERS is neurotoxic and confirmed it in Cu/Zn SOD1 L39R-linked familial ALS (fALS) motor neurons derived from patients. This shows that ERS pathology has a key role in ALS (Rajpurohit et al. 2020). The works referred to above have once again reinstated the necessity for tuning ERS pathways for the sake of therapy in degenerative diseases. The approach that involves driving on the protective side of the UPR while avoiding its damaging side by targeting its specific components seems to be very promising for the development of new treatments against these troublesome illnesses.

9.4 Crosstalk Between UPR and Other Cellular Stress Responses

Besides its involvement in protein balance, the unfolded protein response (UPR) communicates with diverse cellular stress responses, thus increasing the load on the cells caused by stress in neurodegenerative diseases. One example of such interaction is oxidative stress where ERS can cause the realization of reactive oxygen species (ROS) which will be lethal for neurons. To illustrate, experiments conducted on models of Alzheimer's disease have demonstrated that ERS not only drives the UPR but also causes an increase in ROS levels, which in turn enhances the oxidation and aggregation of amyloid-beta (Aβ) peptides (Uddin et al. 2020). Similarly, in Parkinson's disease, misfolded proteins such as SNCA accumulate and induce ERS and ROS production, which instigates mitochondrial dysfunction and subsequently increases neuronal apoptosis (Colla 2019). The essence of these data is that the entangled relationship between ERS and oxidative stress leads to the joint contribution of both to the pathogenesis of neurodegenerative diseases.

Moreover, UPR is also involved in the process of mitochondrial dysregulation, which is inflamed and leads to a vicious cycle of cellular stress. Mitochondrial dysfunction is a common symptom of neurodegenerative diseases and its relationship with ERS can greatly affect cellular health. For instance, the gene associated with ERS mutant SOD1 proteins expressed in Amyotrophic Lateral Sclerosis (ALS) sufferers also damage the mitochondria, which results in energy deficits and more cell death (Tadic et al. 2014). Inflammation even makes this situation worse, where ERS can activate pathways such as NF-κB, which in turn cause more neuronal damage. For example, the research presented by Colla (2019) identified that ERS-induced UPR activation in Parkinson's disease models also led to the release of pro-inflammatory cytokines that further contributed to the neuroinflammatory environment. Such stress responses must have synergistic interplay to provide ground for the detailed and more comprehensive design of the disease-modifying therapy for neurodegenerative diseases. To summarize, see Fig. 9.3.

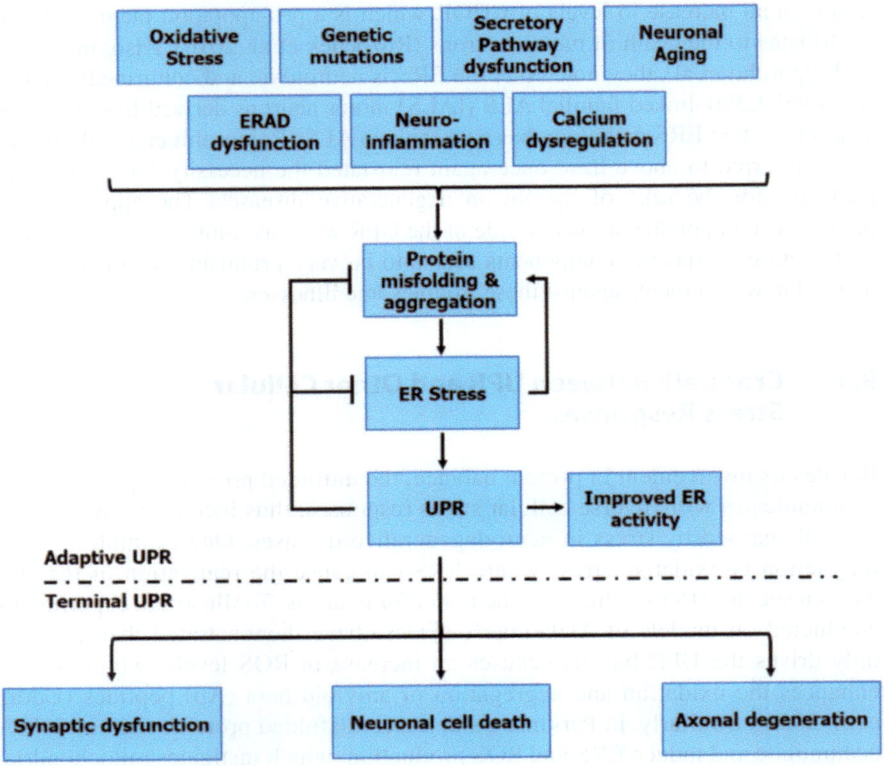

Fig. 9.3 A flowchart illustration depicting the relationship between the possible cues for the ERS and how it gets managed in the cell. The terminal UPR under the overwhelming condition leads to neurodegeneration

9.4.1 Oxidative Stress

Misfolded proteins when ERS is causing NADPH oxidase activation can induce ROS production, the production of excess free radicals, which disrupt normal cellular redox states and exacerbate oxidative stress and neuronal damage.

9.4.2 Mitochondrial Dysfunction

The connection between ERS and mitochondrial dysfunction is through the regulation of calcium homeostasis and the action of CHOP, which promotes mitochondrial outer membrane permeabilization (MOMP), leading to energy deficits and more cell death.

9.4.3 Inflammation

By the release of ERS sensors like IRE1, which activates pro-inflammatory cytokine production, contributing to a neuroinflammatory environment, ERS can trigger inflammatory pathway activation, such as NF-κB.

9.5 The Role of ER Chaperones in Neuroprotection

ER chaperones are integral to the maintenance of homeostasis of proteins within the ER; they are responsible for the achievement of protein folding, assembly, and quality control. Hence, the neuroprotective aspect of ER chaperones is an important detail for the therapeutic targets of ERS in neurodegenerative conditions. ER chaperones such as GRP78/BiP, calnexin, and calreticulin are the main elements in the maintenance of the protein balance and the prevention of protein misfolding. This part discusses their specific functions, GRP78/BiP in the context of neurodegenerative disease, calnexin and calreticulin that have some protecting roles, and finally the potential of modulation. Regarding ER chaperones as possible therapeutic strategies, the effective pathways in ER chaperone research are also highlighted.

9.5.1 GRP78/BiP in Neurodegenerative Diseases

GRP78, or BiP, is a central player in the unfolded protein response (UPR), a signaling cascade that is essential to the maintenance of ER homeostasis during stress conditions. It is recognized that in the neurodegenerative disorders such as Alzheimer's and Parkinson's disease, the altered regulation of GRP78 protein expression is a significant factor in disease progression. GRP78 is the main actor, binding to collapsed or wrongly folded proteins, to prevent their aggregation and to help them refold correctly. When the ERS is on, that is the case with a lot of the unfolded protein load, such as nutrient deprivation or oxidative stress, GRP78/BiP is expressed more to reduce the load of misfolded proteins and start the UPR (Wang et al. 2009). Some studies noted that the augmentation of GRP78 could have effects that protect neurons because of the ability of the ER to fold and process proteins/misfolded proteins correctly that may cause cell death (McGrath 2018).

In contrast, an insufficient level of GRP78 worsens the neuronal vulnerability to ER stress and thus causes defective protein folding and leads to increased cell death through apoptosis. This connection, thus, gives the clear message that GRP78 is involved not only in protein folding but also in the modulation of cell survival under stress. Thus, novel therapeutic approaches are under consideration to select the target for GRP78 and to induce its neuroprotective effect in the setting of neurodegeneration. For example, pharmacological or genetic induction of GRP78 expression has shown promise in summer research projects by minimizing ERS and repairing neuronal damage. Finding out the roles of GRP78 in illnesses of the nervous system in fine detail gives clues to the ways of therapeutic intervention, which in turn, helps cells to be more resistant to ERS. Hence, it may lead to a decrease of the speed of progression of the condition to end up in better conditions for the patients that are affected.

9.5.2 Calnexin and Calreticulin in Neuroprotection

Calnexin and calreticulin are integral components of the ER quality control machinery, specializing in the correct folding of glycoproteins by interacting with their carbohydrate moieties. These ER chaperones collaborate with lectin-like domains and bind transiently to specific glycoproteins to ensure their proper folding and assembly before they are transported to their functional locations. They ensure that only correctly folded proteins are eligible to go down to the secretory pathway, while the misfolded ones should be sequestered for degradation. In a scenario of neurodegenerative diseases where the traditional misplaced proteins and ERS are common, both calnexin and calreticulin can benefit from increased expression or activity (Danilczyk et al. 2000). Research has shown that an increase in the expression of calnexin and calreticulin caused more of the glycoproteins to be correctly folded, and in this way, the misfolded proteins in the ER lumen were reduced. The reduced number of misfolded proteins contributes to the activation of ERS, which has been recognized as a signal for neuron apoptosis under such pathological conditions. Therefore, increasing the levels of calnexin and calreticulin in models of neurodegeneration, thus improving the neuronal survival and proper functioning, is likely to be achieved (Paskevicius et al. 2023).

These results clearly establish that this chaperone system is crucial for maintaining ER function not just under normal physiological conditions but also during states that are pathological or where the capacity of the ER to fold proteins is exceeded (Kozlov and Gehring 2020). Calnexin and calreticulin not only function bilaterally but also cope with managing oxidative stress-induced neurodegenerative disorders. The interventions that are intended to regulate their expression or activity may offer ways for the discovery of new treatments that would reduce the problem of protein misfolding, alleviate ER stress, and retain neuronal integrity in Alzheimer's, Parkinson's, and other conditions characterized by protein aggregation and ER dysfunction.

9.6 Pathological Conditions Leading to Sustained UPR

Many pathological states are associated with prolonged UPR. Neuronal dysfunction, frequently encountered in neurodegenerative diseases, is regarded as a long-term UPR. This is caused by multiple mechanisms that are active at the same time. One such mechanism is age-related ER changes, for example, higher oxidative stress and decreased protein folding capacity, which may induce chronic UPR activation (Cao and Kaufman 2012). Besides, metabolic dysregulation, a typical finding in diabetes and obesity, may further enhance ERS and cause chronic UPR (Yuan et al. 2024). Environmental factors such as infectious agents and toxins can also lead to ERS, consequently opening up the UPR pathway (Corazzari et al. 2017). In addition, genetic mutations that hinder the locals of the ER can lead to misfolding of proteins and, therefore, to chronic UPR. So, these factors illustrate the pathology's complexity before which a person has a good deal of respect and signify the necessity of mechanisms of therapy in unburdening ERS-related diseases. We will discuss the conditions mentioned in detail in this section.

9.6.1 Genetic Mutations and ERS

Genetic mutations serve as the foundation of all protein folding mechanisms in the ER. Neurons are notable as they can suffer from protein homeostasis distortion more than other cells because of their sophisticated wiring. Neurodegenerative disorders are a clear example of conditions that arise from this distortion. Due to the disturbances in the delicate balance of proteins, the folding of proteins happens incorrectly, leading to their accumulation in the ER and consequently initiating the UPR. UPR then reduces the protein synthesis, increases the protein folding, and promotes the degradation of misfolded proteins which all contributes to the overall balance of the ER. In situations of chronic misfolding, however, the persistence of UPR can alternatively act in the pathogen's favor (Russell et al. 2023). Familial Alzheimer's disease (FAD) is a specific type of early-onset Alzheimer's disease that comes due to mutations in the germline. It was mentioned before that mutations in presenilins (PSEN1 and PSEN2) can lead to FAD. Presenilins are integral proteins that mess up with the processing of the amyloid precursor protein (APP) as they are embedded in the cell membrane and are thus involved in the process. The mutations introduce a misfolding of Aβ, and these products initiate the UPR. The beta-amyloid peptide, in this case, is not the only particle of the ER that participates in the inflammation and death of neurons. The by-product of the UPR pathway, which is active for a long time and working against the neurons, will be a high level of cell death, inflammation, and loss of synaptic communication (Sauer 2017).

Research into the molecular causes of ERS is of great importance for developing more effective therapies. Scientists are working on the idea of changing the activity of UPR and promoting the flipping of proteins. Some of the strategies are small molecules that are necessary for the correct protein folding, chaperones that help the protein quality control system, and also gene therapies that are used for treating specific mutations.

9.6.2 Environmental Factors and ERS

Environmental stressors are a significant item in the list of factors that make cellular homeostasis imbalanced. Pesticides, toxins, infections, and traumatic brain injuries, on the other hand, are key factors that have been reported to directly interfere with ER function. The ER can be described as a multi-tool because it is responsible for multiple functions like protein folding, lipid synthesis, and calcium storage. When the ER is overwhelmed with incorrect folding or other stressors, it moves into the state of ERS. The condition is found in the biological response called the unfolded protein response (UPR). UPR works towards offsetting the misfolded protein burden and restoring the normal ER function by increasing the protein folding ability, degrading the misfolded proteins, and curtailing the protein synthesis. Although UPR is protective initially, the failure of the feedback mechanism or uncontrolled activation of UPR can lead to the disease state. Exposure to pesticides has been an important subject of discussion because of its assumed effects on the human body. In reality, several pesticides are implicated in the development of Parkinson's disease (National Institutes of Health 2011). Parkinson's disease is the progressive loss of dopaminergic neurons within the brain over time. How are chemicals and pesticides working on UPR activation and ERS? Rotenone and paraquat, both of which are widely used pesticides, are extensively believed to have adverse impacts on human health. They are both respiratory inhibitors, leading to the enhancement of oxidative stress and the misfolding of proteins in the ER. The pesticides used are the factors that trigger ERS, which then set off UPR. UPR works to repair the ER by increasing proteases and chaperones which are both required for protein quality control. However, excessive chronic UPR can make the condition of neurons worse. Misfolded proteins are present at exorbitant rates which disturb the normal cellular pathways and cause the death of dopaminergic cells (Dovonou et al. 2023).

9.6.3 Metabolic Dysregulation and ERS

Metabolic disorders such as diabetes and obesity disbalance the normal cellular metabolism and can lead to the development of ERS. ER is the most emphasized cellular entity required for protein synthesis, lipid metabolism, and calcium storage. Mutations and stressors (e.g., hypoxia, oxidative stress) are known to trigger ER dysfunction by the accumulation of misfolded proteins, while normal conditions will result in only correctly folded proteins being secreted. In the context of neurodegenerative diseases, metabolic dysregulation exacerbates neuronal ERS. Alzheimer's disease affects the brain by causing cognitive impairment and memory loss and is therefore connected with ERS. Insulin resistance and hyperglycemia, which are characteristics of diabetes, also promote ERS in those who contract Alzheimer's. The imbalance in glucose metabolism exerts a negative effect on neuronal functions and emphasizes the accumulation of misfolded proteins within the ER. For this reason, the UPR is set in motion. But as the time goes by, the continued stress can end up with changes in the CNS, which can lead to neuronal cell death (Yuan et al. 2024).

9.6.4 Age-Related Changes in ER Function

One of the major risk factors that a person can be exposed to is aging, as it is known to be a significant risk factor for neurodegenerative diseases and is correlated with a diminished ER function. With age-related changes in ER chaperone levels, protein folding capacity, and UPR activation, neurons will be more susceptible to ERS (Brown and Naidoo 2012). The investigation of the age-related ER dysfunction mechanisms may provide new avenues for the treatment and prevention of age-associated neurodegenerative diseases.

9.6.5 Sustained UPR and Neuronal Dysfunction

Chronic UPR can result in maladaptative mechanisms that further contribute to neurodegeneration and cell death. Long-term UPR could lead to sustained protein synthesis suppression, depletion of calcium from the ER, and initiation of pro-apoptotic pathways. There are some therapeutic approaches that are designed to control the degree and course of UPR activation by the Galinara gene, which will help in treating the ERS-mediated neurodegeneration (Murao and Nishitoh 2017).

9.7 Impact of Aberrant Protein Accumulation on Cellular Homeostasis

The presence of misfolded proteins in cells causes a huge impact on the balance in cells and even causes them to develop several pathologies. This is especially true in the case of neurodegenerative diseases which are characterized by protein aggregation. In this section, we discuss the complicated interplay between protein aggregation and neurodegenerative diseases. Knowledge on how protein aggregates are generated is a significant part of pharmacological developments of anti-aggregation agents. UPR is one of the stress pathways that maintain the normal metabolic state of cells. It triggers the production of chaperones which are known to mediate protein aggregates in cells that are stressed, as well as autophagy (Yadav et al. 2022), the degradation of misfolded proteins, and other cellular recovery mechanisms. The endoplasmic reticulum-associated degradation (ERAD) pathway is the main factor in the dynamic equilibrium of protein folding by wrong proteins being detected and disposed of, thus keeping them from being deposited in the ER.

9.7.1 Protein Aggregation and Neurodegeneration

The buildup of proteins that are not properly folded is prevalent in neurodegenerative disorders, such as Alzheimer's disease, Parkinson's disease, and Amyotrophic Lateral Sclerosis (ALS). Neurodegenerative disorders are frequently linked to the formation of protein aggregates and similar misfolding proteins that impede the normal balance of cells and consequently lead to neuronal loss. In the case of the aggregation of gamma-carboxyl (GCA) peptides in the brain, the result is the formation of plaques on the outside and neurofibrillary tangles within the neurons. The presence of aggregates disturbs the synapse, that is, they hamper the communication between neurons and lead them to finally die. In this case, the aggregates of proteins are responsible for the sequestration of crucial cellular components and the alteration of different cellular functions, such as redistribution of organelles, apoptosis, and autophagy. The accumulation of SNCA proteins into Lewy bodies in Parkinson's disease acts like a sponge for proteins and also diminishes the activity of the respiratory chain which is thought to be a crucial factor in the development of the disease (Benskey et al. 2016).

These misfolded proteins that aggregate exercise a significant role in cellular failure since they impose on the integrity of organelles such as the mitochondria (MT) and ER. The ER is the center of protein folding and quality control. However, a buildup of unfolded proteins overwhelms its capacity and leads to ERS. This stress causes activation of the unfolded protein response (UPR) which is a cellular mechanism aimed at restoring normal cell ER function. In ALS, for example, the accumulation of superoxide dismutase 1 (SOD1) aggregates causes oxidative stress and cell death because of inhibited mitochondrial dynamics. The partnership between protein aggregation and mitochondrial dysfunction creates an unbreakable loop which damages the neurons and promotes the faster progression of the disease.

9.7.2 Mechanisms and Cellular Responses to Protein Aggregates

Protein aggregates can occur due to various mechanisms such as genetic variants, post-translational modifications, and environmental stresses. For example, mutations in SOD1 that result in misfolded proteins are seen in familial forms of Amyotrophic Lateral Sclerosis (ALS). The over-misfolding/aggregation of SOD1 leads to cellular dysfunctions. Post-translational modifications such as phosphorylation, ubiquitination, and glycosylation can be done in such a way as to initiate or stabilize protein aggregates (Ramesh et al. 2020). One process that occurs frequently in Alzheimer's is tau hyperphosphorylation that promotes the aggregation of tau which is a component of neurofibrillary tangles that impair neuronal function. Environmental agents, such as oxidative stress, metal ion imbalance, and exposure to toxins, also increase the effects of misfolding on the proteins. The aggregates may acquire different arrangements, e.g., amyloid fibrils, and soluble oligomers, each of which may have distinct toxic characteristics. Amyloid fibrils are ordered and very stable, while smaller, soluble oligomers are created that usually are more toxic to cells since they are able to attach and interrupt cellular membranes (Verma et al. 2015).

Clarifying the pathways implicated in the formation of aggregates and their spread will be the key to producing efficient medication. The cells have accomplished this through their efforts on stopping the aggregation of misfolded proteins.

Molecular chaperones are the proteins that facilitate the correct conformation of nascent proteins and thus prevent their aggregation by either facilitating the proper folding of the proteins or refolding the misfolded ones. The endoplasmic reticulum-associated degradation (ERAD) pathway identifies the misfolded proteins tangled in the ER and subjects them to ubiquitination and proteolytic degradation afterwards. The autophagy is a process which removes larger protein aggregates and damaged organelles from the cellular environment. The organelles involved are lysosomes and the aggregation of protein material inside the autophagosomes (Kenney and Benarroch 2015). However, neurodegenerative diseases can overwhelm these protective actions. To illustrate, in Parkinson's disease, the aggregation of SNCA into Lewy bodies may impair the autophagic-lysosome pathway and lead to more dysfunctional organelles and more protein aggregation (Navarro-Romero et al. 2020). This leads to a cycle of cellular stress and neuronal damage that goes round and round.

The improvement of the efficiency of the cellular quality control systems may become a way of reducing the impact of protein aggregates on neuronal functioning. A heat shock molecular chaperone that is targeted to raise its open activity could help cure Huntington's disease. Pharmacological methods activating the ERAD and/or increasing the flux of the autophagic pathway could lead to the elimination of accumulated aggregates. For instance, small molecules like rapamycin have been reported to trigger autophagy while also decreasing the buildup of protein aggregates in Huntington's disease models (Yin et al. 2017). Furthermore, the alteration of the UPR unfolded protein response would balance the processes of protein synthesis, folding, and degradation, respectively. Reducing chronic ER stress and enhancing adaptive UPR signaling can help in reducing the toxic effects of protein aggregates. By boosting the cellular defense mechanisms, one can hope to bring protein homeostasis back and save neurons from the damaging effects of protein aggregation. Therapy for neurodegenerative diseases opens up a new pathway through this.

9.7.3 Role of ERAD in Protein Homeostasis

The endoplasmic reticulum-associated protein degradation (ERAD) pathway is an indispensable cellular mechanism that targets proteins that are not folded correctly in the ER for degradation. This is a three-step process with the steps being recognition, ubiquitination, and retro-translocation. The recognition of misshaped or mutated proteins is based on the identification of substructures in proteins such as exposed hydrophobic regions, unpaired cysteine residues, or immature glycans (Gariballa and Ali 2020). For example, a mammalian cell mechanism named glycan processing is available, in which calnexin/calreticulin assists immature glycoproteins to attain a native conformation. Nevertheless, in the end, only terribly misfolded proteins need to be extracted from calnexin/calreticulin, and this is accomplished by members of the EDEM (ER degradation-enhancing α-mannosidase-like protein) family.

The second step, ubiquitination, is the attaching of the protein molecules ubiquitin to the misfolded proteins, thus tagging them for degradation. This is mediated through an enzymatic reaction cascade, which is initiated by the E1 enzyme. The last phase, retro-translocation, includes the transport of the ubiquitinated proteins from the ER back into the cytosol. It is thought this is facilitated by Hrd1 E3 ubiquitin-protein ligase, which can act as the retro-translocon or dislocon (Penke et al. 2018). The cytosolic proteins are then degraded by proteasome, which is a complex responsible for breaking down the proteins.

Broken pathways of the ERAD have been associated with various neurodegenerative diseases. For these diseases, accumulation of misfolded proteins and ER stress consequently causes neuronal dysfunction. Furthermore, equal neuroglial cells that suffer ERS activate and carry out the damaged inflammatory responses that help produce the neurotoxic milieu. For example, hereditary hemorrhagic telangiectasia types 1 and 2 (HHT1 and HHT2) and familial pulmonary arterial hypertension (FPAH), which are associated with mutations in Endoglin, ALK1, and BMPR2, respectively, have been linked to the loss of function phenotype through the ER retention of some of their disease-causing variants (Abdalla and Letarte 2006). This has led in some cases to premature protein degradation via the proteasomal pathway. Thus, maintaining normal ERAD functioning is a necessity for healthy neurons.

9.8 Natural Compounds as Modulators of ERS

The ER is not only the protein constructor but also the polypeptide assembly plant and polypeptide operational center of eukaryotic cells. It is also sometimes called the calnexin cycle. ER homeostasis is disrupted when there is a misfolding challenge occurring due to endogenous (genetic mutation, oxidative damage, and metabolic imbalances) or exogenous factors. As a result, the defective polysomes accumulate in the ER, and some of them are devoid of functional proteins. In the last few years, natural compounds have been the focus of many researchers who study bio-protectors of ERS. These substances can be either plant-derived or are produced in nature, have many pharmacological effects, and are effective in diseases associated with ERS. This section summarizes the modulation of ERS by phytochemicals, herbal extracts, and other natural compounds. Their mechanisms of action will be elaborated, along with a focus on preclinical and clinical studies demonstrating their therapeutic value. Using nature's technology, these compounds provide a new approach in the treatment of ERS-associated diseases.

9.8.1 Overview of Natural Compounds

Natural compounds, including phytochemicals and herbal extracts, have recently become known as significant modulators of endoplasmic reticulum (ER) stress and unfolded protein response (UPR) pathways. These compounds, sourced from a variety of natural organisms like plants and marine life, show various bioactive properties that alter the response of cells to stress. An outstanding example is curcumin, a polyphenolic substance from turmeric. Curcumin has been the subject of numerous studies on ERS, particularly in the case of Alzheimer's disease and Parkinson's disease (Turer and Sanlier 2025). It has been demonstrated that curcumin can limit ERS and UPR activation, which are a crucial part of the development of these diseases. The protective effect is realized through its antioxidant and anti-inflammatory properties, whereby it alleviates cellular stress that leads to neurodegeneration (Genchi et al. 2024). Model organisms have shown that curcumin treatment leads to less accumulation of misfolded proteins in the ER, which in turn, helps to decrease the activation of UPR pathways that cause cell death (Esmaealzadeh et al. 2024). This offers new hope for curcumin as a therapeutic agent to treat the conditions related to ERS.

Aside from curcumin, herbal extracts such as *Ginkgo biloba* and *Panax ginseng* are also known to modulate ERS and UPR pathways. *Ginkgo biloba*, a common herbal remedy, is composed of a mix of triterpenes and flavonoids which have been found to be beneficial for the brain (Barbalho et al. 2022). Evidence indicates that the *Ginkgo biloba* extract has an effect on ERS through the reduction of oxidative stress and modulation of the UPR, which can be of help in curing neurodegenerative diseases (Achete de Souza et al. 2020). *Panax ginseng*, another famous herbal extract, is traditionally used in folk medicine for its potential neuroprotective properties. The ginsenosides, which are the active ingredients, have an impact on ERS pathways and thus are a factor in the improvement of neurodegenerative symptoms (Ong et al. 2015; Hou et al. 2020; Yang et al. 2024). Both *Ginkgo biloba* and *Panax ginseng* present a case of traditional herbal medicines involving diverse bioactive components that are being explored for their ability to modulate ERS and thereby improve cellular health with neurodegenerative diseases as an example.

9.8.2 Mechanisms of Action of Natural Compounds

Natural compounds can modulate ERS through a variety of mechanisms, including the enhancement of ER chaperone activity, inhibition of oxidative stress, and regulation of UPR signaling pathways. ER chaperones, such as BiP (binding immunoglobulin protein) and GRP94 (glucose-regulated protein 94), play a crucial role in protein folding and quality control within the ER. Certain natural compounds can increase the expression or activity of these chaperones, thereby improving the cell's ability to handle misfolded proteins and alleviate ERS. For instance, compounds like resveratrol, a polyphenol found in grapes, have been shown to upregulate ER chaperones, enhancing their protective effects and aiding in the restoration of ER homeostasis (Azargoonjahromi and Abutalebian 2024). This mechanism is particularly important in conditions where protein misfolding and aggregation are prominent, such as neurodegenerative diseases.

In addition to enhancing chaperone activity, natural compounds can also combat ERS by inhibiting oxidative stress. Oxidative stress contributes significantly to ER dysfunction by promoting protein misfolding and accumulation of reactive oxygen species (ROS) (Ma et al. 2020). Antioxidants such as quercetin, found in many fruits and vegetables, have been shown to reduce oxidative stress and subsequently decrease ERS. By scavenging ROS and protecting cellular components, these compounds help maintain ER function and mitigate the activation of UPR pathways (Costa et al. 2016). Furthermore, natural compounds can directly influence UPR signaling pathways. For example, the flavonoid epigallocatechin gallate (EGCG) from green tea has been observed to modulate UPR signaling by affecting key sensors such as IRE1 and PERK (Weinreb et al. 2009). This modulation can help fine-tune the cellular stress response, promoting cell survival and function under stressful conditions. Understanding these specific mechanisms is crucial for optimizing the therapeutic potential of natural compounds, allowing for targeted interventions that harness their benefits in treating diseases associated with ERS.

9.8.3 Preclinical and Clinical Studies of Natural Compounds

Curcumin, a polyphenol obtained from turmeric, has become popular with researchers interested in exploring its potential therapeutic properties for Alzheimer's disease (AD). Preclinical trials have stressed the importance of addressing ER stress through curcumin modulation. In AD models, curcumin has been found to diminish ER stress markers such as GRP78/BiP which is one of the primary chaperone proteins involved in protein folding in the ER. With this upregulation of the stress markers, it seems curcumin has a consequence of downregulating the ER burden. Besides, curcumin is also implicated in the alteration of UPR important signalings like IRE1α, PERK, and ATF6 which eventually result in the inhibition of the ERS signaling pathway. Also, it acts as a positive autophagy regulator by the mechanism of removing misfolded proteins and protein aggregates (Forouzanfar et al. 2020). Enabling neurons to overcome ERS-induced apoptosis becomes curcumin's major role through its multifaceted antioxidant, anti-inflammatory, and protein-folding actions.

Resveratrol, which is considerably plentiful in grapes and red wine, has additionally been a focus of research relative to its actions in Parkinson's disease (PD) and the connection with ERS. In preclinical models of PD, resveratrol is the factor that has proved most beneficial for the proper functioning of ERS. It has a great impact on the reduction of protein aggregation which is one of the leading causes of ERS-associated neuron dysfunction (Dos Santos et al. 2022). Resveratrol by the activation of ER chaperone proteins, such as BiP/GRP78, greatly assists in proper protein folding, reducing the subsequent stress of misfolding (Derf et al. 2019). Similar to curcumin, resveratrol stimulates the autophagy process, thereby freeing the cell of malfunctioning proteins, and this might lead to less damage in neurons. Resveratrol can also exert beneficial effects on dopaminergic neurons by modulating the pathway of ERS which, in turn, may result in the halting of PD progression. The neuroprotective potential of resveratrol has been evaluated through an in vitro study of human stem cell-derived neuronal cells that uncovered its ability to reduce the neurotoxicity of monocrotophos, a highly toxic organophosphate insecticide (Jahan et al. 2018).

At present, trials are being carried out to investigate whether the use of curcumin and resveratrol as well as other natural compounds in the treatment of patients with neurodegenerative diseases is safe and effective. The focus of these trials is to explore the safety profile of these compounds, ascertain the best doses, and identify unwanted effects. The outcomes of these studies will be necessary to appreciate what makes these compounds useful in controlling ERS and improving neuronal health (Andrade et al. 2023). As far as this goes, the experiments are instrumental in paving the way to new forms of therapies and, in the long run, serve as natural remedies that may contribute to the reform of neurodegenerative disease management.

9.9 Antioxidant Properties of Phytochemicals in ERS Mitigation

Oxidative stress is caused by the surplus of reactive oxygen species (ROS) and insufficient antioxidant capacity of the body. Among ROS are hydrogen peroxide (H_2O_2), hydroxyl radicals (OH^-), singlet oxygen ($^1O_2$), and superoxide (O_2^-) species that are generated by numerous cellular processes. When levels of ROS cross the limits of the body's neutralization capacity, they can trigger damage to cellular components including proteins, lipids, and DNA. This kind of oxidative injury may result in a cellular stress response, particularly in the ER (Di Meo et al. 2016). The resulting dysfunction of ER is one of the main players in triggering the UPR, a cellular mechanism that specifically restores the ER to a steady state and eases oxidative stress. Antioxidant phytochemicals, which are bioactive compounds found in plants, have garnered attention for their potential to combat oxidative stress and its associated disorders. These phytochemicals, such as flavonoids, polyphenols, and carotenoids, play a vital role in preventing and mitigating the effects of oxidative damage (Akbari et al. 2022). Antioxidants are those substances which help in the process of scavenging ROS and neutralizing free radicals, consequently reducing oxidative damage and, in turn, ERS. These compounds not only have antioxidant activity but also have anti-inflammatory effects, which play a pivotal role in regulating oxidative stress-induced inflammation and cellular equilibrium (Leyane et al. 2022). Their neuroprotective activities are noteworthy because they protect neurons from oxidative damage and possibly prevent the occurrence of neurodegenerative diseases.

9.9.1 Mechanism of Antioxidant Phytochemicals

Antioxidant phytochemicals are bioactive compounds found in plants that possess the capacity to neutralize reactive oxygen species (ROS) and oxidative stress, thereby eliminating endoplasmic reticulum (ER) stress. Some of the predominant antioxidant phytochemicals are flavonoids, polyphenols, and carotenoids.

Flavonoids such as quercetin are mostly found in fruits and vegetables, which are famous for their effectiveness in the removal of free radicals. Quercetin acts as an antioxidant by neutralizing ROS, reducing lipid peroxidation, and promoting the stability of membranes in cells directly. Polyphenols, such as resveratrol that is present in grapes and red wine, show similar results and are also powerful antioxidants. Resveratrol enhances the activity of some of the antioxidant enzymes, which are the superoxide dismutase (SOD) and the catalase that the body produces on its own; these enzymes are involved in the detoxification of ROS and oxidative stress shield (Jia et al. 2019). Carotenoids are basically astaxanthin that is present in aquatic forms, and these carotenoids are remarkable antioxidants as they can directly change the single oxygen state which protects cellular parts from oxidative stress (Brotosudarmo et al. 2020). On the whole, the joint action of these antioxidant phytochemicals averts ER malfunctioning by counteracting the oxidative stress and thus ensuring cellular integrity, particularly in neurons that are often attacked by this oxidative stress.

Antioxidant phytochemicals have protective action against free radical activity through various intracellular mechanisms. The main pattern is to scavenge ROS directly; these compounds can degrade ROS and prevent them from causing cellular damage. Quercetin and astaxanthin are two of the compounds which can donate electrons to free radicals, and as a result of this, they remain harmless and do not cause oxidative damage to lipids, proteins, and DNA (Costa et al. 2016). The next step is endogenous antioxidant defense upregulation. Resveratrol can activate the nuclear factor erythroid 2-related factor 2 (Nrf2) pathway, leading to the increase in the expression of antioxidant enzymes such as SOD, catalase, and glutathione peroxidase (Jia et al. 2019). This upregulation enhances the cell's intrinsic ability to counteract oxidative stress. In addition, antioxidant phytochemicals can inhibit the pro-oxidant enzymes such as NADPH oxidase, which is also engaged in ROS generation (Stevenson and Hurst 2007). The action of compounds like curcumin to block these enzymes allows the body to limit the amount of ROS produced and eliminates oxidative stress, which in turn helps in the recovery from ERS and other associated functional cell issues.

The collaborative actions of different antioxidant phytochemicals greatly enhanced their protective properties, making this approach the most promising route. For instance, the mixing of curcumin and resveratrol was documented to confer more significant protection against ERS and oxidative injury than either compound alone (Bisht et al. 2010; Son et al. 2013). This synergistic effect is associated with the two compounds' actions on different systems. Curcumin inhibitors with strong anti-inflammatory and protein-modulating properties together with resveratrol's potent antioxidant and gene-regulating effects provide a complete protective atmosphere against oxidative stress and ER dysfunction. By examining such combinations, breakthroughs will be made for more effective disease management for conditions characterized by oxidative stress and ERS, such as neurodegenerative disorders. By using this synergistic potential of antioxidant phytochemicals, researchers can develop treatments that will have maximum neuroprotection and improve the strength of cells to withstand stress-induced damage.

9.9.2 Clinical Applications of Antioxidant Phytochemicals

Antioxidant phytochemicals which are organic compounds of plants have become one of the most popular topics in the medical field due to their promise of curing many illnesses. These bioactive substances including flavonoids, polyphenols, and carotenoids will do the trick for you as they have such a stunning capability of deactivating reactive oxygen species (ROS) and lessening oxidative stress. From neurodegenerative illnesses to skin disease, the clinical applications of antioxidant phytochemicals are plenty and broad (Prakash and Gupta 2009; Zhang et al. 2015). The following bullet points show the remarkable role and action of these substances in promoting health and combating illness, including specific examples and the cellular pathways involved.

9.10 Herbal Extracts and Their Role in Inducing ER Chaperones

Herbal extracts, which are complex mixtures of bioactive compounds that are derived from plants, have been used as primary medicine for many years in traditional medicine. Their ample and diverse therapeutic properties, such as the ability to regulate endoplasmic reticulum (ER) stress, as along with ER chaperones proteins that assist with protein folding and overall cellular homeostasis, are among the highest achievements of these extracts (Chang et al. 2016). The current topic discusses the range of herbal extracts that influence ERS, particularly using *Ginkgo biloba* and *Panax ginseng* as examples.

9.10.1 *Ginkgo biloba* Extract

Ginkgo biloba extract (GBE) is noted for its high content of flavonoids and terpenoids, which are the compounds that offer significant neuroprotective properties. GBE has been investigated a lot for its capacity to activate ER chaperones and decrease ERS (Lai et al. 2011; de Souza et al. 2020). A remarkable examination that occurred in 2019 showed that GBE could decrease the levels of ERS in rat neurons and therefore acts as a protective factor against apoptosis (Tian et al. 2019). This kind of neuroprotective property is highly relevant in the case of neurodegenerative diseases such as Alzheimer's disease. Clinical trials have produced data showing the positive effect of GBE on memory and cognitive functions in Alzheimer's patients (Huang et al. 2020). An example is a double-blind, placebo-controlled, randomized clinical trial that found that GBE has a significant impact on variables manifesting cognitive functions in patients with mild to moderate stages of Alzheimer's disease (Dodge et al. 2008). The UPR pathway is activated, and the ER chaperones are expressed, thus decreasing ERS. On top of that, GBE has antioxidative properties which add to its neuroprotective effects by lowering the oxidative effect, which often causes the ER to malfunction.

9.10.2 *Panax ginseng* Extract

Panax ginseng extract (PGE) is yet another very powerful herbal extract which gets its potential from ginsenosides, the compounds that modulate ERS and augment the expression of ER chaperones. The results of preclinical studies confirm the ability of PGE to protect cells from apoptosis, caused by ERS acquisition, and show that PGE improves cognitive function in various neurodegenerative disease models (Feng et al. 2022). A good example is a study involving mice that showed that PGE could considerably enhance their skills in spatial learning and memory (Kim et al. 2017). The ginsenosides that are contained in PGE have many ways to affect the body including the modulation of the UPR, the enhancement of antioxidant defenses, and the regulation of gene expression. The compounds activate the PERK, IRE1, and ATF6 pathways, which are the main components of the UPR. Consequently, the expression of ER chaperones get increased, thereby improving the protein folding capacity of the ER (Liu et al. 2016). Moreover, PGE's anti-inflammatory properties also play a role in its overall neuroprotective effects, and thus, PGE is viewed as a candidate for the treatment of neurodegenerative diseases.

9.10.3 Clinical Evidence and Therapeutic Potential

Clinical evidence of testing herbs has emerged to provide the best assurance of safety and effectiveness in regulating ERS and for neuronal health (Dsouza et al. 2022). These extracts are also proposed candidates for neurodegenerative diseases, and they might also be used alongside conventional treatments as supporting therapies. For instance, in one of the studies among Alzheimer patients, the *Ginkgo biloba* extract was shown to have a significant positive effect on the cognitive function of patients. The *Panax ginseng* extract also proved its worth by diminishing the symptoms of patients with cognitive impairment and at the same time by enhancing cognitive function. Nevertheless, more studies remain necessary to prove the efficacy of the mentioned therapeutic remedies and the active compounds responsible for their potential. This means that the bioactive components are isolated, attention is given to pharmacokinetics and pharmacodynamics, and immense clinical trials are organized to give affirmative proof of these extracts in terms of efficacy and safety. The above work is, of course, a starting point for developing effective solutions for these diseases and thus fully leveraging the recovery potential of plants.

9.11 Therapeutic Potential of Natural Compounds in Neuroinflammation

The introduction of natural compounds in treating neuroinflammation is seen as a potential solution, especially in age-related ailments such as Alzheimer's and Parkinson's. Neuroinflammation is the major issue to overcome in these diseases, which is commonly caused by oxidative stress and inflammation, leading to neuronal injury and finally the consequent acceleration of the disease's pathology. The chronic state of ERS causes the activation of inflammatory signaling pathways NF-κB and JNK that results in the production of pro-inflammatory cytokines TNF-α, IL-1β, and IL-6, which make the damage to neural cells worse (Chipurupalli et al. 2021). The thorough understanding of the dialectic between ERS and neuroinflammation is like a missing puzzle piece in creating new therapeutic strategies that would function on both pathways to either stop or slow down the disease.

Natural compounds like curcumin and resveratrol, which possess anti-inflammatory properties, are a great solution to the dual problem of neuroinflammation and excessive ERS, by negatively affecting the processes that cause these ailments. The active components are the antioxidants that curcumin and resveratrol operate through to tackle inflammation. For instance, curcumin disables NF-κB, the critical transcription factor in an inflammatory response, and this in turn reduces the pro-inflammatory genes (Shishodia 2013). Similarly, resveratrol via the activation of SIRT1 pathway with the concomitant improvement of antioxidant activity demonstrated neuroprotection (Gerszon et al. 2014). EGCG is a polyphenol of green tea which possesses multiple mechanisms, i.e., inhibition of COX-2 and iNOS which are the enzymes contributing to inflammation (Mokra et al. 2022). The combined actions of these compound mechanisms lead to low levels of neuroinflammation, thus protecting neuron cells, and therefore, they have the potential to delay neurodegenerative processes.

The findings from preclinical studies on the use of natural agents to reduce neuroinflammation and protection of neurons from ERS-induced damage rates have been promising. Animal models of neurodegenerative disorders such as cognitive function tests have reported considerable progress after therapy with curcumin, resveratrol, and EGCG (Madhubala et al. 2024). The ongoing clinical studies on patient populations aim to evaluate the efficacy and safety of these compounds as well (Hroudová and Fišar 2024). The results so far have been very promising, showing that these natural agents can significantly improve cognitive abilities by lowering markers of inflammation and enhancing overall life quality. Nonetheless, their full therapeutic effect, target dosing, and long-term effects still await exploration.

9.11.1 Therapeutic Targeting of UPR in Neurodegenerative Diseases

In neurodegenerative diseases, therapeutic targeting is developing rapidly and increasingly becoming more sophisticated, with a focus on restorating ER protein homeostasis. This strategy consists of a spectrum of avant-garde measures, including small molecule modulators and gene therapy to biologics and personalized medicine. The purpose of these interventions is to strengthen the adaptive UPR, reduce adverse responses, and finally ensure the survival of neurons. A recent study on Alzheimer s disease (AD) by Li et al. (2024) examined the effect of small molecule inhibitors of PERK and IRE1 on human-derived neuronal cell cultures and found that these inhibitors were able to significantly lower the levels of ERS and apoptosis (Li et al. 2024). The effects of gene editing technologies, like CRISPR/Cas9, are accordingly applied to make such specific UPR component interventions possible (Haney 2018). Such cutting-edge strategies will be the basis for the advent of new and effective therapies against Alzheimer's disease, which will target the mitigation of ERS and its detrimental effects on neuronal health.

As for Amyotrophic Lateral Sclerosis (ALS), Matus et al. (2013) have investigated the case of BiP/GRP78 overexpression on the subcellular dynamics of ER chaperones in different ALS models, where they found that it lowered ERS markers and thus resulted in delayed motor neuron degeneration. Besides, small molecule modulators that induce the overexpression of BiP/GRP78 chaperones are being studied (Matus et al. 2013). Furthermore, another molecule, Sephin1, acts by selectively inhibiting phosphatase PPP1R15A, leading to enhancement of the UPR and protection against misfolding of the proteins in ALS (Das et al. 2015). On the contrary, these small molecules can help fine-tune UPR signaling, which will lead to the preference of cell survival over apoptosis. Meanwhile, these approaches suggest that changing UPR pathways can be a valid remedy for ERS, neuronal apoptosis, and just plain slowing of ALS.

Chemical chaperones, which are mostly small molecules, such as 4-phenylbutyric acid (4-PBA) and taur ursodeoxycholic acid (TUDCA), have also been proven effective in preclinical models by upregulating the expression of key ER chaperones like GRP78, calnexin, and calreticulin. Through the stabilization of intermediates of protein folding by reducing the accumulation of misfolded proteins, these compounds have been helpful in alleviating ER stress in neurons (Patel et al. 2023). Chemical chaperones by augmenting the ER's ability to cope with protein folding problems not only decrease neuronal apoptosis but also promote cellular resilience which might be of therapeutic benefit in diseases characterized by protein aggregation and ER dysfunction.

Gene therapy modalities also come up with a new approach to modulating UPR components for neuroprotection. New avenues of gene education are being pursued, the establishment of which is expected to guard even more nerve cells. As an example, protective UPR components such as XBP1s and ATF6, believed to be involved in adaptive stress responses, are transported in these strategies. The study indicates that the overexpression of XBP1s in a transgenic mouse model for Parkinson's disease (PD) led to a decrease in dopaminergic neuron loss and an improvement in motor function (Valdés et al. 2014). Furthermore, compounds like ER-OASIS (Oxygenase-Antioxidant Synthetic Isopeptide), either targeted to the ER or not, are being developed to reduce ER oxidative stress directly, leading to less protein misfolding and enhanced neuronal function (Jiang et al. 2016). Therefore, these gene therapy and antioxidant strategies aim to restore the affected ER's caloric equilibrium, alleviate neuronal stress, and indeed promote the survival of cells; hence, the hope of their application in successful treatments for neurodegenerative disorders is born.

PERK inhibitors such as GSK2606414 have vast potential in the area of reducing ER stress that is out of bounds in preclinical models. The molecules act on the PERK pathway that is vital in the modulation of protein synthesis and stress response; therefore, it is the target that is required (Rozpędek-Kamińska et al. 2020). IRE1 modulators like 4μ8C are capable of rearranging the IRE1/XBP1 path and thus managing ER stress and inflammation (Siwecka et al. 2021). ATF6 activators, such as Ceapin-A7, boosts the adaptive UPR response; therefore, cellular repair and survival also increases (Gallagher and Walter 2016). These small molecules are instrumental in the selective weathering of the ER stress, while improving neuronal function, even though their clinical efficacy and safety require further investigation.

Another burgeoning area of development is the practice of personalized medicine which is designed for individually selected treatment plans that will be more effective and with fewer side effects. The personalized medicine approach addresses the multifaceted pathophysiology of neurodegenerative diseases by offering individualized therapies that are more effective. The identification of unique misfolded proteins or differential UPR signatures guides the choice of appropriate small molecule modulators or biologics (Swindley 2021). By focusing on the actual causes that affect everyone, the personalized medicine sector aims to deliver more appropriate treatments that will improve the end results and cut down the side effects. The research that validates these biomarkers and those that develop the specific needs for each individual are ongoing (Ahmad et al. 2023). The thorough approach has the promise of improving not just the effectiveness but also the quality of life of patients with untreated diseases. Monoclonal antibody and peptide-based therapies represent a top-notch modality to knock down unwanted proteins involved in ER stress and UPR pathways. Abnormal proteins that suffer aggregation due to ER stress are being targeted by new biologics, which will bind and neutralize these proteins to stop their accumulation, repairing cellular damage as a result. For example, antibodies targeting tau or amyloid-beta proteins are under investigation for their potential to modulate neuroinflammation and improve neuronal health (Valera et al. 2016; Ceyzériat et al. 2020). Likewise, peptide-based therapies mimic or intensify the action of natural ER chaperones; thus, proteotoxicity conditions are prevented (Rajan et al. 2021). Certainly, biologics and peptides point to a modern way of neurodegenerative disease therapy, with ongoing research and clinical trials paving the way.

Protein aggregation diseases, which include Alzheimer's disease and systemic amyloidosis, are marked by the abnormal buildup of protein aggregates that can occur in various cellular compartments including the extracellular space, cytoplasm, and nucleus. These conditions are characterized by the formation of highly structured regions known as amyloids. Amyloids are non-disintegrating, filamentary aggregates that overpower the ingenuity of cellular quality control mechanisms and, as a result, are associated with severe cellular damage. The proteins that form amyloids are constantly in a state of equilibrium, being either monomers that are soluble or in the form of various insoluble aggregated states. However, several determinants largely govern this equilibrium, including the concentration of proteins, their interactions with other proteins, and the condition of the cellular environment. Pathological changes in the equilibrium can lead to pathological aggregation and are a part of the cause and progress in the indicated diseases (Hoseki et al. 2010).

The therapeutic options for reducing protein aggregation are different, but one of the methods is through the consumption of small molecules which block protein aggregation by targeting specific steps in the amyloid formation process (Eisele et al. 2015). These compounds may stabilize native protein structures or prevent the aggregation of proteins by disrupting regions that are prone to aggregation. Another strategy that has been put forward concentrates on enhancing chaperone activity. Chaperones are proteins that aid in the folding of other proteins and prevent misfolding. Our approach is to stimulate the activity of chaperones that would then properly fold the newly formed proteins and thus alleviate cellular stress.

9.12 Conclusion and Future Perspectives

Researchers are making significant strides in developing therapeutic strategies to address ERS and its impact on neurodegenerative diseases. Metabolic pathways are being targeted as an important approach, since treatments that increase insulin sensitivity and regulate glucose metabolism have been shown to lower ERS frequency as a side effect, thus improving neuronal health and slowing down the disease (Yuan et al. 2024). Their homeostasis goal is primarily restoration where they contribute to the biology of protein degradation, such as the ubiquitin-proteasome system (UPS), and promote autophagy. These systems help clean the toxic protein aggregates which are dangerous to neurons. The small molecule CLR01 inhibits major protein aggregates, such as amyloid precursors and alpha-synuclein (Calabrese et al. 2022). Others strategies, such as enhancing the molecular chaperones GRP78 and calnexin, which aid in proper protein folding and assist their function, present a suitable approach to mitigating ERS and restoring cellular balance.

Applications of molecular biology technologies such as CRISPR and RNA interference in understanding the functions of ER chaperones and their role in the pathogenesis of neurodegenerative diseases are improving rapidly (Chehelgerdi et al. 2024; Kumari et al. 2024). High-throughput screening provides a technique for the identification of small molecules that are involved in the modulation of chaperone activity, while proteomics and transcriptomics indicate the interaction of chaperones with autophagy and oxidative stress pathways (Chang et al. 2008). Natural compounds and exosomes from human stem cells that carry bioactive molecules to modulate the neuroinflammatory pathway and neuron survival are also the focus of therapies (Jahan et al. 2023; Kumar 2023; Nayeem et al. 2023). The researchers are forging ahead with the plan to restore cellular systems to their normal operation, conserve neuronal functionality, and delay the onset of neurodegenerative disorders by means of a patchwork of these different methods, thus introducing a whole new way of treating patients.

References

Abdalla SA, Letarte M (2006) Hereditary haemorrhagic telangiectasia: current views on genetics and mechanisms of disease. J Med Genet 43(2):97–110

Achete de Souza G, de Marqui SV, Matias JN, Guiguer EL, Barbalho SM (2020) Effects of Ginkgo biloba on diseases related to oxidative stress. Planta Med 86(6):376–386

Adav SS, Sze SK (2020) Hypoxia-induced degenerative protein modifications associated with aging and age-associated disorders. Aging Dis 11(2):341–364

Ahmad A, Imran M, Ahsan H (2023) Biomarkers as biomedical bioindicators: approaches and techniques for the detection, analysis, and validation of novel biomarkers of diseases. Pharmaceutics 15(6):1630

Ajoolabady A, Lindholm D, Ren J, Pratico D (2022) ER stress and UPR in Alzheimer's disease: mechanisms, pathogenesis, treatments. Cell Death Dis 13(8):706

Akbari B, Baghaei-Yazdi N, Bahmaie M, Mahdavi Abhari F (2022) The role of plant-derived natural antioxidants in reduction of oxidative stress. Biofactors 48(3):611–633

Andrade S, Nunes D, Dabur M, Ramalho MJ, Pereira MC, Loureiro JA (2023) Therapeutic potential of natural compounds in neurodegenerative diseases: insights from clinical trials. Pharmaceutics 15(1):212

Azargoonjahromi A, Abutalebian F (2024) Unraveling the therapeutic efficacy of resveratrol in Alzheimer's disease: an umbrella review of systematic evidence. Nutr Metab (Lond) 21(1):15

Barbalho SM, Direito R, Laurindo LF, Marton LT, Guiguer EL, Goulart RA, Tofano RJ, Carvalho ACA, Flato UAP, Capelluppi Tofano VA, Detregiachi CRP, Bueno PCS, Girio RSJ, Araujo AC (2022) Ginkgo biloba in the aging process: a narrative review. Antioxidants (Basel) 11(3)

Benskey MJ, Perez RG, Manfredsson FP (2016) The contribution of alpha synuclein to neuronal survival and function – implications for Parkinson's disease. J Neurochem 137(3):331–359

Bisht K, Wagner K-H, Bulmer AC (2010) Curcumin, resveratrol and flavonoids as anti-inflammatory, cyto-and DNA-protective dietary compounds. Toxicology 278(1):88–100

Brotosudarmo THP, Limantara L, Setiyono E, Heriyanto (2020) Structures of astaxanthin and their consequences for therapeutic application. Int J Food Sci 2020(1):2156582

Brown MK, Naidoo N (2012) The endoplasmic reticulum stress response in aging and age-related diseases. Front Physiol 3:263

Calabrese G, Molzahn C, Mayor T (2022) Protein interaction networks in neurodegenerative diseases: from physiological function to aggregation. J Biol Chem 298(7):102062

Cao SS (2015) Endoplasmic reticulum stress and unfolded protein response in inflammatory bowel disease. Inflamm Bowel Dis 21(3):636–644

Cao SS, Kaufman RJ (2012) Unfolded protein response. Curr Biol 22(16):R622–R626

Ceyzériat K, Zilli T, Millet P, Frisoni GB, Garibotto V, Tournier BB (2020) Learning from the past: a review of clinical trials targeting amyloid, tau and neuroinflammation in Alzheimer's disease. Curr Alzheimer Res 17(2):112–125

Chang K-H, Chen I-C, Lin H-Y, Chen H-C, Lin C-H, Lin T-H, Weng Y-T, Chao C-Y, Wu Y-R, Lin J-Y (2016) The aqueous extract of Glycyrrhiza inflata can upregulate unfolded protein response-mediated chaperones to reduce tau misfolding in cell models of Alzheimer's disease. In: Drug design, development and therapy, pp 885–896

Chang L, Bertelsen EB, Wisén S, Larsen EM, Zuiderweg ER, Gestwicki JE (2008) High-throughput screen for small molecules that modulate the ATPase activity of the molecular chaperone DnaK. Anal Biochem 372(2):167–176

Chehelgerdi M, Chehelgerdi M, Khorramian-Ghahfarokhi M, Shafieizadeh M, Mahmoudi E, Eskandari F, Rashidi M, Arshi A, Mokhtari-Farsani A (2024) Comprehensive review of CRISPR-based gene editing: mechanisms, challenges, and applications in cancer therapy. Mol Cancer 23(1):9

Chipurupalli S, Samavedam U, Robinson N (2021) Crosstalk between ER stress, autophagy and inflammation. Front Med 8:758311

Colla E (2019) Linking the endoplasmic reticulum to Parkinson's disease and alpha-synucleinopathy. Front Neurosci 13:560
Corazzari M, Gagliardi M, Fimia GM, Piacentini M (2017) Endoplasmic reticulum stress, unfolded protein response, and cancer cell fate. Front Oncol 7:78
Costa LG, Garrick JM, Roque PJ, Pellacani C (2016) Mechanisms of neuroprotection by quercetin: counteracting oxidative stress and more. Oxid Med Cell Longev 2016:2986796
Credle JJ, Forcelli PA, Delannoy M, Oaks AW, Permaul E, Berry DL, Duka V, Wills J, Sidhu A (2015) α-Synuclein-mediated inhibition of ATF6 processing into COPII vesicles disrupts UPR signaling in Parkinson's disease. Neurobiol Dis 76:112–125
Cubillos-Ruiz JR, Bettigole SE, Glimcher LH (2017) Tumorigenic and immunosuppressive effects of endoplasmic reticulum stress in cancer. Cell 168(4):692–706
Danilczyk UG, Cohen-Doyle MF, Williams DB (2000) Functional relationship between calreticulin, calnexin, and the endoplasmic reticulum luminal domain of calnexin. J Biol Chem 275(17):13089–13097
Das I, Krzyzosiak A, Schneider K, Wrabetz L, D'Antonio M, Barry N, Sigurdardottir A, Bertolotti A (2015) Preventing proteostasis diseases by selective inhibition of a phosphatase regulatory subunit. Science 348(6231):239–242
Dawood AA, Altobje MA (2020) Inhibition of N-linked glycosylation by tunicamycin may contribute to the treatment of SARS-CoV-2. Microb Pathog 149:104586
de Souza GA, de Marqui SV, Matias JN, Guiguer EL, Barbalho SM (2020) Effects of Ginkgo biloba on diseases related to oxidative stress. Planta Med 86(06):376–386
Derf A, Mudududdla R, Bharate SB, Chaudhuri B (2019) Inhibitors of Aβ42-induced endoplasmic reticular unfolded protein response (UPRER), in yeast, also rescue yeast cells from Aβ42-mediated apoptosis. Eur J Pharm Sci 128:118–127
Di Meo S, Reed TT, Venditti P, Victor VM (2016) Role of ROS and RNS sources in physiological and pathological conditions. Oxidative Med Cell Longev 2016(1):1245049
Dodge H, Zitzelberger T, Oken B, Howieson D, Kaye J (2008) A randomized placebo-controlled trial of Ginkgo biloba for the prevention of cognitive decline. Neurology 70(19_part_2):1809–1817
Dos Santos MG, Schimith LE, Andre-Miral C, Muccillo-Baisch AL, Arbo BD, Hort MA (2022) Neuroprotective effects of resveratrol in in vivo and in vitro experimental models of Parkinson's disease: a systematic review. Neurotox Res 40(1):319–345
Dovonou A, Bolduc C, Soto Linan V, Gora C, Peralta MR III, Lévesque M (2023) Animal models of Parkinson's disease: bridging the gap between disease hallmarks and research questions. Transl Neurodegener 12(1):36
Dsouza VL, Shivakumar AB, Kulal N, Gangadharan G, Kumar D, Kabekkodu SP (2022) Phytochemical based modulation of endoplasmic reticulum stress in Alzheimer's disease. Curr Top Med Chem 22(22):1880–1896
Eisele YS, Monteiro C, Fearns C, Encalada SE, Wiseman RL, Powers ET, Kelly JW (2015) Targeting protein aggregation for the treatment of degenerative diseases. Nat Rev Drug Discov 14(11):759–780
Ekundayo BE, Obafemi TO, Adewale OB, Obafemi BA, Oyinloye BE, Ekundayo SK (2024) Oxidative stress, endoplasmic reticulum stress and apoptosis in the pathology of Alzheimer's disease. Cell Biochem Biophys 82:1–21
Esmaealzadeh N, Miri MS, Mavaddat H, Peyrovinasab A, Ghasemi Zargar S, Sirous Kabiri S, Razavi SM, Abdolghaffari AH (2024) The regulating effect of curcumin on NF-kappaB pathway in neurodegenerative diseases: a review of the underlying mechanisms. Inflammopharmacology 32:2125
Feng H, Xue M, Deng H, Cheng S, Hu Y, Zhou C (2022) Ginsenoside and its therapeutic potential for cognitive impairment. Biomolecules 12(9):1310
Forouzanfar F, Read MI, Barreto GE, Sahebkar A (2020) Neuroprotective effects of curcumin through autophagy modulation. IUBMB Life 72(4):652–664
Gallagher CM, Walter P (2016) Ceapins inhibit ATF6α signaling by selectively preventing transport of ATF6α to the Golgi apparatus during ER stress. elife 5:e11880

Gariballa N, Ali BR (2020) Endoplasmic reticulum associated protein degradation (ERAD) in the pathology of diseases related to TGFbeta signaling pathway: future therapeutic perspectives. Front Mol Biosci 7:575608

Genchi G, Lauria G, Catalano A, Carocci A, Sinicropi MS (2024) Neuroprotective effects of curcumin in neurodegenerative diseases. Foods 13(11)

Gerszon J, Rodacka A, Puchała M (2014) Antioxidant properties of resveratrol and its protective effects in neurodegenerative diseases. Med J Cell Biol 4(2):97–117

Hamilton S, Terentyev D (2018) Proarrhythmic remodeling of calcium homeostasis in cardiac disease; implications for diabetes and obesity. Front Physiol 9:1517

Hamilton S, Terentyev D (2022) ER stress and calcium-dependent arrhythmias. Front Physiol 13:1041940

Haney M (2018) Identification of genetic modifiers of neurodegenerative diseases using CRISPR-Cas9 screens. Stanford University

Hetz C, Mollereau B (2014) Disturbance of endoplasmic reticulum proteostasis in neurodegenerative diseases. Nat Rev Neurosci 15(4):233–249

Hoseki J, Ushioda R, Nagata K (2010) Mechanism and components of endoplasmic reticulum-associated degradation. J Biochem 147(1):19–25

Hou W, Wang Y, Zheng P, Cui R (2020) Effects of ginseng on neurological disorders. Front Cell Neurosci 14:55

Hroudová J, Fišar Z (2024) Alzheimer's disease approaches-focusing on pathology, biomarkers and clinical trial candidates. Prog Neuro-Psychopharmacol Biol Psychiatry 111069:111069

Huang L-K, Chao S-P, Hu C-J (2020) Clinical trials of new drugs for Alzheimer disease. J Biomed Sci 27:1–13

Jahan S, Kumar D, Mukherjee S, Roy K, Siddiqui AJ, Khan A, Rajpurohit CS, Aldosari S (2023) Stem cells, derived exosomes, and associated signaling molecules in neuroprotection. In: Applications of stem cells and derived exosomes in neurodegenerative disorders. Springer, pp 259–290

Jahan S, Kumar D, Singh S, Kumar V, Srivastava A, Pandey A, Rajpurohit C, Khanna V, Pant A (2018) Resveratrol prevents the cellular damages induced by monocrotophos via PI3K signaling pathway in human cord blood mesenchymal stem cells. Mol Neurobiol 55(11):8278–8292

Jia R, Li Y, Cao L, Du J, Zheng T, Qian H, Gu Z, Jeney G, Xu P, Yin G (2019) Antioxidative, anti-inflammatory and hepatoprotective effects of resveratrol on oxidative stress-induced liver damage in tilapia (Oreochromis niloticus). Comp Biochem Physiol C Toxicol Pharmacol 215:56–66

Jiang T, Sun Q, Chen S (2016) Oxidative stress: a major pathogenesis and potential therapeutic target of antioxidative agents in Parkinson's disease and Alzheimer's disease. Prog Neurobiol 147:1–19

Kenney DL, Benarroch EE (2015) The autophagy-lysosomal pathway: general concepts and clinical implications. Neurology 85(7):634–645

Kim J-I, Jeon SG, Kim KA, Kim J-J, Song EJ, Jeon Y, Kim E, Lee KB, Kwak JH, Moon M (2017) Platycodon grandiflorus root extract improves learning and memory by enhancing synaptogenesis in mice hippocampus. Nutrients 9(7):794

Kim S, Kim DK, Jeong S, Lee J (2022) The common cellular events in the neurodegenerative diseases and the associated role of endoplasmic reticulum stress. Int J Mol Sci 23(11):5894

Kozlov G, Gehring K (2020) Calnexin cycle–structural features of the ER chaperone system. FEBS J 287(20):4322–4340

Kumar D (2023) iPSCs and their role in amelioration of neurodegenerative disorders. In: Jahan S, Siddiqui AJ (eds) Applications of stem cells and derived exosomes in neurodegenerative disorders. Springer, Singapore, pp 111–137

Kumari S, Kamiya A, Karnik SS, Rohilla S, Dubey SK, Taliyan R (2024) Novel gene therapy approaches for targeting neurodegenerative disorders: focusing on delivering neurotrophic genes. Mol Neurobiol:1–26

Lai H-C, Yeh Y-C, Ting C-T, Lee W-L, Lee H-W, Wang L-C, Chen W-T, Lin S-Y, Wu A, Liu T-J (2011) Ginkgo biloba extract 761 prevents hypoxia-triggered cardiomyocyte apoptosis through inhibiting mitochondrial and ER stress-induced apoptotic signaling. Biomed Prev Nutr 1(4):282–293

Leyane TS, Jere SW, Houreld NN (2022) Oxidative stress in ageing and chronic degenerative pathologies: molecular mechanisms involved in counteracting oxidative stress and chronic inflammation. Int J Mol Sci 23(13):7273

Li F, Zhu X, Xu X, Zhou J, Lu R, Wang S, Xing G, Ye Y (2024) Dibromoacetonitrile induced autophagy by mediating the PERK signalling pathway and ROS interaction in HT22 cell. Toxicology 501:153698

Liu H, Yang J, Li L, Shi W, Yuan X, Wu L (2016) The natural occurring compounds targeting endoplasmic reticulum stress. Evid Based Complement Alternat Med 2016(1):7831282

Liu Z, Lv Y, Zhao N, Guan G, Wang J (2015) Protein kinase R-like ER kinase and its role in endoplasmic reticulum stress-decided cell fate. Cell Death Dis 6(7):e1822

Lukasiak A, Zajac M (2021) The distribution and role of the CFTR protein in the intracellular compartments. Membranes (Basel) 11(11)

Ma R-H, Ni Z-J, Thakur K, Zhang F, Zhang Y-Y, Zhang J-G, Wei Z-J (2020) Natural compounds play therapeutic roles in various human pathologies via regulating endoplasmic reticulum pathway. Med Drug Discov 8:100065

Madhubala D, Patra A, Khan MR, Mukherjee AK (2024) Phytomedicine for neurodegenerative diseases: the road ahead. Phytother Res 38:2993

Matus S, Valenzuela V, Medinas DB, Hetz C (2013) ER dysfunction and protein folding stress in ALS. Int J Cell Biol 2013(1):674751

McGrath EP (2018) Investigating the role of IRE1α RNase activity in breast cancer. School of Natural Sciences, National University of Ireland, Galway

Mercado G, Castillo V, Soto P, López N, Axten JM, Sardi SP, Hoozemans JJ, Hetz C (2018) Targeting PERK signaling with the small molecule GSK2606414 prevents neurodegeneration in a model of Parkinson's disease. Neurobiol Dis 112:136–148

Mokra D, Joskova M, Mokry J (2022) Therapeutic effects of green tea polyphenol (–)-Epigallocatechin-3-Gallate (EGCG) in relation to molecular pathways controlling inflammation, oxidative stress, and apoptosis. Int J Mol Sci 24(1):340

Murao N, Nishitoh H (2017) Role of the unfolded protein response in the development of central nervous system. J Biochem 162(3):155–162

National Institutes of Health (2011) NIH study finds two pesticides associated with Parkinson's disease [news release], February 11

Navarro-Romero A, Montpeyo M, Martinez-Vicente M (2020) The emerging role of the lysosome in Parkinson's disease. Cells 9(11)

Nayeem U, Ahmed S, Kumar D, Jahan S, Khan A, Rather G, Siddiqui AJ (2023) Role of stem cells as a protective agent against neurological complications. In: Applications of stem cells and derived exosomes in neurodegenerative disorders. Springer, pp 69–90

Negroni A, Prete E, Vitali R, Cesi V, Aloi M, Civitelli F, Cucchiara S, Stronati L (2014) Endoplasmic reticulum stress and unfolded protein response are involved in paediatric inflammatory bowel disease. Dig Liver Dis 46(9):788–794

Ong WY, Farooqui T, Koh HL, Farooqui AA, Ling EA (2015) Protective effects of ginseng on neurological disorders. Front Aging Neurosci 7:129

Paskevicius T, Farraj RA, Michalak M, Agellon LB (2023) Calnexin, more than just a molecular chaperone. Cells 12(3):403

Patel S, Pangarkar A, Mahajan S, Majumdar A (2023) Therapeutic potential of endoplasmic reticulum stress inhibitors in the treatment of diabetic peripheral neuropathy. Metab Brain Dis 38(6):1841–1856

Penke B, Bogar F, Crul T, Santha M, Toth ME, Vigh L (2018) Heat shock proteins and autophagy pathways in neuroprotection: from molecular bases to pharmacological interventions. Int J Mol Sci 19(1)

Petersen MC, Shulman GI (2018) Mechanisms of insulin action and insulin resistance. Physiol Rev 98(4):2133–2223

Prakash D, Gupta K (2009) The antioxidant phytochemicals of nutraceutical importance. Open Nutr J 2(1):20

Radford H, Moreno JA, Verity N, Halliday M, Mallucci GR (2015) PERK inhibition prevents tau-mediated neurodegeneration in a mouse model of frontotemporal dementia. Thermochim Acta Neuropathol 130:633–642

Rajan R, Ahmed S, Sharma N, Kumar N, Debas A, Matsumura K (2021) Review of the current state of protein aggregation inhibition from a materials chemistry perspective: special focus on polymeric materials. Mater Adv 2(4):1139–1176

Rajpurohit CS, Kumar V, Cheffer A, Oliveira D, Ulrich H, Okamoto OK, Zatz M, Ansari UA, Khanna VK, Pant AB (2020) Mechanistic insights of astrocyte-mediated hyperactive autophagy and loss of motor neuron function in SOD1 L39R linked amyotrophic lateral sclerosis. Mol Neurobiol 57:4117–4133

Ramesh M, Gopinath P, Govindaraju T (2020) Role of post-translational modifications in Alzheimer's disease. Chembiochem 21(8):1052–1079

Rios-Ocampo WA, Navas MC, Buist-Homan M, Faber KN, Daemen T, Moshage H (2020) Hepatitis C virus proteins core and NS5A are highly sensitive to oxidative stress-induced degradation after eIF2alpha/ATF4 pathway activation. Viruses 12(4)

Rozpędek-Kamińska W, Siwecka N, Wawrzynkiewicz A, Wojtczak R, Pytel D, Diehl JA, Majsterek I (2020) The PERK-dependent molecular mechanisms as a novel therapeutic target for neurodegenerative diseases. Int J Mol Sci 21(6):2108

Rozpedek W, Pytel D, Mucha B, Leszczynska H, Diehl JA, Majsterek I (2016) The role of the PERK/eIF2α/ATF4/CHOP signaling pathway in tumor progression during endoplasmic reticulum stress. Curr Mol Med 16(6):533–544

Russell ND, Jorde LB, Chow CY (2023) Characterizing genetic variation in the regulation of the ER stress response through computational and cis-eQTL analyses. G3: Genes Genomes Genet 13(12):jkad229

Sauer A (2017) What you need to know about familial Alzheimer's disease. Alzheimers. net

Shishodia S (2013) Molecular mechanisms of curcumin action: gene expression. Biofactors 39(1):37–55

Siwecka N, Rozpędek-Kamińska W, Wawrzynkiewicz A, Pytel D, Diehl JA, Majsterek I (2021) The structure, activation and signaling of IRE1 and its role in determining cell fate. Biomedicine 9(2):156

Son Y, Lee JH, Chung H-T, Pae H-O (2013) Therapeutic roles of heme oxygenase-1 in metabolic diseases: curcumin and resveratrol analogues as possible inducers of heme oxygenase-1. Oxid Med Cell Longev 2013(1):639541

Stauffer WT, Blackwood EA, Azizi K, Kaufman RJ, Glembotski CC (2020) The ER unfolded protein response effector, ATF6, reduces cardiac fibrosis and decreases activation of cardiac fibroblasts. Int J Mol Sci 21(4)

Stevenson DE, Hurst RD (2007) Polyphenolic phytochemicals–just antioxidants or much more? Cell Mol Life Sci 64(22):2900–2916

Swindley O (2021) Manipulation of endoplasmic reticulum folding and assembly machinery for improved production of complex biotherapeutics. University of Sheffield

Tadic V, Prell T, Lautenschlaeger J, Grosskreutz J (2014) The ER mitochondria calcium cycle and ER stress response as therapeutic targets in amyotrophic lateral sclerosis. Front Cell Neurosci 8:147

Tian J, Popal MS, Liu Y, Gao R, Lyu S, Chen K, Liu Y (2019) Ginkgo Biloba leaf extract attenuates atherosclerosis in streptozotocin-induced diabetic ApoE-/-mice by inhibiting endoplasmic reticulum stress via restoration of autophagy through the mTOR signaling pathway. Oxid Med Cell Longev 2019(1):8134678

Turer BY, Sanlier N (2025) Relationship of curcumin with aging and Alzheimer and Parkinson disease, the most prevalent age-related neurodegenerative diseases: a narrative review. Nutr Rev 83(3):e1243–e1258

Uddin MS, Tewari D, Sharma G, Kabir MT, Barreto GE, Bin-Jumah MN, Perveen A, Abdel-Daim MM, Ashraf GM (2020) Molecular mechanisms of ER stress and UPR in the pathogenesis of Alzheimer's disease. Mol Neurobiol 57:2902–2919

Valdés P, Mercado G, Vidal RL, Molina C, Parsons G, Court FA, Martinez A, Galleguillos D, Armentano D, Schneider BL (2014) Control of dopaminergic neuron survival by the unfolded protein response transcription factor XBP1. Proc Natl Acad Sci 111(18):6804–6809

Valera E, Spencer B, Masliah E (2016) Immunotherapeutic approaches targeting amyloid-β, α-synuclein, and tau for the treatment of neurodegenerative disorders. Neurotherapeutics 13(1):179–189

Verma M, Vats A, Taneja V (2015) Toxic species in amyloid disorders: oligomers or mature fibrils. Ann Indian Acad Neurol 18(2):138–145

Wang M, Wey S, Zhang Y, Ye R, Lee AS (2009) Role of the unfolded protein response regulator GRP78/BiP in development, cancer, and neurological disorders. Antioxid Redox Signal 11(9):2307–2316

Weinreb O, Amit T, Mandel S, Youdim MB (2009) Neuroprotective molecular mechanisms of (-)-epigallocatechin-3-gallate: a reflective outcome of its antioxidant, iron chelating and neuritogenic properties. Genes Nutr 4(4):283–296

Wu D, Huang LF, Chen XC, Huang XR, Li HY, An N, Tang JX, Liu HF, Yang C (2023) Research progress on endoplasmic reticulum homeostasis in kidney diseases. Cell Death Dis 14(7):473

Yadav VK, Awasthi P, Behl R, Kumar A (2022) HSc70 interactome reveal major role of macroautophagy and minor role of chaperone mediated autophagy in K-Ras G12V cell proliferation and survival. J Proteome 264:104614

Yang Y, Chen W, Lin Z, Wu Y, Li Y, Xia X (2024) Panax notoginseng saponins prevent dementia and oxidative stress in brains of SAMP8 mice by enhancing mitophagy. BMC Complement Med Ther 24(1):144

Ye H, Chen C, Wu H, Zheng K, Martín-Adrados B, Caparros E, Francés R, Nelson LJ, Gómez Del Moral M, Asensio I (2022) Genetic and pharmacological inhibition of XBP1 protects against APAP hepatotoxicity through the activation of autophagy. Cell Death Dis 13(2):143

Yin Y, Sun G, Li E, Kiselyov K, Sun D (2017) ER stress and impaired autophagy flux in neuronal degeneration and brain injury. Ageing Res Rev 34:3–14

Yong J, Johnson JD, Arvan P, Han J, Kaufman RJ (2021) Therapeutic opportunities for pancreatic beta-cell ER stress in diabetes mellitus. Nat Rev Endocrinol 17(8):455–467

Yoon H, Shaw JL, Haigis MC, Greka A (2021) Lipid metabolism in sickness and in health: emerging regulators of lipotoxicity. Mol Cell 81(18):3708–3730

Yuan S, She D, Jiang S, Deng N, Peng J, Ma L (2024) Endoplasmic reticulum stress and therapeutic strategies in metabolic, neurodegenerative diseases and cancer. Mol Med 30(1):40

Zatterale F, Longo M, Naderi J, Raciti GA, Desiderio A, Miele C, Beguinot F (2019) Chronic adipose tissue inflammation linking obesity to insulin resistance and type 2 diabetes. Front Physiol 10:1607

Zhang Y-J, Gan R-Y, Li S, Zhou Y, Li A-N, Xu D-P, Li H-B (2015) Antioxidant phytochemicals for the prevention and treatment of chronic diseases. Molecules 20(12):21138–21156

Translational Applications of Natural Compounds in Combatting Neurodegenerative Diseases and Exploring Future Prospects

10

Arshya Hashim and Faria Fatima

Abstract

Neurodegenerative diseases are a group of disorders characterized by the progressive degeneration and dysfunction of the structure and function of the nervous system. These neurological illnesses remain incurable, and current medicines only alleviate the symptoms. The translational application of natural compounds is gaining momentum against neurodegenerative diseases which involves taking findings from basic research and applying them in clinical settings for therapeutic purposes. Several natural compounds have shown promise in preclinical studies for their potential neuroprotective effects, such as curcumin which is found in turmeric and has antioxidant and anti-inflammatory properties that may be neuroprotective against diseases like Alzheimer's. Developing curcumin-based supplements or pharmaceuticals for clinical trials could help evaluate their effectiveness in preventing or slowing neurodegenerative processes. Resveratrol, present in red grapes and wine, has been studied for its potential neuroprotective effects and ability to activate sirtuins, proteins associated with longevity. Polyphenols in green tea, such as epigallocatechin gallate (EGCG) and also omega-3 fatty acids, found in fish oil, have anti-inflammatory and neuroprotective properties which can be supplements on cognitive function and disease progression in neurodegenerative disorders. The future prospects of translational applications involving natural compounds against neurodegenerative diseases hold great promise and involve several key areas of development

A. Hashim (✉)
Department of Biotechnology, Dr. D. Y. Patil Arts, Commerce and Science, Pune, Maharashtra, India
e-mail: arshya.hashim@dypvp.edu.in

F. Fatima
Department of Agriculture, Integral Institute of Agricultural Science and Technology, Integral University, Lucknow, Uttar Pradesh, India

S. Jahan et al. (eds.), *A Mechanistic Exploration of Natural Compounds for Neuronal Health*, https://doi.org/10.1007/978-981-95-0125-0_10

like precision medicine, advanced formulations via nanotechnology, biomarker discovery, targeting multiple pathways, clinical trial design, regulatory pathways, and patient-centric approach. Continued translational research is crucial for advancing our understanding of the therapeutic potential of natural compounds against neurodegenerative diseases. This involves moving beyond preclinical studies to well-designed clinical trials that provide evidence of safety, efficacy, and optimal dosages for these natural interventions. Integrating natural compounds into conventional treatment approaches or developing novel therapeutics based on these compounds holds promise for the future of neurodegenerative disease management.

Keywords

Natural product · Dietary supplement · Cognitive function · Nanotechnology · Biomarkers

10.1 Introduction

Neurodegenerative diseases are a group of disorders characterized by the progressive degeneration and death of nerve cells (neurons) in the brain and nervous system. This leads to a gradual decline in cognitive, motor, and sometimes sensory functions. These diseases can affect different parts of the brain and nervous system, resulting in a variety of symptoms depending on the specific condition. The exact causes of neurodegenerative diseases are not fully understood, but they likely involve a combination of genetic, environmental, and lifestyle factors (Livingston et al. 2020) shown in Fig. 10.1.

The demography of neurodegenerative diseases varies globally and is influenced by factors such as age distribution, genetics, environmental exposures, and healthcare infrastructure. There is a higher prevalence in developed countries due to longer life expectancy and better diagnostic practices. However, rates are rapidly increasing in developing countries as their populations age. Approximately 50 million people worldwide are living with dementia, with Alzheimer's disease accounting for 60–70% of these cases. It primarily affects older adults, and the risk increases significantly with age, especially after 65 years (Ali et al. 2023). Parkinson's disease (PD) affects about ten million people globally. It is most commonly diagnosed in individuals over 60 years old, though early-onset Parkinson's can occur before 50. Huntington's disease (HD) affects approximately 3–7 per 100,000 people of European ancestry and is less common in Asian and African populations (Livingston et al. 2020). Symptoms typically appear between 30 and 50 years old, though juvenile forms can occur. Amyotrophic lateral sclerosis (ALS) affects about 2–3 per 100,000 people worldwide. It is most commonly diagnosed between ages 55 and 75. Multiple sclerosis (MS) is most commonly diagnosed between ages 20 and 50, affecting 2.8 million people worldwide. It is more common in regions farther from the equator, such as North America, Europe, and Australasia (De Pedro-Cuesta et al. 2015; Livingston et al. 2020).

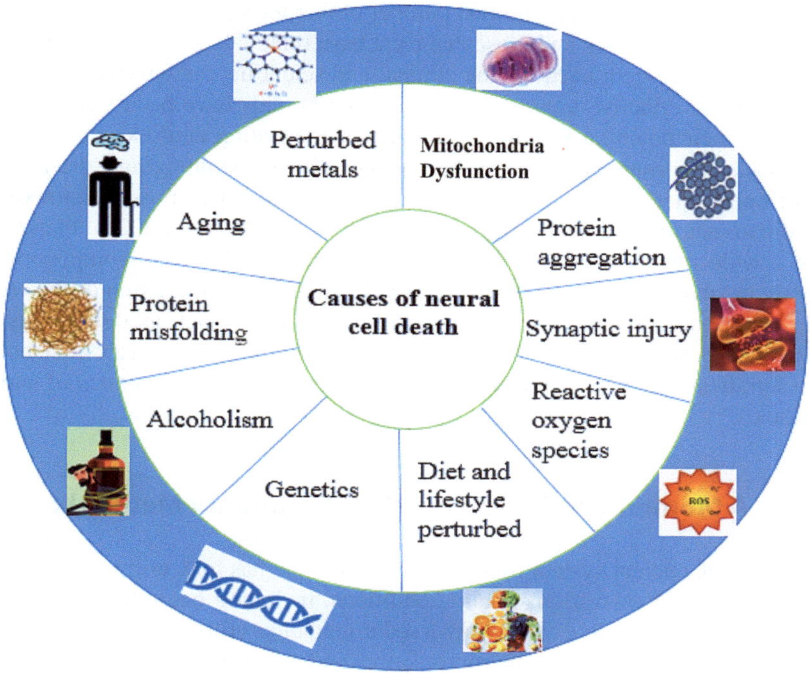

Fig. 10.1 Diagrammatic representation of causes of neural cell death

The exact mechanisms of neurodegeneration are complex and not fully understood, but several key processes that contribute are protein misfolding and aggregation, abnormal folding, and accumulation of specific proteins which can form toxic aggregates, such as amyloid plaques and tau tangles in Alzheimer's disease, Lewy bodies in Parkinson's disease, and huntingtin protein aggregates in Huntington's disease (Keshavan et al. 2021). Oxidative stress is an imbalance between the production of reactive oxygen species (free radicals) and the body's ability to detoxify them and can also damage cellular components, including DNA, proteins, and lipids (Oo et al. 2023: Nayeem et al. 2023). Mitochondrial dysfunction can lead to energy deficits and increased production of reactive oxygen species. Impaired mitochondrial function is observed in many neurodegenerative diseases (Alqahtani et al. 2023). Neuro-inflammation is the chronic activation of the immune response in the brain, involving glial cells (microglia and astrocytes). Neuroinflammation can contribute to neuronal damage and disease progression (Patani et al. 2023). Genetic mutations in specific genes can lead to hereditary forms of neurodegenerative diseases. Examples include mutations in the amyloid protein precursor (APP), presenilin-1 (PSEN1), and presenilin-2 (PSEN2) genes in Alzheimer's disease, and the huntingtin (HTT) gene in Huntington's disease (Akçimen et al. 2023).

Currently, there are only a limited number of therapies available for neurodegenerative diseases (NDDs), and those that exist have relatively low success rates. This underscores the urgent need for the clinical and scientific communities to develop new and more effective treatments. Natural compounds have been studied extensively for their potential benefits in treating or preventing neurodegenerative diseases such as Alzheimer's disease, Parkinson's disease, Huntington's disease, and amyotrophic lateral sclerosis (ALS). These compounds often possess antioxidant, anti-inflammatory, and neuroprotective properties (Jamal et al., 2020). Plants produce a wide range of secondary metabolites, which are organic compounds not directly involved in the normal growth, development, or reproduction of the organism. These compounds often serve ecological functions such as defense against herbivores, pathogens, and competition. However, many secondary metabolites have significant pharmacological activities and are used in the treatment of various diseases.

10.2 A Momentary Review on Plant Secondary Metabolites

Ayurveda, the ancient system of medicine from India, has long utilized plant-based treatments to address various health conditions, including neurodegenerative diseases. As Gregory et al. explains, Ayurvedic medicine utilizes secondary metabolites from plants that offer neuroprotection and possess anti-inflammatory and antioxidant properties. Jack et al. defines secondary metabolites as organic compounds that are not directly associated with the growth, development, or reproduction of plants, fungi, or microorganisms. These metabolites are much more useful as they play vital roles in signaling, defense, and interacting with the environment. Plants have different types of secondary metabolites which include terpenoids, alkaloids, and phenylpropanoids or phenolic compounds. As shown in Fig. 10.2, these metabolites can be broadly divided into two groups, which are metabolites containing nitrogen and those without nitrogen.

10.3 Ethnopharmacological History of Complementary Drugs for Neurodegenerative Diseases

Humans have throughout history used natural resources to treat various neurological disorders. Current-day science is proving ancient treatments from traditional Chinese medicine (TCM), Ayurveda, and ethnic medicines that are gaining familiarity all over the world. Neurodegenerative disorders such as Alzheimer's, Parkinson's, Huntington's, and amyotrophic lateral sclerosis (ALS) can be particularly troublesome due to the absence of effective treatments and the progressive nature of these diseases. The natural resources have the ability to decelerate the progress of the disease and mitigate the symptoms by serving neuroprotective, antioxidant, and anti-inflammatory functions (Gregory et al. 2021; Jack Jr et al. 2019).

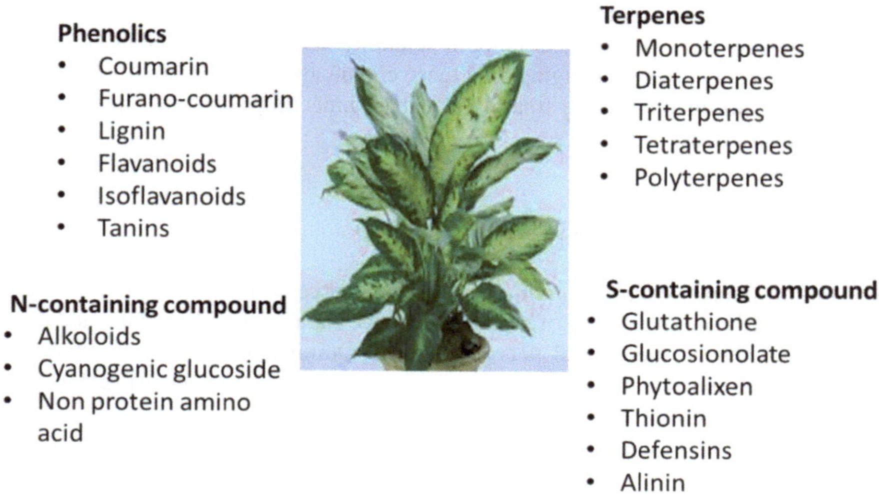

Fig. 10.2 Categories of secondary metabolites

For example, Ashwagandha (*Withania somnifera*) is an ancient Ayurvedic herb for mental peace and sharpness. Evidence suggests it may improve cognitive abilities and reduce the rate of amyloid plaque buildup (Mamatha and Baragur 2024). Similar is the case with *Ginkgo biloba* which is used in TCM for enhancing memory and circulation. It has also been shown to improve other aspects of cognition and even some symptoms of dementia (Yu et al. 2023). Huperzine A, derived from *Huperzia serrata*, has been studied for its memory-enhancing effects. It acts as an acetylcholinesterase inhibitor, an enzyme that advances Alzheimer's (Tripathi et al. 2024).

In advances of traditional medicine, such well-known herbs like *Echinacea*, which is recognized for amping up immunity, or willow bark used for pain serve as synonyms of our understanding of plant-based medicines. Current research seems fixated on a lesser-studied array of plants such as *Alchornea laxiflora*, *Acanthus montanus*, *Ficus platyphylla*, *Sutherlandia frutescens*, *Gladiolus dalenii*, and *Voacanga africana*, specifically noting the plants for their potential use in treating neurodegenerative diseases (Islam et al. 2023). Modern scientific discoveries paired with ancient wisdom are unlocking new dynamic approaches to achieve brain health.

10.4 Traditional Plant-Based Treatments for Neurodegeneration: Historical Perspectives and Modern Implications

The historical background of natural compounds in the treatment of neurodegenerative diseases confirms to power of natural substances to offer new therapeutic possibilities and the importance of traditional knowledge. Natural compounds provide hope in the combat against crippling neurodegenerative diseases from ancient Ayurvedic and TCM practices to modern scientific study (Mukherjee et al. 2023). Investigations are always changing, and these chemicals might eventually be essential parts of all-encompassing treatment plans connecting conventional medicine with modern science.

10.4.1 Early Discoveries (4500–1500 B.C)

The opium poppy (*Papaver somniferum*) is the source of morphine, which established the basis for knowledge of natural substance capacity to relieve pain and their possible effects on the central nervous system. Quinine, known for antimalarial qualities, is isolated from Cinchona bark and prompted more investigation into compounds derived from plants (Buccellato et al. 2023).

10.4.2 Mid-Twentieth Century

L-Dopa was discovered in the 1960s from the plant *Mucuna pruriens* and has become a key player in treating Parkinson's disease by helping boost dopamine levels in the brain. On the other hand, reserpine, extracted from *Rauwolfia serpentina*, has been used to tackle hypertension and various psychiatric conditions, revealing the fascinating effects that natural alkaloids can have on the central nervous system (Sharif-Rad.M et al., 2020; Gregory et al. 2021).

10.4.3 Late Twentieth and Early Twenty-First Centuries

In the latest era, extensive research into the protective effects of turmeric (*Curcuma longa*) demonstrates its effectiveness for Alzheimer's therapy due to its anti-inflammatory and antioxidant curcumin properties (Zia et al. 2021). Research continued to study EGCG from green tea as an agent capable of stopping amyloid plaque generation while controlling oxidative stress in degenerative neurological conditions (Payne et al. 2022). Many phytocannabinoids originate from *Cannabis sativa* through which cannabidiol (CBD) demonstrates promising neuroprotective properties while reducing inflammation and disability symptoms for patients with multiple sclerosis and ALS according to Xiang et al. (2023).

10.5 Pharmacological Impact of Secondary Metabolites on the Treatment of Neurodegenerative Diseases

In accordance with recent studies, a group of secondary metabolites that are effective against progression of neurodegenerative diseases are flavonoids, phenolic acids and stilbenes, alkaloids, carotenoids, catechins, and terpenes (Fig. 10.2)

(Zhang et al. 2024). Secondary metabolites, derived from various plants, have harvested significant attention for their primary effectiveness in addressing neurodegenerative diseases. These compounds exhibit a diverse array of pharmacological activities that contribute to their neuroprotective effects. Continued research into these natural compounds may lead to more effective treatments for NDDs, harnessing the potential of traditional medicine and modern pharmacology.

10.5.1 Preventing Protein Misfolding and Aggregation

Proteins are like the tiny machines of our cells, and for them to work properly, they need to fold into very specific three-dimensional shapes. But sometimes, it does not occur due to genetic mutations, environmental stress, or simple mistakes during their production, and the proteins do not fold the way they were supposed to. When this phenomenon occurs, they can take on abnormal shapes, exposing sticky, water-repelling (hydrophobic) parts that should be hidden inside. These sticky bits cause misfolded proteins to clump together, forming aggregates (Agorogiannis et al. 2004). These clumps can show up as small groups called oligomers, long strands known as fibrils, or even larger, more complex structures. They tend to pile up in neurons, interfering with how brain cells work and communicate. This process of protein misfolding and aggregation is a key contributor to many neurodegenerative diseases, such as Alzheimer's, Huntington's, and Parkinson's (Mehdi 2023). In a healthy brain, proteins fold perfectly, but when misfolding occurs, it sets off a cascade of damage that the brain often struggles to control.

Promising methods to prevent protein misfolding, a crucial step in the development of neurodegenerative diseases, are offered by natural substances. Plant-based products perform especially well when it comes to proteins since they prevent them from misfolding and forming harmful clusters which disable brain cells. Turmeric contains a compound known as curcumin. Research asserts that curcumin is proficient at inhibiting the development of alpha-synuclein fibrils in Parkinson's and inhibiting aggregation of amyloid-beta in Alzheimer's (Zakariya et al. 2022). Resveratrol, which is found in peanuts, berries, and grapes, also possesses antioxidant properties and affects how proteins are folded. It has been demonstrated that it does not only inhibit the formation of fibrous amyloid-beta, but also facilitates the lower expression of amyloid-beta (Meng et al. 2020).

One of the primary components of green tea, epigallocatechin gallate, is well-known in the scientific community for its ability to modify protein aggregates. Research has shown it can restructure destructive aggregates of alpha-synuclein and amyloid-beta, potentially averting their transformation to more pernicious forms (Ashley et al. 2022). Another natural shield is quercetin, a flavonoid abundantly available in fruits, vegetables, and grains. In addition to stabilizing protein structures and preventing their aggregation, quercetin provides strong antioxidant protection which may benefit general brain health (Costa et al. 2016). Together, these natural compounds offer a great promise, not as cures, but as effective, non-invasive herbal tools capable of halting or slowing the mechanisms of neurodegenerative diseases (Zakariya et al. 2022). Additionally, resveratrol from peanuts, berries, and grapes helps regulate protein folding while providing antioxidant benefits. It has been shown to prevent the formation of amyloid-beta fibrils and even assist in their removal (Meng et al. 2020).

Additionally, the olive oil extract oleuropein possesses both anti-inflammatory and antioxidant properties when extracted from *Olea europaea* L. Scientific research indicates that oleuropein decreases the accumulation of tau protein tangles while inhibiting amyloid-beta fibril formation in Alzheimer's disease patients (Masood et al. 2021). Ginkgolides represent terpenes extracted from *Ginkgo biloba* trees, and scientists aim to obtain them to protect nerve cells by modifying protein folding and aggregation pathways (Liu et al., 2022; Cai et al. 2022). Ashwagandha properties include withanolides that interfere with protein aggregation events responsible for different neurodegenerative conditions according to scientific reports. The compounds found in these plants have potential applications in contemporary medical practice for misfolding and erroneous protein folding disorders (Paul et al. 2021).

10.5.2 Antioxidant Activities

The appearance and advancement of neurodegenerative diseases are significantly influenced by oxidative stress. When body immunity is weak and the production of reactive oxygen species (ROS) is out of balance, the result is oxidative stress. This imbalance can cause destruction in the brain by harming cells, impacting mitochondrial function, and eventually culminating in neuron death.

According to scientific understanding, mitochondria serve as the main producers of ROS in the cell, which are commonly referred to as powerhouses. The destruction of cell membranes or proteins alongside genomic alteration impedes the energy production mechanism which starts the cell death procedures. The brain needs ROS molecules for signaling and balance performance; however, certain types of the molecules trigger significant medical complications. In Alzheimer's disease, amyloid-beta (Aβ) peptide deposition produces ROS that leads to damage of neurons. The aggregation of tau protein becomes probable during oxidative stress conditions where the protein creates damaging tangles that kill cells and affect cell functionality (Naseri et al. 2019; Simunkova et al. 2019). Parkinson's disease allows the important protein α-synuclein to play a vital role in generating oxidative stress which leads the protein to form aggressive Lewy body clusters as it aggregates. Dopamine-producing cells of the substantia nigra specifically suffer damage from this disease process, as they are already at risk because dopamine metabolism naturally produces ROS. Similar to this, mutations in the antioxidant enzyme SOD1 in amyotrophic lateral sclerosis (ALS) impair its ability to detoxify superoxide radicals, which causes ROS to accumulate and ultimately results in the death of motor neurons. Further progression of nerve cell damage involves mitochondrial issues and increased ROS, which are also prevalent in ALS (Naseri et al. 2019; Ionescu-Tucker and Cotman 2021).

Resveratrol, a polyphenol present in grapes and red wine, activates pathways like SIRT1 and Nrf2 that enhance the cell's antioxidant defenses. It has been shown to protect neurons by reducing ROS levels and inhibiting inflammation (Kung et al. 2021). Epigallocatechin gallate is a major component of green tea. EGCG exerts strong antioxidant effects by neutralizing free radicals and upregulating antioxidant enzymes such as glutathione peroxidase and superoxide dismutase (Payne et al. 2022). Quercetin, found in apples, onions, and berries, is a flavonoid that has been shown to reduce oxidative stress by scavenging free radicals and chelating metal ions that catalyze ROS production (Khan et al. 2019). *Ginkgo biloba* extract contains flavonoids and terpenoids that possess antioxidant properties. It enhances mitochondrial function and reduces oxidative damage in neurons (Luthra and Roy 2022). Naturally occurring in spinach, broccoli, and potatoes, alpha-lipoic acid is a powerful antioxidant that can regenerate other antioxidants, such as vitamins C and E, and directly scavenge ROS. N-Acetylcysteine (NAC), a precursor to glutathione, helps replenish intracellular levels of this critical antioxidant, enhancing the cell's ability to neutralize ROS and repair oxidative damage (Barbalho et al. 2022).

Vitamin E is a lipid-soluble antioxidant that shields cell membranes from lipid peroxidation brought on by reactive oxygen species (ROS). Vitamin C is majorly present in citrus fruits, berries, nuts, seeds, and leafy greens, is a water-soluble antioxidant that scavenges free radicals in the aqueous compartments of the cell, and regenerates oxidized vitamin E (Icer et al. 2021). Coenzyme Q10 (ubiquinone), found in meat, fish, and whole grains, is involved in the mitochondrial electron transport chain and acts as an antioxidant, reducing ROS production and protecting mitochondrial function (Jimenez-Jimenez et al. 2023). These natural compounds offer a promising complementary approach to traditional therapies for neurodegenerative diseases, potentially mitigating the damaging effects of oxidative stress and improving neuronal health.

10.5.3 Anti-inflammatory Activities

An inflammation in the brain and spinal cord is a highly intricate response that plays a major role in many neurodegenerative diseases. The resident immune cells in the brain, known as microglia, are the main actors here. Microglia slowly observe the environment of the brain under normal circumstances, but they go into alert mode in response to an injury or illness. They can sense signals of danger like pathogens or decaying cells through specialized surface detectors called pattern recognition receptors (or Toll-like receptors, TLRs). Upon activation, microglia change their morphology and start to produce inflammatory substances such as reactive oxygen species (ROS), chemokines, and cytokines including IL-1β, TNF-α, and IL-6. This reaction is beneficial at the beginning towards the removal of damage; however, when it lasts too long, it becomes harmful, causing chronic inflammation and further cellular injury.

Another critical player is the astrocyte, a star-shaped cell that normally supports neurons and maintains the blood-brain barrier (BBB), a barrier that controls what gets into the brain. In addition to swelling and proliferation (a process known as astrogliosis), reactive astrocytes release inflammatory chemicals and reactive oxygen species (ROS) that worsen the situation. This same reactivity can result in permeabilization of the blood-brain barrier, enabling harmful immune cells to traverse through the blood into the brain and engender more inflammation. In acuteness, inflammation is essential to protect the brain, whereas in prolonged situations, it possibly transforms right into a wicked cycle of injury, severely contributing to neurodegenerative diseases like Parkinson's and Alzheimer's (Subhramanyam et al. 2019).

Various plant compounds have emerged as promising agents for relieving brain inflammation and protecting neurons. For example, the quercetin flavonoid, which you can find in, among others, apples, onions, and berries, is more than just fighting oxidative stress as an antioxidant. It also suppresses the NF-κB pathway, a key driver of inflammation. As a result, harmful cytokines decrease and motor as well as cognitive function improves in different models of the disease (Chiang et al. 2023). Likewise, the *Ginkgo biloba* extract, rich in flavonoids and terpenoids, has been shown to enhance memory, decrease brain inflammation, and provide neuroprotection in Alzheimer's and Parkinson's disease models through similar mechanisms that includes suppression of NF-κB as well as upregulation of antioxidant enzymes. Another dominant compound is alpha-lipoic acid (ALA), which secures cells by activating the Nrf2 pathway, the master switch for the body's antioxidant defense. It also inhibits inflammation by reducing NF-κB activity, making it beneficial for both Alzheimer's and Parkinson's (Dos Santos et al. 2019). N-Acetylcysteine (NAC), a supplement version of an amino acid found in foods like poultry and yogurt, helps the brain replenish glutathione, one of its most vital antioxidants. It lowers reactive oxygen species (ROS) and inflammation markers, supporting neuron survival (Morgan et al. 2021).

Omega-3 fatty acids, especially EPA and DHA, found in fish oil, are also known to reduce inflammation by blocking the NF-κB pathway. They improve the fluidity and function of cell membranes and have been shown to enhance memory and slow the progression of neurodegenerative diseases (Grant and Blake 2023). Lastly, Boswellic acids from Boswellia (frankincense) offer a natural anti-inflammatory effect by targeting 5-lipoxygenase (5-LOX) and NF-κB, helping to calm neuroinflammation and support cognitive health, especially in Alzheimer's disease models (Khaafi and Javadi 2023). Altogether, these plant-based compounds offer exciting potential for managing and possibly slowing down neurodegenerative diseases by addressing two major culprits: stress and chronic inflammation.

10.5.4 Anti-apoptotic and Neurotrophic Activities

Apoptosis, or programmed cell death, is a carefully controlled process that helps get rid of old, damaged, or potentially dangerous cells without triggering inflammation or harming surrounding tissues. This process is vital for keeping our bodies balanced, supporting proper development, protecting us from disease, and keeping the immune system in check (Zhang et al. 2024). Apoptosis is carried out by a group of enzymes known as caspases. These are like molecular scissors. Firstly, initiator caspases (such as caspase-8 and caspase-9) get activated and then turn on effector caspases (like caspase-3 and caspase-7), which go on to cut up specific parts of the cell, leading to its breakdown (Van Opdenbosch and Lamkanfi 2019). There is also a group of proteins called the Bcl-2 family that act like the gatekeepers of cell survival. Some of them are Bax, Bak, and Bid which encourage apoptosis by making the mitochondria (the cell's energy producers) more "leaky," which helps trigger cell death. Similarly, Bcl-2, Bcl-xL, and Mcl-1 functions to stop that from happening, helping the cell hang on by protecting mitochondrial integrity and blocking the release of cytochrome c, a key player in the death process (Zhang et al. 2024).

Then there are the IAPs (inhibitors of apoptosis proteins), which can block apoptosis by binding to and disabling caspases. When apoptosis goes off track, it can cause problems. For example, if harmful or self-attacking cells are not cleared out properly, it may lead to autoimmune diseases where the body starts attacking itself. Remarkably, nature offers some powerful tools to influence apoptosis. Curcumin, a natural compound found in turmeric, has been shown to protect cells by blocking or modifying the pathways that lead to cell death. It does this by interacting with key signaling routes like NF-κB, Akt, and p53, all of which play roles in whether a cell lives or dies.

Several natural compounds have shown powerful protective effects against cell death, particularly in the brain, by helping cells survive under stress. These compounds work by blocking harmful signals that trigger apoptosis (programmed cell death) and boosting the expression of protective proteins like Bcl-2 and Bcl-xL, which help keep cells alive (Antonio and Giuseppina 2019), Resveratrol, found in grapes and red wine, activates the SIRT1 pathway, known for promoting longevity and cell survival. It also supports the PI3K/Akt pathway, which keeps mitochondria—the cell's energy centers—healthy and prevents the release of molecules that trigger cell death (Sireesh and Phanithi 2021). Also, ginsenosides (particularly Rg1 and Rb1 from ginseng) also activate the PI3K/Akt and ERK1/2 pathways, increasing protective proteins while reducing those that lead to apoptosis (Nian et al. 2023). Epigallocatechin gallate is a powerful antioxidant from green tea, and it helps stabilize mitochondria and blocks enzymes called caspases that drive cell death by influencing the JNK and p38 MAPK pathways (Priscila et al. 2021). Similarly, compounds such as bacosides from *Bacopa monnieri* are known to enhance brain-supportive factors like BDNF, which helps with the growth and survival of nerve cells. Additionally, withanolides from Ashwagandha play a role in regenerating nerve connections, while Huperzine A boosts neurotransmitter levels and offers protection to neurons. Ferulic acid and quercetin provide both antioxidant and anti-apoptotic advantages by increasing neurotrophic factor expression and soothing stress-related signals. Also, kaempferol aids in cell survival by balancing crucial pathways like PI3K/Akt and NF-κB, which ultimately helps reduce cell damage and fosters resilience. All in all, these natural compounds showcase the promising potential of plant-based medicine in safeguarding brain health and supporting cognitive well-being over the long draw.

10.5.5 Acetylcholinesterase Inhibition Activity

Acetylcholinesterase (AChE) participates as one of the crucial enzymes within the brain, regulating the communication between nerve cells. It hydrolyses acetylcholine (ACh), which serves as an integral component of the neurotransmitter system responsible for memory and learning, following the delivery of its message. AChE operates by removing acetylcholine from the synaptic cleft, or the space situated between neurons, thereby resetting the system in preparation for the subsequent signal. This sequence of actions is crucial for the brain and its functions, and in other words, cognition. However, it is also linked with Alzheimer's disease (AD). Studies suggest that AChE can bind to sticky forms of another protein amyloid-beta (Aβ) that clusters into plaques typically located within the brain of Alzheimer patients. AChE appears to enhance this process of plaque formation, escalating the damage of brain cells and memory loss (Seyyed et al. 2023). Fortunately, there are some natural compounds which seem to offer promise in this context. This includes resveratrol, a polyphenol found in red grapes and wine, which is said to inhibit AChE (Chandrasekaran et al., 2023).

Additionally, it activates the SIRT1 pathway, which is essential for preserving the health and longevity of cells (Simunkova et al. 2019). Galantamine, an AChE blocker alkaloid that enhances acetylcholine's effects on specific brain receptors, is also well-known. In, Lazarova et al. (2021) highlighted Huperzine A, a fascinating natural substance that is extracted from a particular kind of moss and functions as a potent and reversible AChE inhibitor (Mak et al. 2021).

10.5.6 Monoamine Oxidase Inhibitors (MAOIs)

Monoamine oxidase (MAO) enzymes reside on the external layer of mitochondria, the power plants of cells, where they rapidly connect with neurotransmitters that transmit messages between nerve cells. The MAO enzymes activate neurotransmitter destruction through the mitochondria, resulting in a stabilized neurotransmitter concentration. At the same time, MAO generates by-products such as aldehydes, ammonia, and hydrogen peroxide (H_2O_2). These chemical by-products create cellular oxidation that is linked to multiple brain abnormalities. Dopamine levels rise in Parkinson's disease (PD), and here is where some drugs called MAO-B inhibitors, such as selegiline and rasagiline, are the main players. Studies indicate that these medications reduce oxidative stress together with movement-related symptom relief and may slow down the progression of the disease (Mazumder et al. 2020). The method of molecular docking enables researchers to better understand how different compounds bind with MAO-A. The research determined that multiple flavonoids including chrysin, myricetin, and genistein from soybeans exhibit strong binding capacity to MAO-A while minimizing its functional activity (Grabska-Kobyłecka et al. 2023). Medical studies using docking analysis have confirmed that baicalein and decursin which are derived from *Scutellaria baicalensis* roots show potential as MAO-A inhibitor substances (Devi et al. 2024).

10.6 Conclusion

Natural compounds have shown promise in preclinical studies for their potential neuroprotective effects. For example, curcumin, found in turmeric, has antioxidant and anti-inflammatory properties that may be neuroprotective against diseases like Alzheimer's. Curcumin-based supplements or pharmaceuticals are being developed for clinical trials to evaluate their effectiveness in preventing or slowing neurodegenerative processes. Resveratrol, present in red grapes and wine, has been studied for its potential neuroprotective effects and ability to activate sirtuins, proteins associated with longevity. Polyphenols in green tea, such as epigallocatechin gallate (EGCG), and also omega-3 fatty acids, found in fish oil, have anti-inflammatory and neuroprotective properties that can support cognitive function and slow disease progression in neurodegenerative disorders.

The future prospects of translational applications involving natural compounds against neurodegenerative diseases hold great promise and involve several key areas of development like precision medicine, advanced formulations via nanotechnology, biomarker discovery, targeting multiple pathways, clinical trial design, regulatory pathways, and patient-centric approach. Continued translational research is crucial for advancing our understanding of the therapeutic potential of natural compounds against neurodegenerative diseases. This involves moving beyond preclinical studies to well-designed clinical trials that provide evidence of safety, efficacy, and optimal dosages for these natural interventions. Integrating natural compounds into conventional treatment approaches or developing novel therapeutics based on these compounds holds promise for the future of neurodegenerative disease management.

References

Agorogiannis EI, Agorogiannis GI, Papadimitriou A, Hadjigeorgiou GM (2004) Protein misfolding in neurodegenerative diseases. Neuropathol Appl Neurobiol 30(3):215–224

Akçimen F, Lopez ER, Landers JE, Nath A, Chiò A, Chia R, Traynor BJ (2023) Amyotrophic lateral sclerosis: translating genetic discoveries into therapies. Nat Rev Genet 24(9):642–658

Ali S, Mukherjee S, Goel D, Ghosh A, Faruq M (2023) Role of stem cells and derived exosomes as a novel therapeutic agent against Alzheimer's and Parkinson's disease. In: Jahan S, Siddiqui AJ (eds) Applications of stem cells and derived exosomes in neurodegenerative disorders. Springer, Singapore

Alqahtani T, Deore SL, Kide AA, Shende BA, Sharma R, Dadarao Chakole R, Nemade LS, Kishor Kale N et al (2023) Mitochondrial dysfunction and oxidative stress in Alzheimer's disease, and Parkinson's disease, Huntington's disease and Amyotrophic Lateral Sclerosis -An updated review. Mitochondrion 71:83–92

Antonio G, Giuseppina T (2019) Curcumin and cancer. Nutrients 11(10):2376

Ashley P, Nahashon S, Taka E, Adinew GM, Karam FA (2022) Soliman Epigallocatechin-3-Gallate (EGCG): new therapeutic perspectives for neuroprotection, aging, and neuroinflammation for the modern age. Biomolecules 12(3):371

Barbalho SM, Direito R, Laurindo LF, Marton LT, Guiguer EL, Goulart RA, Tofano RJ, Carvalho ACA, Flato UAP, Capelluppi Tofano VA, Detregiachi CRP, Bueno PCS, Girio RSJ, Araújo AC (2022) Ginkgo biloba in the aging process: a narrative review. Antioxidants (Basel) 11(3):525

Buccellato FR, D'Anca M, Tartaglia GM, Del Fabbro M, Scarpini E, Galimberti D (2023) Treatment of Alzheimer's disease: beyond symptomatic therapies. Int J Mol Sci 24(18):13900
Cai D, Luo Z, Su J, Gan H, Wang Z, Liu X, Li S, Wu J, Chen J, Ma R, Huang M, Zhong G (2022) Exposure-response analysis and mechanism of Ginkgolide B's neuroprotective effect in acute cerebral ischemia/reperfusion stage in rat. Biol Pharm Bull 45(4):409–420
Chandrasekaran V, Hediyal TA, Anand N, Kendaganna PH, Gorantla VR, Mahalakshmi AM, Ghanekar RK, Yang J, Sakharkar MK, Chidambaram SB (2023) Polyphenols, autophagy and neurodegenerative diseases: a review. Biomolecules 13(8):1196
Chiang MC, Tsai TY, Wang CJ (2023) The potential benefits of quercetin for brain health: a review of anti-inflammatory and neuroprotective mechanisms. Int J Mol Sci 24(7):6328
Costa LG, Garrick JM, Roquè PJ, Pellacani C (2016) Mechanisms of neuroprotection by quercetin: counteracting oxidative stress and more. Oxidative Med Cell Longev 2016:2986796
De Pedro-Cuesta J, Rábano A, Martínez-Martín P, Ruiz-Tovar M, Alcalde-Cabero E, Almazan-Isla J et al (2015) Comparative incidence of conformational, neurodegenerative disorders. PLoS One 10(9)
Devi SNS, Premkumar T, Vino S, Radhika R, Shoba G, Deepa M (2024) Insights into dietary phytochemicals targeting Parkinson's disease key genes and pathways: a network pharmacology approach. Comput Biol Med 172:108195
Dos Santos SM, Romeiro CFR, Rodrigues CA, Cerqueira ARL, Monteiro MC (2019) Mitochondrial dysfunction and alpha-lipoic acid: beneficial or harmful in Alzheimer's Disease? Oxid Med Cell Longev 2019:8409329
Grabska-Kobyłecka I, Szpakowski P, Król A, Książek-Winiarek D, Kobyłecki A, Głąbiński A, Nowak D (2023) Polyphenols and their impact on the prevention of neurodegenerative diseases and development. Nutrients 15(15):3454
Grant WB, Blake SM (2023) Diet's role in modifying risk of Alzheimer's disease: history and present understanding. J Alzheimer's Dis 96(4):1353–1382
Gregory J, Vengalasetti YV, Bredesen DE, Rao RV (2021) Neuroprotective herbs for the management of Alzheimer's disease. Biomolecules 11(4):543
Icer MA, Arslan N, Gezmen-Karadag M (2021) Effects of vitamin E on neurodegenerative diseases: an update. Acta Neurobiol Exp (Wars) 81(1):21–33
Ionescu-Tucker A, Cotman CW (2021) Emerging roles of oxidative stress in brain aging and Alzheimer's disease. Neurobiol Aging 107:86–95
Islam A, Mishra A, Ahsan R, Fareha S (2023) Phytopharmaceuticals and herbal approaches to target neurodegenerative disorders. Drug Res (Stuttg) 73(7):388–407
Jack CR Jr, Therneau TM, Weigand SD, Wiste HJ, Knopman DS, Vemuri P et al (2019) Prevalence of biologically versus clinically defined Alzheimer spectrum entities using the National Institute on Aging-Alzheimer's Association research framework. JAMA Neurol 76(10):1174–1183
Jamal QMS, Siddiqui MU, Alharbi AH, Albejaidi F, Akhtar S, Alzohairy MA, Kamal MA, Kesari KK (2020) A computational study of natural compounds from Bacopa monnieri in the treatment of Alzheimer's disease. Curr Pharm Des 26(7):790–800
Jimenez-Jimenez FJ, Alonso-Navarro H, García-Martín E, Agúndez JAG (2023) Coenzyme Q10 and dementia: a systematic review. Antioxidants (Basel). 12(2):533
Keshavan A, Pannee J, Karikari TK, Rodriguez JL, Ashton NJ, Nicholas JM et al (2021) Population-based blood screening for preclinical Alzheimer's disease in a British birth cohort at age 70. Brain 144(2):434–449
Khaafi F, Javadi B (2023) Molecular targets underlying the neuroprotective effects of boswellic acid: a systematic review. Mini Rev Med Chem 23(19):1912–1925
Khan H, Ullah H, Aschner M, Cheang WS, Akkol EK (2019) Neuroprotective effects of quercetin in Alzheimer's Disease. Biomolecules. 10(1):59
Kung HC, Lin KJ, Kung CT, Lin TK (2021) Oxidative stress, mitochondrial dysfunction, and neuroprotection of polyphenols with respect to resveratrol in Parkinson's disease. Biomedicine 9(8):918
Lazarova MI, Tsekova DS, Tancheva LP, Kirilov KT, Uzunova DN, Vezenkov LT, Tsvetanova ER, Alexandrova AV, Georgieva AP, Gavrilova PT, Dragomanova ST, Papazova MG, Handzhiyski

YS, Kalfin RE (2021) New galantamine derivatives with inhibitory effect on acetylcholinesterase activity. J Alzheimers Dis 83(3):1211–1220
Liu M, Peng Y, Che Y, Zhou M, Bai Y, Tang W, Huang S, Zhang B, Deng S, Wang C, Yu Z (2022) MiR-146b-5p/TRAF6 axis is essential for Ginkgo biloba L. extract GBE to attenuate LPS-induced neuroinflammation. Front Pharmacol 13:978587
Livingston G, Huntley J, Sommerlad A, Ames D, Ballard C, Banerjee S et al (2020) Dementia prevention, intervention, and care: 2020 report of the Lancet Commission. Lancet 396(10248):413–446
Luthra R, Roy A (2022) Role of medicinal plants against neurodegenerative diseases. Curr Pharm Biotechnol 23(1):123–139
Mak S, Li W, Fu H, Luo J, Cui W, Hu S, Pang Y, Carlier PR, Tsim KW, Pi R, Han Y (2021) Promising tacrine/huperzine A-based dimeric acetylcholinesterase inhibitors for neurodegenerative disorders: from relieving symptoms to modifying diseases through multitarget. J Neurochem 158(6):1381–1393
Mamatha NM, Baragur VS (2024) Ashwagandha- Withania somnifera (L.) Dunal as a multipotent neuroprotective remedy for genetically induced motor dysfunction and cellular toxicity in human neurodegenerative disease models of drosophila. J Ethnopharmacol 10:318
Masood SB, Tariq U, Iahtisham-Ul-Haq AN, Rizwan M (2021) Neuroprotective effects of oleuropein: recent developments and contemporary research. J Food Biochemist 45(12):e13967
Mazumder MK, Borah A, Choudhury S (2020) Inhibitory potential of plant secondary metabolites on anti-parkinsonian drug targets: relevance to pathophysiology, and motor and non-motor behavioural abnormalities. Med Hypotheses 137:109544
Mehdi AR (2023) Protein Misfolding and Aggregation in Proteinopathies: Causes, Mechanism and Cellular Response. Diseases, 11(1):30
Meng X, Zhou J, Zhao CN, Gan RY, Li HB (2020) Health benefits and molecular mechanisms of resveratrol: a narrative review. Foods 9:340
Morgan AM, Hassanen EI, Ogaly HA, Al Dulmani SA, Al-Zahrani FAM, Galal MK, Kamel S, Rashad MM, Ibrahim MA, Hussien AM (2021) The ameliorative effect of N-acetylcysteine against penconazole induced neurodegenerative and neuroinflammatory disorders in rats. J Biochem Mol Toxicol 35(10):e22884
Mukherjee S, Ali S, Hashmi S, Jahan S (2023) History, origin and types of neurological disorders. In: Applications of stem cells and derived exosomes in neurodegenerative disorders. Springer Link, pp 1–32
Naseri NN, Wang H, Guo J, Sharma M, Luo W (2019) The complexity of tau in Alzheimer's disease. Neurosci Lett 705:183–194
Nayeem U, Ahmed S, Kumar D, Jahan S, Rather G, Siddiqui AJ (2023) Role of stem cells as a protective agent against neurological complications. In: Applications of stem cells and derived exosomes in neurodegenerative disorders. Springer Nature Link, pp 69–90
Nian Z, Mao F, Cheng S (2023) Mechanism research and application for ginsenosides in the treatment of hepatocellular carcinoma. Biomed Res Int 2023:7214037
Oo EO, Gerke-Duncan MB, Damian Holsinger RM (2023) Oxidative stress and antioxidants in neurodegenerative disorders antioxidants. Baseline 12(2):517
Patani R, Hardingham GE, Liddelow SA (2023) Functional roles of reactive astrocytes in neuroinflammation and neurodegeneration. Nat Rev Neurol 19(7):395–409
Paul S, Chakraborty S, Anand U, Dey S, Nandy S, Ghorai M, Saha SC, Patil MT, Kandimalla R, Proćków J, Dey A (2021) Withania somnifera (L.) Dunal (Ashwagandha): a comprehensive review on ethnopharmacology, pharmacotherapeutics, biomedicinal and toxicological aspects. Biomed Pharmacother 143:22–45
Payne A, Nahashon S, Taka E, Adinew GM, Soliman KFA (2022) Epigallocatechin-3-Gallate (EGCG): new therapeutic perspectives for neuroprotection, aging, and neuroinflammation for the modern age. Biomolecules 12(3):371
Priscila BG, Ana CRS, Yraima C (2021) Green Tea Epigallocatechin-3-gallate (EGCG) targeting protein misfolding in drug discovery for neurodegenerative diseases. Biomolecules 11(5):767

Seyyed MB, Mohammad A, Alireza M (2023) Acetylcholinesterase inhibitory activity of Ferula plants and their potential for treatment of Alzheimer's disease. J Complement Integr Med 21:1172

Simunkova M, Alwasel SH, Alhazza IM, Jomova K, Kollar V, Rusko M, Valko M (2019) Management of oxidative stress and other pathologies in Alzheimer's disease. Arch Toxicol 93(9):2491–2513

Sireesh KT, Phanithi PB (2021) Differential role of SIRT1/MAPK pathway during cerebral ischemia in rats and humans. Sci Rep 11:6339

S-Rad M, Chintha L, Dias DA, Anca OD, Mohamad FM et al (2020) Impact of natural compounds on neurodegenerative disorders: from preclinical to pharmacotherapeutics. J Clin Med 9(4):1061

Subhramanyam CS, Wang C, Hu Q, Dheen ST (2019) Microglia-mediated neuroinflammation in neurodegenerative diseases. Semin Cell Dev Biol 94:112–120

Tripathi PN, Lodhi A, Rai SN, Nandi NK, Dumoga S, Yadav P, Tiwari AK, Singh SK, El-Shorbagi AA, Chaudhary S (2024) Review of pharmacotherapeutic targets in Alzheimer's disease and its management using traditional medicinal plants. Degener Neurol Neuromuscul Dis 14:47–74

Van Opdenbosch N, Lamkanfi M (2019) Caspases in cell death, inflammation, and disease. Immunity 50(6):1352–1364

Xiang L, Wang Y, Liu S, Liu B, Jin X, Cao X (2023) Targeting protein aggregates with natural products: an optional strategy for neurodegenerative diseases. Int J Mol Sci 24(14):11275

Yu T, Xing Y, Gao Q, Wang D, Chen H, Wang H, Zhang Y (2023) Ginkgo biloba extract drives gut flora and microbial metabolism variation in a mouse model of Alzheimer's disease. Pharmaceutics 15(12):2746

Zakariya SM, Zehra A, Khan RH (2022) Biophysical insight into protein folding, aggregate formation and its inhibition strategies. Protein Pept Lett 29:22–36

Zhang W, Zhuang S, Guan H, Li F, Zou H, Li DJ (2024) New insights into the anti-apoptotic mechanism of natural polyphenols in complex with Bax protein. Biomol Struct Dyn 42(6):3081–3093

Zia A, Farkhondeh T, Pourbagher-Shahri AM, Samarghandian S (2021) The role of curcumin in aging and senescence: molecular mechanisms. Biomed Pharmacother 134:111119

Neurovascular Coupling Dysfunction: Natural Products as a Remedy to Manage Neurodegenerative Disorders

11

Renu Negi, Shweta Sharma, and Bhaskar Sharma

Abstract

Neurovascular coupling (NVC) is a fundamental physiological process that ensures an adequate blood supply to activate neurons in brain regions to meet their metabolic demands. It plays a pivotal role in sustaining cerebral blood flow (CBF) within the neurovascular unit (NVU), which possesses pericytes, endothelial cells, glial cells, and neurons. The NVU plays a crucial role in maintaining the stability of the blood-brain barrier (BBB) and CBF, which are vital for sustaining normal brain function. Altered NVC can result in long-term impairment of neuron activity, which is associated with neurodegenerative disorders such as Alzheimer's, Parkinson's, amyotrophic lateral sclerosis (ALS), and frontotemporal dementia (FTD). Recently, natural compounds have come to light as propitious candidates for preventing and treating neurodegenerative disorders due to their diverse bioactivities and lower side effect profiles than synthetic drugs. The potential therapeutic effect of these compounds is gaining attention in clinical trials and translational research. This chapter explores the underlying pathophysiology of NVC that contributes to neurodegenerative disorders (NDDs). Furthermore, it highlights the therapeutic promise of natural products to restore

R. Negi (✉)
Systems Toxicology Group, CSIR-Indian Institute of Toxicology Research Vishvigyan Bhavan, Lucknow, Uttar Pradesh, India

Academy of Scientific and Innovative Research (AcSIR), Ghaziabad, Uttar Pradesh, India

S. Sharma
School of Biotechnology, Jawaharlal Nehru University, New Delhi, India

B. Sharma
Neurobiology Laboratory, Department of Anatomy, All India Institute of Medical Sciences, New Delhi, India

S. Jahan et al. (eds.), *A Mechanistic Exploration of Natural Compounds for Neuronal Health*, https://doi.org/10.1007/978-981-95-0125-0_11

and manage NVC function in NDDs, as they possess anti-inflammatory, antioxidant, and neuroprotective characteristics.

Keywords

Neurovascular coupling (NVC) · Neurodegenerative disorders · NVC dysfunction · Natural compounds · BBB

11.1 Introduction

Interestingly, the human brain accounts for only 2% of our body weight, but it consumes about 20% of the body's oxygen and glucose at rest. Another interesting fact is that all energy is obtained from the oxidative metabolism of glucose (Dienel and Hertz 2001). Additionally, the brain requires a significant substrate supply and metabolic end-product clearance, which is compromised by its restricted ability to store intracellular energy. These traits, coupled with the critical function compared to other organs, create a condition where it is necessary for blood flow to be precisely aligned with changes in neuronal activity. To achieve this, the brain has extensive vascularization and sophisticated regulation of blood flow to meet its metabolic demands (Attwell et al. 2010; Dienel and Hertz 2001). One crucial element in maintaining an adequate blood supply is neurovascular coupling (NVC), which regulates and adjusts the regional cerebral blood flow network based on neuronal activity and metabolism. The brain has unique physiology, as it is not able to store glucose in the form of fat. It requires instant energy in the form of glucose in its active regions. To meet this metabolic need, neurovascular coupling (NVC) is vital to provide the brain with sufficient oxygen and nutrients. Understanding the function of NVC is essential for developing therapeutic interventions for various disease conditions. Alteration in NVC leads to reduction in cerebral blood flow, resulting in hypoxia. In addition, the BBB disruption in the brain also inhibits the entry of glucose and other required nutrients and the removal of unwanted byproduct material. Additionally, decreased production of trophic factors by neurovascular unit (NVU) cells can increase the susceptibility of neurons and glial cells to diseases. Disruptions in protein clearance processes can lead to the deposition of protein molecules like Aβ and tau, which contribute to proteinopathies. Several studies have reported that dysfunction in NVU is associated with NDDs including AD (Toledo et al. 2013), FTD (Janelidze et al. 2017), ALS (Murphy et al. 2012), PD, and dementia with Lewy bodies (Janelidze et al. 2017).

Natural products hold significant promise in mitigating neurovascular dysfunction as they possess anti-inflammatory, antioxidant, and neuroprotective characteristics. These chemical compounds target various mechanisms that maintain BBB integrity, reduce oxidative stress, and enhance mitochondrial function, thereby improving neurovascular health and overall brain function (Hamel et al. 2008; Kamitani and Tong 2005). These compounds, including flavonoids, curcumin, resveratrol, and omega-3 fatty acids, contain strong anti-inflammatory and antioxidant properties that reduce oxidative stress and inflammation, which are major causes of neurovascular dysfunction. Additionally, natural products such as *Ginkgo biloba* extract can enhance CBF, thus improving the communication between neurons and blood vessels. They also offer neuroprotective benefits, safeguarding neurons from damage and apoptosis. Moreover, certain natural compounds, such as polyphenols in green tea, help prevent the accumulation of proteins associated with neurodegenerative diseases like amyloid-beta and tau. Overall, natural products support the health and functionality of the neurovascular unit, making them promising candidates for therapeutic strategies against neurovascular coupling dysfunction. This chapter explores the mechanisms of NVC, its dysfunction in relation to various diseases, especially neurodegenerative diseases, and the alleviation of these dysfunctions using natural products.

11.2 Overview of Neurovascular Unit and Its Components

The brain possesses a highly specialized vascular system that is distinct in structure and function from the vasculature of peripheral organs. Importantly, these processes work in harmony to regulate brain homeostasis and ensure optimal functioning, a role that is fulfilled by the NVU. It is a highly integrated and dynamic system composed of neurons, glial cells (astrocytes, microglia), and vascular components, including vascular smooth muscle cells (VSMCs), pericytes, and endothelial cells (as described in Fig. 11.1). The dynamic interaction among these components assists in maintaining blood flow in the brain, supplying neurons with energy substrates, and removing waste products (Iadecola 2017).

Neurons are vital components of the NVU, serving as the primary drivers that trigger elevated CBF during metabolic hyperemia and as the brain's main energy consumers (Howarth et al. 2021). They impact the NVU responses by interacting with astrocytes, microglia, and pericytes and alter the CBF (Attwell et al. 2010). Neurons regulate blood flow in two distinct ways: direct and indirect. In the direct way, it secretes the vaso-modulatory molecules that affect the arteries. In the indirect way, it activates the cells like astrocytes and pericytes, which then release chemical substances to modulate the capillaries (Biesecker et al. 2016; Mishra et al. 2016). Nitric oxide (NO) and prostanoids are vasoactive molecules of neuronal origin and play an important role in NVC. In addition, astrocytes also participate importantly in NVC by activating cells, neurons, and pericytes. Studies have illustrated that neuronal activity stimulates astrocytes primarily through glutamatergic signaling pathways (Fergus and Lee 1997; Zonta et al. 2003). For instance, Carmignoto's group showed that activation of metabotropic glutamate receptors (mGluRs) on astrocytes increases intracellular $Ca2^{+}$ concentration, which results in arteriolar dilation. This is mediated through a cyclooxygenase product, probably prostaglandin E2 (PGE2), of arachidonic acid (AA) metabolism that accounts for astrocyte-mediated arteriole dilation. Furthermore, Mishra et al. (2016) illustrated that neural activity increments raise calcium in astrocytes by the action of P2X1 receptor activation. The elevated calcium results in the synthesis of PGE2, which

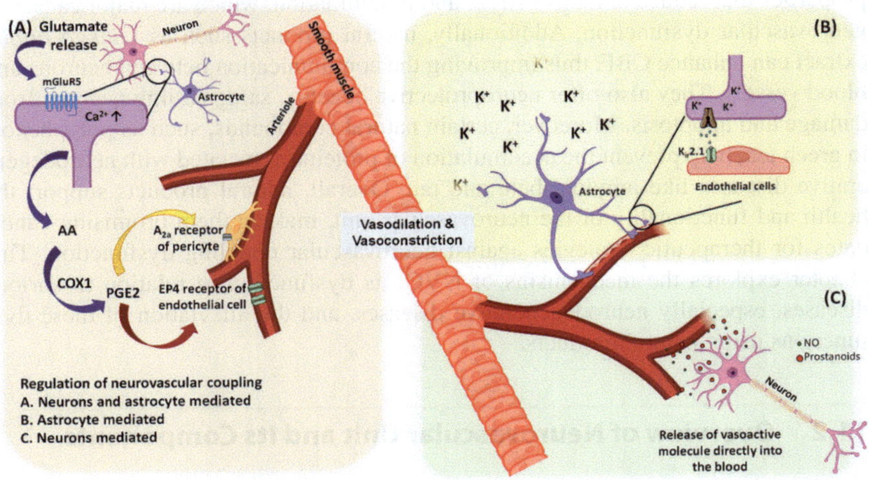

Fig. 11.1 The diagram illustrates the components of the neurovascular unit, including vascular smooth muscle cells (VSMCs), pericytes, and endothelial cells, and their interactions in regulating blood flow. Neurovascular coupling mechanism: (**a**) Indirect (astrocyte-mediated): Neurons release chemical compounds that activate astrocytes, leading them to release vasoactive molecules, resulting in vasoconstriction or vasodilation. (**b**, **c**) Direct (astrocytes and neurons): Neurons and astrocytes release vasoactive molecules that are directly secreted into the bloodstream, modulating cerebral blood flow

gets released from astrocytic end feet. PGE2 subsequently activates EP4 receptors on pericytes, causing capillary dilation (Mishra et al. 2016). Beyond PGE2, astrocytic activation results in the production and secretion of other chemical mediators, for instance, NO (Carmignoto and Gómez-Gonzalo 2010; Chisari et al. 2004; De Labra et al. 2009; Li et al. 2003; Metea and Newman 2006; Rancillac et al. 2006), epoxyeicosatrienoic acids (EETs) (Anderson et al. 2004; Carmignoto and Gómez-Gonzalo 2010; Koehler et al. 2009), glutamate, adenosine, and ATP, all of which can modulate arterioles. This complex signaling network highlights the integral role of astrocytes in controlling CBF in response to neuronal activity.

The capillaries and arteriole play a critical role in initiation, conduction, and regulation of NVCs (Longden et al. 2017; Mishra et al. 2016). Longden et al. (2017) highlighted the role of endothelial cells (capillary) in these processes, demonstrating that these cells are key in initiating and conducting NVC signals. They proposed that the Kir2.1 channel in capillary endothelial cells is activated by the extracellular potassium flux during an action potential, producing a hyperpolarizing electrical signal. The dilatation of penetrating arterioles and pial arteries caused by this signal's subsequent upstream transmission increases blood flow to the signal's source (Chen et al. 2011; Longden et al. 2017; Tian et al. 2010).

11.2.1 Neurovascular Coupling Mechanisms

Neurovascular coupling is the connection between neurons and vascular capillaries that ensures the adequate supply of nutrients required for neuronal function (Sokoloff 1996). Cerebral blood flow at rest exhibits regional variation in proportion to the metabolic demands of the brain. Consequently, regions with higher energy utilization, such as the inferior colliculus, exhibit higher blood flow, while areas with minimal energy consumption, such as the white matter, show lower blood flow. This dynamic process is driven by potent vascular responses to vasoactive signals released during periods of heightened neuronal activity, which leads to elevated CBF—a phenomenon referred to as functional hyperemia (Chaigneau et al. 2003; Cox et al. 1993; Freygang and Sokoloff 2013). NVC is regulated by three primary mechanisms (Phillips et al. 2016; Spronck et al. 2012). First, the feedback mechanism is where metabolic activity produces metabolites such as CO_2 and lactate, which causes blood vessels to dilate and modulates blood flow (Hosford and Gourine 2019; Iadecola 2017; Phillips et al. 2016; Spronck et al. 2012; Yang et al. 2017). Second, a feedforward mechanism suggests that neuronal signaling triggers NVC. This process occurs through the release of mediators such as potassium ions (K^+), NO, and prostanoids during synaptic activity and also induces other cells to secrete vasoactive molecules to modulate cerebral blood flow (Yang et al. 2017). Third, astrocyte-mediated regulation responds to local changes in oxygen, CO_2, lactate, pH, and neurotransmitter release resulting from metabolic activity. These astrocytes serve as intermediaries between tissue pressure and neural activity through the use of mechanosensors [38, 40]. As blood flow increases in blood vessels, it triggers shear stress on the vessel walls, which, in turn, causes the release of vasoactive molecules such as nitric oxide, leading to vasodilation (Spronck et al. 2012). Chemical mediators, including prostaglandin, potassium, and NO, are released by astrocytes in response to glutamate stimulation. These mediators act on vascular smooth muscles, resulting in vasodilation (Huneau et al. 2015; Iadecola 2017; Metea and Newman 2006) (see Fig. 11.1). Additionally, the autonomic nervous system regulates the constriction and dilation of arteriolar smooth muscles by modulating sympathetic and parasympathetic tone (Phillips et al. 2016; Spronck et al. 2012; Yang et al. 2017). The NVU is a vital component of the BBB that regulates homeostasis and meticulously controls the movement of immune cells and other molecules within the brain. Reduced CBF or impaired neurovascular coupling could deprive neurons of essential energy and create a toxic environment due to inefficient clearance of metabolites. Multiple sclerosis (MS), spinal cord injury, traumatic brain injury, stroke, and dementia, including Alzheimer's disease (AD) and AD-related mixed dementia, are among the neurological conditions that commonly exhibit these symptoms (Joo Turoni et al. 2011; Lin et al. 2011; Sachdeva et al. 2020; Sharma et al. 2020; Stickland et al. 2019; Tarantini et al. 2017). Understanding the biological and cellular mechanisms in NVC is crucial for developing therapeutic interventions to address imbalances between metabolic demand and supply, as observed in various ailments, which are briefly discussed below.

11.2.2 Neurovascular Coupling Dysfunction: Neurodegenerative Disorders

Neurovascular coupling plays a pivotal role in sustaining the BBB and managing cerebral blood flow (CBF) (Liu et al. 2019; Sweeney et al. 2018). Neurovascular coupling (NVC) dysfunction is significantly correlated with the pathogenesis of neurodegenerative diseases (NDDs), particularly AD, PD, and ALS (Rajkumar 2024). Studies suggest that the BBB disruption, linked with NVC impairment, is a pathophysiological characteristic of NDDs (Sweeney et al. 2018). Disruption of NVC has been linked to PD. Similarly, BBB disruption has also been reported in chemically induced Parkinson's disease (PD) mouse models (Carvey et al. 2005; Chen et al. 2008). Collins et al. (2012) suggested that the high ROS level and inflammatory cytokines secreted by activated microglia and astrocytes lead to rearrangement of tight junction proteins within the BBB in PD patients (Collins et al. 2012). In addition, the reduced expression of the transporter protein P-glycoprotein (P-gp) in the midbrain of PD patients is associated with BBB dysfunction (Kortekaas et al. 2005). The reduction in P-gp is thought to lead to the accumulation of α-synuclein which is implicated in the pathogenesis of PD (Kortekaas et al. 2005).

In addition, the impaired NVC (BBB) function is a significant pathological characteristic in ALS. Mice with ALS-related SOD1 mutations exhibit reduced levels of tight junction proteins, including occludin, ZO-1, and claudin-5, which impair barrier function (Zhong et al. 2008). Likewise, ALS patients' lumbar spinal cords showed reductions in ZO-1 and occludin (Henkel et al. 2009). Reduced amounts of perivascular occludin, collagen IV, and astrocyte end-feet have been found in the spinal cords of ALS patients after their deaths. The blood-brain barrier is compromised by the deterioration of tight junctions, endothelial cells, and astrocytic end-feet, which leads to vascular leakage and entry of toxic substances into the central nervous system (Miyazaki et al. 2011). In ALS, the expression of GLUT-1 and CD146, crucial for endothelial cell function, is diminished. Even before clinical symptoms manifest, mice with SOD1 mutations have a 30–45% decrease in spinal cord blood flow in the cervical and lumbar regions (Garbuzova-Davis et al. 2008). Overall, the impairment in the NVU starts before the appearance of clinical symptoms and is critical to ALS pathogenesis (Garbuzova-Davis et al. 2011).

Amyloid-β (Aβ) buildup in the brain is also linked to NVU dysfunction, most likely as a result of decreased cerebral blood flow (CBF) and increased BBB permeability (Sweeney et al. 2018). Endothelial cells in Alzheimer's disease patients exhibit elevated levels of the receptor for advanced glycation end products (RAGE) and decreased expression of P-glycoprotein (P-gp) and low-density lipoprotein receptor-related protein 1 (LRP1) (Yamazaki and Kanekiyo 2017; Zenaro et al. 2017). The movement of Aβ across the blood-brain barrier depends on these proteins. On the abluminal side, LRP1 facilitates the internalization of Aβ, which is then transported to lysosomes for degradation (Candela et al. 2015; Pflanzner et al. 2011). An ATP-driven P-gp transporter on the luminal surface of endothelial cells aids in the removal of Aβ, and a decrease in its expression causes Aβ to accumulate (Cirrito et al. 2005; Schinkel 1999). RAGE is a transporter that transports Aβ from peripheral circulation across the BBB. Its increased expression results in greater Aβ influx, contributing to AD (Kook et al. 2012; Marco and Skaper 2006; Wan et al. 2015). Additionally, occludins, claudins, and ZO-1 are tight junction proteins, and alterations in their expression affect BBB permeability (Kook et al. 2012; Marco and Skaper 2006; Wan et al. 2015).One study reported that Aβ1-42 oligomers target tight junction proteins and increase BBB permeability by reducing the expression of occludin, claudin-5, and ZO-1 in endothelial cells (Kook et al. 2012; Wan et al. 2015). Thus, alterations in NVC are associated with the etiology of NDDs and understanding its function could aid in therapeutic intervention.

11.3 Neurovascular Coupling Dysfunction: Can Natural Products Help?

Natural products have evolved over the course of millions years and have long been a cornerstone of medical chemistry and pharmacology because of their diverse biological activities and drug-like characteristics (Atanasov et al. 2021; Oyebode et al. 2016). They hold significant promise in preventing neurovascular dysfunction due to their cytoprotective properties. These products directly and indirectly improve neurovascular coupling, which helps in preserving the BBB by alleviating oxidative stress and enhancing mitochondrial function, thereby improving neurovascular health and overall brain function (Hamel et al. 2008; Tong et al. 2005). A critical factor in neurovascular coupling impairment, associated with numerous neurological disorders, is the dysfunction of the BBB (Yu et al. 2020a). ROS play a major role in disrupting the BBB by triggering the signaling pathway that causes deterioration of tight junction (TJ) proteins and the dysregulation of adherens junction (AJ) proteins (Pun et al. 2009). Accordingly, antioxidants have been proposed as potential neuroprotective agents in conditions that lead to BBB damage. Elevated ROS levels can activate several signaling molecules, such as HIF-1, NF-κB, and MMPs, causing a loss of BBB integrity and functional impairment (Kim et al. 2022).

This section discusses natural products and their mechanisms to manage or improve NVC dysfunction. Various studies have illustrated the beneficial effects of flavonoids and polyphenols in mitigating BBB disruption and alleviating neurodegeneration (Cheng et al. 2018; Jang et al. 2014; Jin et al. 2019). These flavonoids demonstrate a variety of antioxidant effects, such as inhibiting ROS production, neutralizing ROS precursors, chelating metals, and scavenging ROS (Cheng et al. 2018; Kumar and Pandey 2013). The mechanism by which these flavonoids exert their effects is summarized in Table 11.1 and detailed in the following discussion.

Quercetin is a flavonoid derived from plants and is prevalent in various fruits and vegetables. In human brain endothelial cells, quercetin mitigated fibrillar Aβ-induced BBB damage by reducing ROS overproduction and preserving SOD activity, thereby decreasing BBB permeability (Li et al. 2015). Yu et al. reported that quercetin preserves neurovascular integrity by modulating the SIRT1/Nrf2/HO-1 pathway in AD (Yu et al. 2020b). The cerebroprotective response of quercetin was reported in HT22 hippocampal neurons by stimulating PI3K/AKT which

Table 11.1 List of phytochemicals and their mechanisms of action in neurovascular coupling disorders

S. no	Natural compound	Source	Mechanism of action	Disorders with NVC dysfunction	Ref.
1	Quercetin	Apples, wolfberries, strawberries, apricots, grapes, sprouts, tomatoes, cauliflower, onions, lettuce, and broccoli	Neuroprotective effects by regulating the SIRT1/Nrf2/HO-1 pathway, decreases BBB permeability, improves cognitive memory by increasing the expression of the NRF2/ARE pathway	AD	Li et al. (2015), Yu et al. (2020b)
			Upregulates ZO-1 and claudin-5, inhibits MMP-9 levels via the Sirt1/Nrf2/HO-1 signaling pathway	Cerebral ischemia-reperfusion (I/R) injury	Selvakumar et al. (2013), Yang et al. (2022)
			Regulates the SIRT1/HO-1/Nrf2 pathway to reduce oxidative damage, restores mitochondrial homeostasis by modulating ARE/PINK1 expression	Parkinson's disease	Sriraksa et al. (2012), Gureev and Popov (2019), Thiruvengadam et al. (2021)
2	Kaempferol	Tea, broccoli, apples, strawberries, and beans	Mitigates BBB disruption by alleviating phosphorylation of NFκB, p65	Cerebral ischemia-reperfusion (I/R) injury	Li et al. (2019)
			Maintains BBB integrity by suppressing the HMGB1/TLR4-mediated inflammatory pathway	Parkinson's disease	Yang et al. (2019)

3	Rutin	Orange peels, tobacco leaves, buckwheat flowers, tomatoes, and *Sophora japonica*	Enhances cognitive memory and protects against Aβ neurotoxicity by activating the BDNF and MAPK pathways. Inhibits neuroinflammation by downregulating the NFκB pathway	AD	Enogieru et al. (2018), Moghbelinejad et al. (2014), Sun et al. (2021)
4	Pinocembrin	Propolis, honey, ginger	Neurovascular protection via the ERK-CREB-BDNF pathway. Protects the neurovascular unit (NVU) from fibrillar Aβ1-42, regulating neurovascular RAGE pathways	AD	Liu et al. (2014)
5	Apigenin	Apples, grapes, onions, oranges, basil, thyme, chamomile tea, wine, and beer	Modulation of redox pathways Nrf2, NFκB, P13/Akt, and MAPK, Inhibition of oxidant enzymes (COX-2, NOS, NO)	AD	Liu et al. (2011)
			Attenuation of upregulated NFκB expression, downregulation of the proinflammatory enzyme (iNOS-1) and proinflammatory cytokines (TNF-α, IL-6), and dopamine D2 receptor (D2 R) upregulation	Parkinson’s disease	Charrière et al. (2024)

(continued)

Table 11.1 (continued)

S. no	Natural compound	Source	Mechanism of action	Disorders with NVC dysfunction	Ref.
6	Curcumin	Constituent of turmeric (*Curcuma longa*)	Anti-inflammatory response by inhibiting the NFκB pathway, attenuating TNF-α, IL-1β, cyclooxygenase-2 (COX-2), and inducible nitric oxide synthase (iNOS)	AD	Liu et al. (2016), Wu et al. (2022)
			Neuroprotective effects via BDNF/tyrosine kinase B (TrkB)-MAPK/ PI-3 K-cyclic AMP response element binding protein (CREB) signaling cascade	Cerebral ischemia-reperfusion (I/R) injury	Sun et al. (2020)
7	Resveratrol	Blackberries, blueberries, peanuts, and red wine	Anti-inflammatory effects by attenuating NF-κB activation, inhibiting p38 MAPK phosphorylation, and reducing COX-2, iNOS, TNF-α, IL-1β, and IL-6 expression. Rescues cortical neurovascular coupling responses via downregulating NADPH oxidase-derived reactive oxygen species (ROS) production	AD	Cicero et al. (2019)

8	Saffron	Derived from the flower of *Crocus sativus*	Protects against brain injury by upregulating the expressions of VEGFR-2 and SRF. Induces autophagy via the AMPK pathway, NF-κB pathway inhibition, suppression of p53 expression in the hippocampus, and improves cognitive abilities	AD	Bie et al. (2011), Zhang et al. (2018)
9	*Ginkgo biloba*	Ginkgo tree	Decreases malondialdehyde (MDA) levels, downregulates pro-inflammatory cytokines (TNF-α, IL-1β), and augments anti-inflammatory cytokine IL-10 expression	Cerebral ischemia-reperfusion (I/R) injury	Owens et al. (2024)
			Neuroprotective effects through the inhibition of NF-κB signaling, supporting anti-apoptotic and anti-inflammatory processes. Promotes neurite growth and supports neuronal plasticity via the PI3K/Akt/mTOR signaling pathway	AD	Lejri et al. (2019)

(continued)

Table 11.1 (continued)

S. no	Natural compound	Source	Mechanism of action	Disorders with NVC dysfunction	Ref.
10	Epigallocatechin gallate (EGCG)	Red wine, fruits such as blueberries, strawberries, apples, grapes, kiwi, gooseberries, green tea, beer, liquor, and chocolate	Modulates endothelial nitric oxide synthase (eNOS) activity. Inhibits neuronal cell death by reducing Bax and caspase-3 levels while increasing Bcl-2 l. Neuroprotective effects via activating the CREB/BDNF/TrkB-PI3K/Akt signaling pathway	Parkinson's disease	Ding et al. (2017), Mohd Sabri et al. (2022), Nan et al. (2018)
11	Cinnamaldehyde	Cinnamon	TRPA1 agonist. Reduces ischemic brain damage by recruiting IKCa/SKCa channels in a calcium-dependent manner	Ischemic stroke	Aubdool et al. (2016), Behringer and Hakim (2019), Guerra et al. (2018)
12	Allyl isothiocyanate	Mustard oil	Exert their activity by covalently modifying the cysteine residues on the intracellular N-terminal region of TRPA1		Guerra et al. (2018), Hinman et al. (2006)
13	Allicin and diallyl disulfide	Garlic			
14	Methyl salicylate	Wintergreen oil			

15	Naringenin	Citrus fruits such as lemons, oranges, tangerines, and grapefruits	Inhibits oxidative stress and inflammation via the SIRT1/FOXO1 signaling pathway. Suppression of oxidative stress by inhibiting the MAPK signaling cascade	Cerebral ischemia/reperfusion (I/R) injury	Zhao et al. (2023)
			Alleviated locomotor impairments and mitigates BBB dysfunction by downregulating matrix metalloproteinases (MMPs) and inducing glial fibrillary acidic protein (GFAP)	Subarachnoid hemorrhage	Adetunji et al. (2023)
16	Cocoa	Cocoa beans, dark chocolate	Improves cognitive ability by activating signaling pathways such MAPK, ERK, and PI3-kinase/Akt cascades. Regulates the gene expression and protein production of eNOS and suppresses the production of proinflammatory cytokines IL-1β, IL-2, and IL-8	AD	Martín et al. (2020), Sorond et al. (2008), Zeli et al. (2022)
17	Catechin	Green tea, black tea, berries, apples, dark chocolate, cocoa powder, red wine	Restores NOS activity and improves endothelial function, enhancing cerebrovascular flow-mediated dilation and learning abilities	Atherosclerosis	Drouin et al. (2011)

(continued)

Table 11.1 (continued)

S. no	Natural compound	Source	Mechanism of action	Disorders with NVC dysfunction	Ref.
18	Icariin	Present in the plant *Epimedium* commonly referred to as horny goat weed or yin yang huo	Activates the NO/cGMP pathway in the hippocampus and cortex of mice, improving learning and memory	AD	Jin et al. (2014)
19	Taurisolo	Aglianico grapes	Vasodilation in the pial arteriolar plexus of the corpus callosum. Prevents the formation of reactive oxygen species (ROS)	Cerebral ischemia/reperfusion (I/R) injury	Lapi et al. (2020)
20	Ginseng	Dried root of *Panax ginseng*	Reduces pro-inflammatory cytokines IL-6 and IL-1β, enhances SOD and iNOS and anti-apoptotic activity, decreases caspase-3 and BAX, and upregulates Bcl-2 expression	Traumatic brain injury	Rastogi et al. (2015)

downregulates GSK-3β activity (Zaplatic et al. 2019), consequently mitigating neuronal death caused by oxidative stress. In a chronic PD mouse model, quercetin reduced oxidative damage by modulating the SIRT1/HO-1/Nrf2 pathway, resulting in the survival of neurons (Sriraksa et al. 2012). According to reports, quercetin activated the NRF2/ARE pathway in rat brains, improving memory function (Thiruvengadam et al. 2021), potentially by mitigating the neurovascular impairments associated with chronic neurodegenerative conditions. Quercetin increases tight junction proteins (ZO-1, claudin-5) and inhibits MMP-9 to improve blood-brain barrier integrity in a rat ischemia/reperfusion model (Selvakumar et al. 2013). Rats undergoing middle cerebral artery occlusion (MCAO) experience less cerebral ischemia-reperfusion (I/R) damage, thanks to the activation of the Sirt1/Nrf2/HO-1 signaling pathway, which is responsible for the protective effects seen (Yang et al. 2022).

Kaempferol is a multifarious flavonoid having antibacterial, anticancer, antioxidant, and anti-inflammatory properties. Kaempferol treatment greatly reduced inflammation and preserved the integrity of the blood-brain barrier in rats experiencing cerebral ischemia/reperfusion, thereby enhancing neurological results. This mechanism involves reducing NF-κB phosphorylation and nuclear translocation, which leads to reduced synthesis of pro-inflammatory proteins. This decrease in pro-inflammatory proteins reduces neuroinflammation, hence guarding against BBB disruption (Li et al. 2019). In the striatum of mice exposed to LPS, another study found that kaempferol blocks the HMGB1/TLR4 pathway, hence lowering neuroinflammation and maintaining BBB integrity. Inhibiting the signaling system protects neurons against inflammation and slows the course of neurodegenerative diseases (Yang et al. 2019).

Rutin, comprising the flavonoid quercetin and the disaccharide rutinose—also known as rutoside—is a glycoside. In the rat hippocampus, rutin upregulates CREB, BDNF, and ERK1 expression, fostering neuronal plasticity and survival in the central nervous system and mitigating Aβ neurotoxicity (Enogieru et al. 2018; Moghbelinejad et al. 2014). Rutin suppresses the NF-κB pathway, which causes neuroinflammation, neuronal degeneration, and the development of Alzheimer's disease (AD). Inhibiting the proliferation of stimulated neuroglial cells reduces pro-inflammatory cytokine release, as evidenced by decreased levels of IL-1β and TNF-α. In addition, the downregulation of GFAP and Iba-1, indicating reduced astrogliosis and microgliosis in P301S mice, has been linked to the amelioration of neurovascular coupling (NVC) impairments (Sun et al. 2021).

Pinocembrin (5,7-dihydroxyflavanone), a flavonoid predominantly sourced from propolis, has been demonstrated to prevent cognitive deficits in APP/PS1 transgenic mice, without affecting the deposition of Aβ or oxidative stress (Liu et al. 2014). It promotes neurovascular protection by preserving the neuropil structure, mitigating neuroglial activation and inflammatory responses, supporting microvascular health, and enhancing cholinergic function via the ERK-CREB-BDNF signaling pathway. It also influences RAGE-mediated signaling. Pinocembrin protects the neurovascular unit (NVU) from fibrillar Aβ1-42 in an in vitro model through modulation of neurovascular RAGE pathways (Liu et al. 2014).

Apigenin is a common flavonoid present in many fruits and vegetables, including parsley, onions, oranges, sprouts, and chamomile tea. When given orally to amnesic mice induced by Aβ25-35, it has shown notable neurovascular protective properties (Liu et al. 2011). This treatment enhances cognitive abilities, preserves the integrity of the neurovascular unit, modulates microvascular function, lowers oxidative injury to the neurovascular unit, and improves localized brain blood flow. It reduces oxidative stress and improves the cholinergic system by elevating acetylcholine (ACh) levels and inhibiting the acetylcholinesterase (AChE) activity. Apigenin directly binds to BDNF and enhances its neurotrophic function, potentially reducing the susceptibility to neurodegenerative diseases (Gao et al. 2023). Preclinical studies have shown apigenin's neuroprotective mechanisms in Parkinson's disease, such as suppressing NF-κB gene expression, decreasing α-synuclein protein accumulation, downregulating proinflammatory enzyme iNOS-1 and cytokines (TNF-α, IL-6), and upregulating dopamine D2 receptor (D2R) expression (Charrière et al. 2024).

Curcumin is a polyphenol that has shown a beneficial impact on neurovascular degeneration by protecting the BBB (Wang et al. 2011). Maintaining the vascular integrity and function within the nervous system helps prevent neurovascular degeneration. Curcumin apparently passes the blood-brain barrier rather well, lowering oxidative stress and inflammation and affecting brain neurochemistry (Reddy et al. 2018; Yang et al. 2005; Zhang et al. 2010). Curcumin improves spatial memory deficits in Alzheimer's transgenic mouse models and primary neuronal-glial cultures derived from rats, according to recent studies showing it increases cholinergic neuronal function. These effects are mediated by inhibiting the NF-κB signaling pathway and modulating PPAR activity (Liu et al. 2016). It alleviates Aβ deposition by attenuating inflammation and cell apoptosis through inhibition of NF-κB transcription, thereby suppressing COX-2, IL-1β, TNF-α, and inducible nitric oxide synthase (iNOS) (Wu et al. 2022). Furthermore, Wang et al. (2011) investigated that curcumin provides neuroprotection in ischemia-reperfusion (I/R) injury through the BDNF/TrkB-mediated MAPK/PI3K/CREB signaling cascade. Similarly, another study found that curcumin enhances cognitive function in a rat model of traumatic brain injury (TBI) by influencing the BDNF system. It upregulates BDNF protein expression, promotes TrkB phosphorylation, and aids in neuronal survival (Sun et al. 2020; Wang et al. 2011).

The natural polyphenolic compound *resveratrol* exists mainly in the form of polydatin (Jia et al. 2017; Saha et al. 2012). It has protective effects on the BBB by regulating MMP-9 and neuroinflammation while enhancing adaptive immune responses (Moussa et al. 2017). The mechanism proceeds by lowering nitric oxide (NO) and inflammatory cytokine tumor necrosis factor-alpha (TNF-α). This inhibitory effect results from reduced p38 MAPK phosphorylation together with blockade of NF-κB activation. Resveratrol decreases the expression levels of cyclooxygenase-2 (COX-2) and inducible nitric oxide synthase (iNOS) (Cicero et al. 2019) which function as critical components in inflammatory response pathways. According to Cicero et al. (2019), resveratrol counters Aβ-induced iNOS overexpression while reducing inflammation through TNF-α inhibition along with the suppression of IL-1β and IL-6 (Cicero et al. 2019). The administration of resveratrol improves cerebromicrovascular endothelial function by decreasing NADPH oxidase-derived reactive oxygen species production which leads to the restoration of cortical neurovascular coupling after neuronal activity in aged mice. The brain's microvascular system benefits from resveratrol to help protect cognitive abilities during aging (Toth et al. 2014). Research has shown that resveratrol-loaded fusogenic liposomes strongly improved NVC responses due to elevated nitric oxide (NO)-mediated vasodilation which demonstrated beneficial effects on cerebromicrovascular function. The encapsulation of resveratrol in fusogenic liposomes resulted in greater cellular uptake in aged rats which activated Nrf2-dependent antioxidant pathways against oxidative stress in cerebromicrovascular endothelial cells (Wiedenhoeft et al. 2019). The research findings demonstrate that fusogenic liposomes could serve as a valuable tool for treating brain microvascular aging problems while improving vascular-related cognitive abilities.

Saffron is derived from the ancient flower saffron crocus and contains bioactive compounds such as crocetin and crocin. Due to the inflammatory and anti-oxidative properties of crocetin, neuroprotection is preserved from neurovascular injuries, coupled with the deterioration of function due to degenerative diseases. Saffron extract mitigates the consequences of brain injuries by blocking apoptosis and fostering the formation of blood vessels. Crocetin, which is found in saffron, increases the sensitivity of endothelial cells to VEGF through the upregulation of its receptors and serum response factor (SRF) to these cells (Wang et al. 2011). This helps in angiogenesis and therefore increased microcirculation in the injured cranial tissues (Bie et al. 2011). By lowering proinflammatory cytokines, blocking NF-κB activation and p53 in the hippocampal region, and lowering Aβ plaque accumulation, crocetin given orally to Alzheimer's disease model mice improved cognitive ability (Zhang et al. 2018).

The *Ginkgo biloba* extract EGb 761® is gaining recognition for the possible treatment of neurovascular coupling (NVC) deficits and cognitive dysfunctions associated with COVID-19 (Owens et al. 2024). Rat models of ischemia/reperfusion (I/R)-induced cerebral damage were used to investigate EGb 761®'s neuroprotective effects. Studies have found lower levels of MDA, TNF-α, and IL-1β, as well as higher levels of IL-10 and increased activity of antioxidant enzymes (Zeli et al. 2022). The evidence on cognitive function and the anti-inflammatory, neuroplasticity-enhancing effects of EGb 761® make it a candidate for therapy for patients with cognitive impairment. Additionally, *Ginkgo biloba* extract increases neurotrophic factors like BDNF and promotes neurite growth by triggering the PI3K/AKT/mTOR signaling pathway and inhibiting PTEN and GSK3β, consequently modulating neuroplasticity (Lejri et al. 2019).

Epigallocatechin gallate (EGCG), the predominant and most prevalent compound in the catechin group, is derived from *Camellia sinensis* (Cai et al. 2014). Evidence points to its role in controlling the activity of endothelial nitric oxide synthase (eNOS), hence improving neurovascular function (Sabri et al. 2022). Recent findings indicate that EGCG can successfully cross the BBB (Pervin et al. 2017), and a single oral dose is reported to regulate cerebral blood flow (CBF) parameters in healthy individuals. However, this does not result in changes in cognitive performance (Pompili et al. 2014). Furthermore, EGCG lowers mortality in rat models of middle cerebral artery occlusion and helps neurons survive, thereby minimizing neural damage. The downregulation of Bax and caspase-3 and the upregulation of Bcl-2 expression help to enable this protective effect (Nan et al. 2018). Furthermore, EGCG's neuroprotective benefits are associated with the activation of the CREB/BDNF/TrkB-PI3K/Akt signaling cascade (Ding et al. 2017). Cerebral conditions such as small vessel diseases (SVD), Alzheimer's disease (AD), and traumatic brain injury (TBI) are reported to impair cerebral blood flow (CBF) because of the loss of endothelial ion signaling pathways (Moccia et al. 2022). Such pathological changes within the body lead to diminished function of ion channels and thus disrupt the responsive (neurovascular) coupling (NVC) needed to supply active neurons with adequate blood. This disorder helps explain the prevalence of vascular cognitive impairment and dementia (VCID) in diseases including AD, SVDs, and TBI (Moccia et al. 2022). Important ion channels with endothelial actions are NMDA receptors, transient receptor potential ankyrin 1 (TRPA1), and inward-rectifier potassium (Kir2.1) channels. Studies show that dietary compounds may activate TRP channels in brain microvascular endothelial cells (ECs), thereby restoring endothelium-dependent vasodilation in cerebral diseases (Behringer and Hakim 2019; Guerra et al. 2018). Decreased TRPA1 expression observed in endothelial cells in the case of ischemic stroke consequently impairs the vasomotor response in brain slices. Cinnamaldehyde (from cinnamon) (Aubdool et al. 2016), a selective TRPA1 agonist, is reported to significantly reduce ischemic brain damage when administered systemically, by recruiting IKCa/SKCa channels in a calcium-dependent manner (Thakore et al. 2021). Other dietary compounds including allyl isothiocyanate (from mustard oil), allicin and diallyl disulphide (from garlic), and methyl salicylate (from wintergreen oil) can also activate TRPA1, hence offering still another possibility for dietary interventions to enhance CBF (Hinman et al. 2006; Moccia et al. 2022).

Naringenin (4′,5,7-trihydroxy-flavonone), a flavonoid classified under flavanones, is widely present in citrus fruits such as grapes, lemons, tangerines, and oranges. Its strong anti-inflammatory and antioxidant action is well-known. Research indicates that naringenin lowers oxidative stress and inflammation caused by oxygen-glucose deprivation/reperfusion (OGD/R), shielding against neuronal cell death. SIRT1/ FOXO1 signaling cascade activation coordinates this response. These naringenin preliminary studies show its possible effectiveness in treating cerebral ischemia/reperfusion (I/R) damage (Zhao et al. 2023). Besides, naringenin was found to have a neurotherapeutic effect on stroke-related neurological injury (Adetunji et al. 2023). For example, after inducing post-subarachnoid hemorrhage in rats, naringenin administration inhibited the MAPK signaling pathway, resulting in reduced oxidative stress and apoptosis as well as improved BBB integrity (Han et al. 2017). In addition, naringenin leads to the decrease of matrix metalloproteinases (MMPs) and induces glial fibrillary acidic protein (GFAP) in 3-nitropropionic acid-induced neurodegeneration in rats. This resulted in improved locomotor performance and less BBB disruption (Gopinath and Sudhandiran 2016).

Cocoa flavanols especially epicatechin affect the functioning of the endothelial cells of the blood vessels in the brain by stimulating eNOS and NO production. This increase in NO causes vasodilation and increases blood flow to the brain (Martín et al. 2020). More importantly, though, the effect of cocoa flavanols on endothelial cells is related with the inhibition of pro-inflammatory cytokines (IL-1β, IL-2, and IL-8) (Hwang and Jeong 2010a). Therefore, cocoa flavanols could help offset the age-related drop in endothelial function, benefiting the central nervous system as well as peripheral blood vessels (Sorond et al. 2008). There have been limited human studies focusing on the neurobiological, cognitive, affective, and behavioral effects of cocoa products rich in flavanols. It is suggested that flavanols may impact brain function, learning, memory, and cognition by modulating key cellular signaling pathways, including MAPK, ERK, and PI3K/AKT cascades (Martín et al. 2020). By inducing gene expression and protein synthesis, these signaling cascades play a pivotal role in reinforcing long-term potentiation and developing persistent memories. Catechin, a flavanol, greatly increases cerebrovascular flow-mediated dilation and cognitive performance in atherosclerotic mice by restoring NOS activity and enhancing endothelial function (Drouin et al. 2011). Through activation of the NO/cGMP pathway in the hippocampal and cortical areas, respectively, the flavanol icariin was shown to enhance cognitive ability (Jin et al. 2014). The Taurisolo supplement taken from the Aglianico grape has been found in later research to induce vasodilation in the corpus callosum's pial arteriolar plexus. After an initial rise in eNOS expression, it lowers reactive oxygen species and boosts capillary perfusion in rats, underscoring the part polyphenols play in cerebrovascular function and cognitive performance (Lapi et al. 2020). Furthermore, Taurisolo has been demonstrated to preserve the integrity and function of blood vessels and cells, reducing brain damage caused by cerebral hypoperfusion, a condition related to global health difficulties and ranking third in disability and mortality. Ginseng is a kind of herbal medication widely utilized throughout Asia. It is obtained from the industrious root of *Panax*, which is dried and peeled and is reputed to heal a variety of ailments while also boosting the immune system. It is proposed as a potential remedy for various neurological disorders, including stroke and acute and chronic neurodegenerative diseases.

Ginseng decreases pro-inflammatory cytokines IL-6 and IL-1β, boosts the anti-inflammatory cytokine IL-10, and mitigates oxidative stress induced by traumatic brain injury (TBI) by increasing SOD andiNOS activity, while reducing malondialdehyde levels. Additionally, ginseng exhibits anti-apoptotic effects by attenuating caspase-3 and BAX levels while upregulating Bcl-2 levels (Xia et al. 2012). Furthermore, it inhibits iNOS, reducing NO production and post-ischemic inflammation (Ye et al. 2011). Ginseng also upregulates the ARE pathway for neuroprotection via the Gbeta1/PI3K/Akt-Nrf2 signaling pathway (Hwang and Jeong 2010a) and offers potential benefits via Akt/Nrf2 in SK-N-SH cells (Hwang et al. 2010b). Natural products enhance neurovascular coupling (NVC) and its underlying mechanisms, as shown in Fig. 11.2 and Table 11.1 While lowering malondialdehyde levels, ginseng reduces pro-inflammatory cytokines IL-6 and IL-1β by raising SOD and iNOS activity, mitigating oxidative stress induced by traumatic brain injury

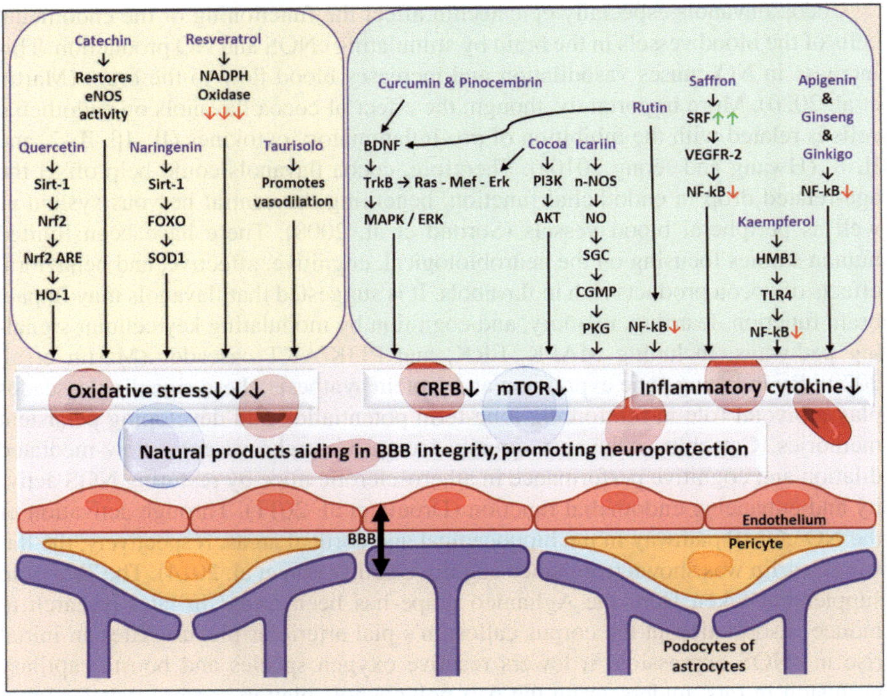

Fig. 11.2 Diagram illustrating how natural substances improve neurovascular coupling by stabilizing inflammatory cytokines, reducing oxidative stress, and activating signaling pathways that provide neuroprotection, thereby preserving the blood-brain barrier (NVU) and supporting overall brain health

(TBI). IL-10 is then raised. Furthermore, ginseng reduces caspase-3 and BAX levels and raises Bcl-2 levels, displaying anti-apoptotic effects (Xia et al. 2012). It also inhibits iNOS, reducing NO production and post-ischemic inflammation (Ye et al. 2011). Ginseng also upregulates the ARE pathway for neuroprotection via the Gbeta1/PI3K/Akt-Nrf2 signaling pathway (Hwang and Jeong 2010a) and potential benefits via Akt/Nrf2 in SK-N-SH cells (Hwang et al. 2010b). As Fig. 11.2 and Table 11.1 illustrate how natural products improve neurovascular coupling (NVC) and their underlying mechanisms.

11.3.1 Preclinical and Clinical Evidence

The clinical trials' primarily goal is to develop numerous therapeutic options for neurodegenerative diseases (NDs), such as cognitive enhancers and anti-amyloid and anti-tau treatments to reduce neuroinflammation. Restoring impaired NVC function is necessary to achieve proper brain functions, and clinical approaches are designed to increase vascular flow and metabolite supply to neurons and thus decrease the impact of NDs.

Research on preclinical resveratrol's effects suggests its possible benefit in treating degenerative diseases, including Parkinson's (PD) and Alzheimer's (AD). Still, there are not enough clinical studies that fairly show its success in human subjects. A phase II clinical trial (NCT01504854) by Turner et al. (2015) on 119 AD patients found mild to moderate effects of resveratrol on cognitive function (Turner et al. 2015). This study demonstrated that resveratrol and its metabolites were present in both plasma and cerebrospinal fluid (CSF), while treatment was safe and tolerated, without any changes to AD biomarkers being observed. Another study (NCT00678431) where AD patients were administered 5 mg of resveratrol for 1 year showed less deterioration in the treatment group, although the results were not statistically significant (Zhu et al. 2018). Huhn et al. (2018) administered 200 mg of resveratrol for 26 weeks in a study (NCT02621554) where participants older than 60 also reported no statistically significant changes in cognitive memory with the treatment (Huhn et al. 2018).

Resveratrol's limited efficacy in these clinical trials results from its poor bioavailability, solubility challenges, first-pass metabolism, and gastrointestinal problems associated with high-dose administration (Ramírez-Garza et al. 2018). Enhancing resveratrol's bioavailability and targeted delivery is crucial for its use in neurodegenerative diseases, and additional research is needed to confirm its efficacy.

Curcumin has shown potential benefits for neurodegenerative diseases (NDs) in preclinical studies, but clinical trials have yielded mixed results. In a 24-week phase-2 clinical trial (NCT000997), mild to moderate Alzheimer's disease (AD) patients' safety and tolerability of curcumin were assessed; nevertheless, no significant improvement in clinical outcomes or biomarkers was observed (Plascencia-Villa and Perry 2023). A 2020 open-label phase-2 trial (NCT04499963) assessed theracurmin (a curcumin colloidal dispersion) for ALS over 6 months, and another ongoing study (NCT04654689) is examining a combination of curcumin, resveratrol, and G04CB02 for ALS, focusing on their potential to improve neurovascular function and overall neuroprotection (ClinicalTrials.gov). Curcumin's low bioavailability and rapid clearance limit its clinical application. Innovative methods like nanotechnology are being explored to enhance its medicinal effects (Sabouni et al. 2023). Nanoformulations of curcumin, namely nanocrystals, polymeric nanoparticles, micelles, liposomes, and nanogels, have shown increased bioavailability in clinical trials (Ringman et al. 2008). More extensive clinical and pharmacokinetic studies are needed to confirm curcumin's potential against NDs by addressing neurovascular dysfunction (Mandal et al. 2020; Wąsik and Antkiewicz-Michaluk 2017).

Few clinical studies have explored quercetin's effectiveness for neurodegenerative diseases, particularly in the context of neurovascular coupling dysfunction. A phase-2 trial (NCT04785300) started in 2022 is investigating quercetin in patients with mild cognitive impairment or Alzheimer's disease (AD), conditions often associated with impaired neurovascular coupling (Cummings et al. 2022). Another pilot study (NCT05422885) is examining whether a regimen of 1250 mg quercetin in every 2 weeks for 12 weeks can improve blood flow, mobility, and cognition in older adults, potentially preventing AD progression by addressing neurovascular coupling dysfunction, with findings anticipated shortly (Grewal et al. 2021). Quercetin's use in AD is limited by poor blood-brain barrier penetrability and low solubility, which are critical factors in enhancing neurovascular coupling. To increase quercetin's bioavailability and effectiveness in neurovascular function restoration, future research should prioritize the development of advanced delivery methods, such as prodrugs and nano-encapsulation. Further evaluations are necessary to assess effective dose and delivery methods for developing a new drug targeting neurodegenerative diseases through improved neurovascular coupling (Madhubala et al. 2024). A recent study (NCT03978052) showed that EGCG could improve cognitive decline and brain connectivity in participants, which might improve neurovascular coupling dysfunction in individuals with cognitive decline (Forcano et al. 2021). Additionally, a phase-2 clinical study (NCT01699711) is evaluating whether EGCG administration could improve cognitive performance and inhibit Alzheimer's-like progression (de la Torre et al. 2016). Another clinical trial (NCT00461942) is investigating the therapeutic potential and safety of EGCG in treating patients with Parkinson's disease. Clinical data on the therapeutic and preventive effects of EGCG on cognitive decline and Alzheimer's disease (AD) are inconsistent (Fernandes et al. 2021; Ide et al. 2014) and cannot be recommended for AD prevention. Robust clinical trials are necessary to identify optimal dosages and ensure targeted delivery (Lange et al. 2022). To assess the influence of American ginseng Cereboost® on cognitive function and mood, particularly regarding neurovascular dysfunction, a double-blind, randomized, placebo-controlled clinical trial was conducted. The study, which included 61 healthy young adults, was preregistered at ClinicalTrials.gov on July 6, 2018 (Identifier: NCT03579095). The results showed that Cereboost® significantly improved cognitive performance, suggesting potential therapeutic benefits for those experiencing neurovascular dysfunction. Cereboost supplementation for short-term memory and cognitive control showed improvements, with even greater enhancements after 2 weeks of daily supplementation. Furthermore, taking this supplement on a daily basis boosted performance accuracy and improved mood while reducing mental exhaustion associated with neurovascular complications (Bell et al. 2022).

Ginkgo biloba extract EGb 761 shows possible therapeutic worth in preclinical studies on patients with mild cognitive impairment (MCI). It has been mentioned that considering the clinical management of MCI, EGb 761 could address some crucial pathological processes like oxidative damage, inflammation, and possibly NVC dysfunction. It was registered on clinicaltrials.gov with ID NCT05594355 (Morató et al. 2023). The study provides strong proof that EGb 761 might drastically lower inflammatory markers including TNF-α, interleukins (IL-6), and C-reactive protein. With low levels of malondialdehyde (MDA) and vice versa, significant improvement was also shown in indicators of oxidative stress including the activity of antioxidant enzymes superoxide dimutase (SOD) and glutathione peroxidase. Through better blood flow and appropriate supply of neurons, EGb 761's anti-inflammatory and antioxidant qualities can help slow down degradation of cognitive performance. Given its safety profile and the mechanistic insights provided by this

Table 11.2 Clinical evaluations of plant-based compounds in the management of neurodegenerative diseases

S. no	Compound	Diseases	Clinical trial ID	Stage	Status
1	Resveratrol	Alzheimer's disease	NCT01504854	Phase II	Completed
			NCT00678431	Phase III	Completed
			NCT02621554	N/A	Completed
			NCT00743743	Phase III	Completed
2	Curcumin	Alzheimer's disease	NCT00099710	Phase II	Completed
		Amyotrophic lateral sclerosis	NCT04499963	Phase II	Completed
		Amyotrophic lateral sclerosis	NCT04654689	Phase II	Completed
3	Quercetin	Alzheimer's disease	NCT04785300	Phase II	Ongoing
			NCT05422885	Phase II	Ongoing
4	EGCG	Cognitive decline	NCT03978052	Phase II	Ongoing
		Cognitive defects in Down's syndrome	NCT01699711	Phase II	Completed
		Parkinson's disease	NCT00461942	Phase II	Completed
5	American ginseng	Cognitive function	NCT03579095	Preliminary study	Completed
6	*Ginkgo biloba* extract (EGb 761)	Mild cognitive impairment	NCT05594355	Phase III	Ongoing

trial, EGb 761 emerges as a promising adjunctive treatment for MCI. Further investigation in larger and longer-term studies is warranted to confirm its efficacy, elucidate its impact on NVC, and determine optimal treatment protocols. (Table 11.2 contains clinical evaluations of plant-based compounds in the treatment of neurodegenerative diseases.)

11.4 Conclusion and Future Perspective

Neurovascular coupling is a critical process that ensures adequate blood flow in active regions of the brain, thus maintaining neuronal heath and function. Dysfunction in NVC has been increasingly associated with the pathogenesis of neurodegenerative diseases such AD, PD, ALS, and vascular dementia. Emerging evidences demonstrate the potential of natural products to modulate NVC and mitigate the progression of these debilitating conditions. Natural products have demonstrated a significant promising impact on preserving the integrity of the BBB, reducing oxidative stress and anti-inflammatory properties. Furthermore, natural products can improve CBF and aid neuronal survival by modulating signaling pathways associated with NVC.

In conclusion, natural products hold great promise as potential therapeutic agents for managing NVC dysfunction and associated neurodegenerative diseases. Continued research in this field will not only deepen our understanding of the pathophysiology of NDDS but also develop novel, effective treatments that harness the power of nature to promote brain health.

11.5 Summary

This chapter explores the function of NVC, a fundamental mechanism that ensures an adequate blood supply to the demands of the active regions of the brain. Since the brain does not store fat to generate instant enery, NVC plays a major role in maintaing its energy demands. The impairment of the NVU, leading to NVC dysfunction, is associated with various NDDs such as AD, PD, and ALS. Natural products play an important role in the prevention of NDDS by modulating NVC dysfunction.

Natural compounds such as flavonoids, ployphenols, and other boactive compounds possess antioxidant, anti-inflamatory, and neuroprotective properties that can improve NVC, maintain BBB integrity, and improve overall brain health. This chaper provides insight into the mechanism by which natural products contribute to improvement of the neurovascular health, and their potential in preventing and treating NDDs. Furthermore, it higlights the significance of ongoing clinical trials and studies to evaluvate the sensitivity and efficacy of natural products against NDDs.

References

Adetunji JA, Fasae KD, Awe AI et al (2023) The protective roles of citrus flavonoids, naringenin, and naringin on endothelial cell dysfunction in diseases. Heliyon 9(6):e17166. https://doi.org/10.1016/j.heliyon.2023.e17166

Anderson CM, Bergher JP, Swanson RA (2004) ATP-induced ATP release from astrocytes. J Neurochem 88(1):246–256. https://doi.org/10.1111/j.1471-4159.2004.02204.x

Atanasov, Atanas G., Sergey B. Zotchev, Verena M. Dirsch, International Natural Product Sciences Taskforce, and Claudiu T. Supuran (2021) Natural products in drug discovery: advances and opportunities. Nat Rev Drug Discov 20(3):200–216. https://doi.org/10.1038/s41573-020-00114-z

Attwell D, Buchan AM, Charpak S, Lauritzen M, Macvicar BA, Newman EA (2010) Glial and neuronal control of brain blood flow. Nature 468(7321):232–243. https://doi.org/10.1038/nature09613

Aubdool AA, Kodji X, Abdul-Kader N et al (2016) TRPA1 activation leads to neurogenic vasodilatation: involvement of reactive oxygen nitrogen species in addition to CGRP and NO. Br J Pharmacol 173(15):2419–2433. https://doi.org/10.1111/bph.13519

Behringer EJ, Hakim MA (2019) Functional interaction among KCa and TRP channels for cardiovascular physiology: modern perspectives on aging and chronic disease. Int J Mol Sci 20(6):1380. https://doi.org/10.3390/ijms20061380

Bell L, Whyte A, Duysburgh C et al (2022) A randomized, placebo-controlled trial investigating the acute and chronic benefits of American ginseng (Cereboost®) on mood and cognition in healthy young adults, including in vitro investigation of gut microbiota changes as a possible mechanism of action. Eur J Nutr 61(1):413–428. https://doi.org/10.1007/s00394-021-02654-5

Bie X, Chen Y, Zheng X, Dai H (2011) The role of crocetin in protection following cerebral contusion and in the enhancement of angiogenesis in rats. Fitoterapia 82(7):997–1002. https://doi.org/10.1016/j.fitote.2011.06.001

Biesecker KR, Srienc AI, Shimoda AM et al (2016) Glial cell calcium signaling mediates capillary regulation of blood flow in the retina. J Neurosci 36(36):9435–9445. https://doi.org/10.1523/JNEUROSCI.1782-16.2016

Cai J, Jing D, Shi M et al (2014) Epigallocatechin Gallate (EGCG) attenuates infrasound-induced neuronal impairment by inhibiting microglia-mediated inflammation. J Nutr Biochem 25(7):716–725. https://doi.org/10.1016/j.jnutbio.2014.02.012

Candela P, Saint-Pol J, Kuntz M et al (2015) In vitro discrimination of the role of LRP1 at the BBB cellular level: focus on brain capillary endothelial cells and brain pericytes. Brain Res 1594:15–26. https://doi.org/10.1016/j.brainres.2014.10.047

Carmignoto G, Gómez-Gonzalo M (2010) The contribution of astrocyte signalling to neurovascular coupling. Brain Res Rev 63(1):138–148. https://doi.org/10.1016/j.brainresrev.2009.11.007

Carvey PM, Zhao CH, Hendey B et al (2005) 6-Hydroxydopamine-induced alterations in blood–brain barrier permeability. Eur J Neurosci 22(5):1158–1168. https://doi.org/10.1111/j.1460-9568.2005.04281.x

Chaigneau E, Oheim M, Audinat E, Charpak S (2003) Two-photon imaging of capillary blood flow in olfactory bulb glomeruli. Proc Natl Acad Sci USA 100(22):13081–13086. https://doi.org/10.1073/pnas.2133652100

Charrière K, Schneider V, Perrignon-Sommet M et al (2024) Exploring the role of apigenin in neuroinflammation: insights and implications. Int J Mol Sci 25(9):5041. https://doi.org/10.3390/ijms25095041

Chen X, Lan X, Roche I, Liu R, Geiger JD (2008) Caffeine protects against MPTP-induced blood-brain barrier dysfunction in mouse striatum. J Neurochem 107(4):1147–1157. https://doi.org/10.1111/j.1471-4159.2008.05697.x

Chen BR, Bouchard MB, McCaslin AFH, Burgess SA, Hillman EMC (2011) High-speed vascular dynamics of the hemodynamic response. NeuroImage 54(2):1021–1030. https://doi.org/10.1016/j.neuroimage.2010.09.036

Cheng X, Yang Y-L, Yang H, Wang Y-H, Guan-Hua D (2018) Kaempferol alleviates LPS-induced neuroinflammation and BBB dysfunction in mice via inhibiting HMGB1 release and down-regulating TLR4/MyD88 pathway. Int Immunopharmacol 56:29–35. https://doi.org/10.1016/j.intimp.2018.01.002

Chisari M, Salomone S, Laureanti F, Copani A, Sortino MA (2004) Modulation of cerebral vascular tone by activated glia: involvement of nitric oxide. J Neurochem 91(5):1171–1179. https://doi.org/10.1111/j.1471-4159.2004.02782.x

Cicero AFG, Ruscica M, Banach M (2019) Resveratrol and cognitive decline: a clinician perspective. Arch Med Sci 15(4):936–943. https://doi.org/10.5114/aoms.2019.85463

Cirrito JR, Deane R, Fagan AM et al (2005) P-glycoprotein deficiency at the blood-brain barrier increases amyloid-beta deposition in an Alzheimer disease mouse model. J Clin Invest 115(11):3285–3290. https://doi.org/10.1172/JCI25247

Collins LM, Toulouse A, Connor TJ, Nolan YM (2012) Contributions of central and systemic inflammation to the pathophysiology of Parkinson's disease. Neuropharmacology 62(7):2154–2168. https://doi.org/10.1016/j.neuropharm.2012.01.028

Cox SB, Woolsey TA, Rovainen CM (1993) Localized dynamic changes in cortical blood flow with whisker stimulation corresponds to matched vascular and neuronal architecture of rat barrels. J Cereb Blood Flow Metab 13(6):899–913. https://doi.org/10.1038/jcbfm.1993.113

Cummings J, Lee G, Nahed P et al (2022) Alzheimer's disease drug development pipeline: 2022. Alzheimer's Dement: Transl Res Clin Interv 8(1):e12295. https://doi.org/10.1002/trc2.12295

de la Torre R, de Sola S, Hernandez G et al (2016) Safety and efficacy of cognitive training plus epigallocatechin-3-gallate in young adults with Down's syndrome (TESDAD): a double-blind, randomised, placebo-controlled, phase 2 trial. Lancet Neurol 15(8):801–810. https://doi.org/10.1016/S1474-4422(16)30034-5

De Labra C, Rivadulla C, Espinosa N, Dasilva M, Cao R, Cudeiro J (2009) Different sources of nitric oxide mediate neurovascular coupling in the lateral geniculate nucleus of the cat. Front Syst Neurosci 3:575. https://doi.org/10.3389/neuro.06.009.2009

Dienel GA, Hertz L (2001) Glucose and lactate metabolism during brain activation. J Neurosci Res 66(5):824–838

Ding M-L, Ma H, Man Y-G, Lv H-Y (2017) Protective effects of a green tea polyphenol, epigallocatechin-3-gallate, against sevoflurane-induced neuronal apoptosis involve regulation of CREB/BDNF/TrkB and PI3K/Akt/mTOR signalling pathways in neonatal mice. Can J Physiol Pharmacol 95(12):1396–1405. https://doi.org/10.1139/cjpp-2016-0333

Drouin A, Bolduc V, Thorin-Trescases N et al (2011) Catechin treatment improves cerebrovascular flow-mediated dilation and learning abilities in atherosclerotic mice. Am J Physiol Heart Circ Physiol 300(3):H1032–H1043. https://doi.org/10.1152/ajpheart.00410.2010

Enogieru AB, Haylett W, Hiss DC, Bardien S, Ekpo OE (2018) Rutin as a potent antioxidant: implications for neurodegenerative disorders. Oxidative Med Cell Longev 2018:6241017. https://doi.org/10.1155/2018/6241017

Fergus A, Lee KS (1997) Regulation of cerebral microvessels by glutamatergic mechanisms. Brain Res 754(1–2):35–45. https://doi.org/10.1016/s0006-8993(97)00040-1

Fernandes L, Cardim-Pires TR, Foguel D, Palhano FL (2021) Green tea polyphenol epigallocatechin-gallate in amyloid aggregation and neurodegenerative diseases. Front Neurosci 15:718188. https://doi.org/10.3389/fnins.2021.718188

Forcano L, Fauria K, Soldevila-Domenech N et al (2021) Prevention of cognitive decline in subjective cognitive decline APOE E4 carriers after EGCG and a multimodal intervention (PENSA): study design. Alzheimer's & Dementia (New York, NY) 7(1):e12155. https://doi.org/10.1002/trc2.12155

Freygang W, Sokoloff L (2013) Quantitative measurement of regional circulation in the central nervous system by the use of radioactive inert gas. Adv Biol Med Phys 6:263–279

Gao AX, Xia TCX, Lin LSY, Dong TTX, Tsim KWK, Therapeutics. (2023) The neurotrophic activities of brain-derived neurotrophic factor are potentiated by binding with apigenin. A common flavone in vegetables, in stimulating the receptor signaling. CNS Neurosci Ther 29(10):2787–2799

Garbuzova-Davis S, Saporta S, Sanberg PRJALS (2008) Implications of blood-brain barrier disruption in ALS. In: Amyotrophic lateral sclerosis, vol 9. Taylor & Francis

Garbuzova-Davis S, Rodrigues MCO, Hernandez-Ontiveros DG et al (2011) Amyotrophic lateral sclerosis: a neurovascular disease. Brain Res 1398:113–125. https://doi.org/10.1016/j.brainres.2011.04.049

Gopinath K, Sudhandiran G (2016) Protective effect of naringin on 3-Nitropropionic acid-induced neurodegeneration through the modulation of matrix metalloproteinases and glial fibrillary acidic protein. Can J Physiol Pharmacol 94(1):65–71. https://doi.org/10.1139/cjpp-2015-0035

Grewal AK, Singh TG, Sharma D et al (2021) Mechanistic insights and perspectives involved in neuroprotective action of quercetin. Biomed Pharmacother 140:111729. https://doi.org/10.1016/j.biopha.2021.111729

Guerra G, Lucariello A, Perna A, Botta L, De Luca A, Moccia F (2018) The role of endothelial Ca2+ signaling in neurovascular coupling: a view from the lumen. Int J Mol Sci 19(4):938. https://doi.org/10.3390/ijms19040938

Gureev AP, Popov VN (2019) Nrf2/ARE pathway as a therapeutic target for the treatment of Parkinson diseases. Neurochem Res 44(10):2273–2279. https://doi.org/10.1007/s11064-018-02711-2

Hamel E, Nicolakakis N, Aboulkassim T, Ongali B, Tong X-K (2008) Oxidative stress and cerebrovascular dysfunction in mouse models of Alzheimer's disease. Exp Physiol 93(1):116–120. https://doi.org/10.1113/expphysiol.2007.038729

Han Y, Jingyuan S, Liu X, Zhao Y, Wang C, Li X (2017) Naringin alleviates early brain injury after experimental subarachnoid hemorrhage by reducing oxidative stress and inhibiting apoptosis. Brain Res Bull 133:42–50. https://doi.org/10.1016/j.brainresbull.2016.12.008

Henkel JS, Beers DR, Wen S, Bowser R, Appel SH (2009) Decreased mRNA expression of tight junction proteins in lumbar spinal cords of patients with ALS. Neurology 72(18):1614–1616. https://doi.org/10.1212/WNL.0b013e3181a41228

Hinman A, Chuang H-H, Bautista DM, Julius D (2006) TRP channel activation by reversible covalent modification. Proc Natl Acad Sci USA 103(51):19564–19568. https://doi.org/10.1073/pnas.0609598103

Hosford PS, Gourine AV (2019) What is the key mediator of the neurovascular coupling response? Neurosci Biobehav Rev 96:174–181. https://doi.org/10.1016/j.neubiorev.2018.11.011

Howarth C, Mishra A, Hall CN (2021) More than just summed neuronal activity: how multiple cell types shape the BOLD response. Philos Trans R Soc Lond B Biol Sci 376(1815):20190630. https://doi.org/10.1098/rstb.2019.0630

Huhn S, Beyer F, Zhang R et al (2018) Effects of resveratrol on memory performance, hippocampus connectivity and microstructure in older adults – a randomized controlled trial. NeuroImage 174:177–190. https://doi.org/10.1016/j.neuroimage.2018.03.023

Huneau C, Benali H, Chabriat H (2015) Investigating human neurovascular coupling using functional neuroimaging: a critical review of dynamic models. Front Neurosci 9:467. https://doi.org/10.3389/fnins.2015.00467

Hwang YP, Jeong HG (2010a) Ginsenoside Rb1 protects against 6-Hydroxydopamine-induced oxidative stress by increasing heme oxygenase-1 expression through an estrogen receptor-related PI3K/Akt/Nrf2-dependent pathway in human dopaminergic cells. Toxicol Appl Pharmacol 242(1):18–28. https://doi.org/10.1016/j.taap.2009.09.009

Hwang YP, Jeong HG (2010b) Ginsenoside Rb1 protects against 6-hydroxydopamine-induced oxidative stress by increasing heme oxygenase-1 expression through an estrogen receptor-related PI3K/Akt/Nrf2-dependent pathway in human dopaminergic cells. Toxicol Appl Pharmacol 242(1):18–28

Iadecola C (2017) The neurovascular unit coming of age: a journey through neurovascular coupling in health and disease. Neuron 96(1):17–42. https://doi.org/10.1016/j.neuron.2017.07.030

Ide K, Yamada H, Takuma N et al (2014) Green tea consumption affects cognitive dysfunction in the elderly: a pilot study. Nutrients 6(10):4032–4042. https://doi.org/10.3390/nu6104032

Janelidze S, Hertze J, Nägga K et al (2017) Increased blood-brain barrier permeability is associated with dementia and diabetes but not amyloid pathology or APOE genotype. Neurobiol Aging 51:104–112. https://doi.org/10.1016/j.neurobiolaging.2016.11.017

Jang J-W, Lee J-K, Hur H, Kim T-W, Joo S-P, Piao M-S (2014) Rutin improves functional outcome via reducing the elevated matrix metalloproteinase-9 level in a photothrombotic focal ischemic model of rats. J Neurol Sci 339(1–2):75–80. https://doi.org/10.1016/j.jns.2014.01.024

Jia Y, Wang N, Liu X (2017) Resveratrol and amyloid-Beta: mechanistic insights. Nutrients 9(10):10. https://doi.org/10.3390/nu9101122

Jin F, Gong Q-H, Xu Y-S et al (2014) Icariin, a phoshphodiesterase-5 inhibitor, improves learning and memory in APP/PS1 transgenic mice by stimulation of NO/cGMP signalling. Int J Neuropsychopharmacol 17(6):871–881

Jin Z, Ke J, Guo P, Wang Y, Wu. (2019) Quercetin improves blood-brain barrier dysfunction in rats with cerebral ischemia reperfusion via Wnt signaling pathway. Am J Transl Res 11(8):4683

Kamitani Y, Tong F (2005) Decoding the visual and subjective contents of the human brain. Nat Neurosci 8(5):679–685. https://doi.org/10.1038/nn1444

Kim Y, Yeon Cho A, Kim HC, Ryu D, Jo SA, Jung Y-S (2022) Effects of natural polyphenols on oxidative stress-mediated blood-brain barrier dysfunction. Antioxidants (Basel, Switzerland) 11(2):197. https://doi.org/10.3390/antiox11020197

Koehler RC, Roman RJ, Harder DR (2009) Astrocytes and the regulation of cerebral blood flow. Trends Neurosci 32(3):160–169. https://doi.org/10.1016/j.tins.2008.11.005

Kook S-Y, Hong HS, Moon M, Ha CM, Chang S, Mook-Jung I (2012) $A\beta_{1-42}$-RAGE interaction disrupts tight junctions of the blood-brain barrier via Ca^{2+}-calcineurin signaling. J Neurosci Off J Soc Neurosci 32(26):8845–8854. https://doi.org/10.1523/JNEUROSCI.6102-11.2012

Kortekaas R, Leenders KL, van Oostrom JCH et al (2005) Blood-brain barrier dysfunction in parkinsonian midbrain in vivo. Ann Neurol 57(2):176–179. https://doi.org/10.1002/ana.20369

Kumar S, Pandey AK (2013) Chemistry and biological activities of flavonoids: an overview. TheScientificWorldJournal 2013:162750. https://doi.org/10.1155/2013/162750

Lange KW, Lange KM, Nakamura Y (2022) Green tea, epigallocatechin gallate and the prevention of Alzheimer's disease: clinical evidence. Food Sci Human Wellness 11(4):765–770. https://doi.org/10.1016/j.fshw.2022.03.002

Lapi D, Stornaiuolo M, Sabatino L et al (2020) The pomace extract taurisolo protects rat brain from ischemia-reperfusion injury. Front Cell Neurosci 14:3. https://doi.org/10.3389/fncel.2020.00003

Lejri I, Grimm A, Eckert A (2019) Ginkgo Biloba extract increases neurite outgrowth and activates the Akt/mTOR pathway. PLoS One 14(12):e0225761. https://doi.org/10.1371/journal.pone.0225761

Li N, Sul J-Y, Haydon PG (2003) A calcium-induced calcium influx factor, nitric oxide, modulates the refilling of calcium stores in astrocytes. J Neurosci Off J Soc Neurosci 23(32):10302–10310. https://doi.org/10.1523/JNEUROSCI.23-32-10302.2003

Li Y, Zhou S, Li J et al (2015) Quercetin protects human brain microvascular endothelial cells from Fibrillar β-Amyloid1-40-induced toxicity. Acta Pharm Sin B 5(1):47–54. https://doi.org/10.1016/j.apsb.2014.12.003

Li W-H, Cheng X, Yang Y-L et al (2019) Kaempferol attenuates neuroinflammation and blood brain barrier dysfunction to improve neurological deficits in cerebral ischemia/reperfusion rats. Brain Res 1722:146361

Lin WH, Hao Q, Rosengarten B, Leung WH, Wong KS (2011) Impaired neurovascular coupling in ischaemic stroke patients with large or small vessel disease. Eur J Neurol 18(5):731–736. https://doi.org/10.1111/j.1468-1331.2010.03262.x

Liu R, Zhang T, Yang H, Lan X, Ying J, Du G (2011) The flavonoid apigenin protects brain neurovascular coupling against amyloid-β 25-35-induced toxicity in mice. J Alzheimer's Dis 24(1):85–100

Liu R, Li JZ, Song JK, Zhou D, Huang C, Bai XY, Xie T et al (2014) Pinocembrin improves cognition and protects the neurovascular unit in Alzheimer related deficits. Neurobiol Aging 35(6):1275–1285

Liu Z-J, Li Z-H, Liu L et al (2016) Curcumin attenuates beta-amyloid-induced neuroinflammation via activation of peroxisome proliferator-activated receptor-gamma function in a rat model of Alzheimer's disease. Front Pharmacol 7:261. https://doi.org/10.3389/fphar.2016.00261

Liu X, Hou D, Lin F et al (2019) The role of neurovascular unit damage in the occurrence and development of Alzheimer's disease. Rev Neurosci 30(5):477–484

Longden TA, Dabertrand F, Koide M et al (2017) Capillary K+ -sensing initiates retrograde hyperpolarization to increase local cerebral blood flow. Nat Neurosci 20(5):717–726. https://doi.org/10.1038/nn.4533

Madhubala D, Patra A, Khan MR, Mukherjee AK (2024) Phytomedicine for neurodegenerative diseases: the road ahead. Phytother Res 38(6):2993–3019. https://doi.org/10.1002/ptr.8192

Mandal M, Jaiswal P, Mishra A (2020) Role of curcumin and its nanoformulations in neurotherapeutics: a comprehensive review. J Biochem Mol Toxicol 34(6):e22478. https://doi.org/10.1002/jbt.22478

Marco S, Skaper SD (2006) Amyloid β-peptide1–42 alters tight junction protein distribution and expression in brain microvessel endothelial cells. Neurosci Lett 401(3):219–224

Martín MA, Goya L, de Pascual-Teresa S (2020) Effect of cocoa and cocoa products on cognitive performance in young adults. Nutrients 12(12):3691. https://doi.org/10.3390/nu12123691

Metea MR, Newman EA (2006) Glial cells dilate and constrict blood vessels: a mechanism of neurovascular coupling. J Neurosci Off J Soc Neurosci 26(11):2862–2870. https://doi.org/10.1523/JNEUROSCI.4048-05.2006

Mishra A, Reynolds JP, Chen Y, Gourine AV, Rusakov DA, Attwell D (2016) Astrocytes mediate neurovascular signaling to capillary pericytes but not to arterioles. Nat Neurosci 19(12):1619–1627. https://doi.org/10.1038/nn.4428

Miyazaki K, Ohta Y, Nagai M et al (2011) Disruption of neurovascular unit prior to motor neuron degeneration in amyotrophic lateral sclerosis. J Neurosci Res 89(5):718–728. https://doi.org/10.1002/jnr.22594

Moccia F, Negri S, Faris P, Angelone T (2022) Targeting endothelial ion signalling to rescue cerebral blood flow in cerebral disorders. Vasc Pharmacol 145:106997. https://doi.org/10.1016/j.vph.2022.106997

Moghbelinejad S, Nassiri-Asl M, Farivar TN et al (2014) Rutin activates the MAPK pathway and BDNF gene expression on beta-amyloid induced neurotoxicity in rats. Toxicol Lett 224(1):108–113. https://doi.org/10.1016/j.toxlet.2013.10.010

Morató X, Marquié M, Tartari JP et al (2023) A randomized, open-label clinical trial in mild cognitive impairment with EGb 761 examining blood markers of inflammation and oxidative stress. Sci Rep 13(1):5406. https://doi.org/10.1038/s41598-023-32515-6

Moussa C, Hebron M, Huang X et al (2017) Resveratrol regulates neuro-inflammation and induces adaptive immunity in Alzheimer's disease. J Neuroinflammation 14(1):1. https://doi.org/10.1186/s12974-016-0779-0

Murphy MJ, Grace GM, Tartaglia MC et al (2012) Widespread cerebral haemodynamics disturbances occur early in amyotrophic lateral sclerosis. Amyotrophic Lateral Sclerosis 13(2):202–209. https://doi.org/10.3109/17482968.2011.625569

Nan W, Zhonghang X, Keyan C, Tongtong L, Wanshu G, Zhongxin X (2018) Epigallocatechin-3-gallate reduces neuronal apoptosis in rats after middle cerebral artery occlusion injury via PI3K/AKT/eNOS signaling pathway. Biomed Res Int 2018:6473580. https://doi.org/10.1155/2018/6473580

Owens CD, Pinto CB, Detwiler S et al (2024) Neurovascular coupling impairment as a mechanism for cognitive deficits in COVID-19. Brain Commun 6(2):fcae080. https://doi.org/10.1093/braincomms/fcae080

Oyebode O, Kandala N-B, Chilton PJ, Lilford RJ (2016) Use of traditional medicine in middle-income countries: a WHO-SAGE study. Health Policy Plan 31(8):984–991. https://doi.org/10.1093/heapol/czw022

Pervin M, Unno K, Nakagawa A et al (2017) Blood brain barrier permeability of (−)-epigallocatechin gallate, its proliferation-enhancing activity of human neuroblastoma SH-SY5Y cells, and its preventive effect on age-related cognitive dysfunction in mice. Biochem Biophys Rep 9:180–186. https://doi.org/10.1016/j.bbrep.2016.12.012

Pflanzner T, Janko MC, André-Dohmen B et al (2011) LRP1 mediates bidirectional transcytosis of amyloid-β across the blood-brain barrier. Neurobiol Aging 32(12):2323

Phillips A, Chan F, Zi MM, Zheng AK, Ainslie P (2016) Neurovascular coupling in humans: physiology, methodological advances and clinical implications. J Cereb Blood Flow Metab 36(4):647–664. https://doi.org/10.1177/0271678X15617954

Plascencia-Villa G, Perry G (2023) Roles of oxidative stress in synaptic dysfunction and neuronal cell death in Alzheimer's disease. Antioxidants (Basel, Switzerland) 12(8):1628. https://doi.org/10.3390/antiox12081628

Pompili M, Innamorati M, Gonda X et al (2014) Pharmacotherapy in bipolar disorders during hospitalization and at discharge predicts clinical and psychosocial functioning at follow-up. Hum Psychopharmacol 29(6):578–588

Pun PBL, Jia L, Moochhala S (2009) Involvement of ROS in BBB dysfunction. Free Radic Res 43(4):348–364. https://doi.org/10.1080/10715760902751902

Rajkumar RP (2024) Revisiting a hypothesis: the neurovascular unit as a link between major depression and neurodegenerative disorders. Front Cell Neurosci 18:1455606. https://doi.org/10.3389/fncel.2024.1455606

Ramírez-Garza SL, Laveriano-Santos EP, Marhuenda-Muñoz M et al (2018) Health effects of resveratrol: results from human intervention trials. Nutrients 10(12):1892. https://doi.org/10.3390/nu10121892

Rancillac A, Rossier J, Guille M et al (2006) Glutamatergic control of microvascular tone by distinct GABA neurons in the cerebellum. J Neurosci 26(26):6997–7006. https://doi.org/10.1523/JNEUROSCI.5515-05.2006

Rastogi V, Santiago-Moreno J, Doré S (2015) Ginseng: a promising neuroprotective strategy in stroke. Front Cell Neurosci 8(2014):1662–5102. https://www.frontiersin.org/journals/cellular-neuroscience/articles/10.3389/fncel.2014.0045. https://doi.org/10.3389/fncel.2014.00457

Reddy PH, Manczak M, Yin X et al (2018) Protective effects of Indian spice curcumin against amyloid-β in Alzheimer's disease. J Alzheimer's Dis 61(3):843–866. https://doi.org/10.3233/JAD-170512

Ringman JM, Younkin SG, Pratico D et al (2008) Biochemical markers in persons with preclinical familial Alzheimer disease. Neurology 71(2):85–92. https://doi.org/10.1212/01.wnl.0000303973.71803.81

Sabouni N, Marzouni HZ, Palizban S et al (2023) Role of curcumin and its nanoformulations in the treatment of neurological diseases through the effects on stem cells. J Drug Target 31(3):243–260. https://doi.org/10.1080/1061186X.2022.2141755

Sabri M, Aiza N, Lee S-K, Murugan DD, Ling WC (2022) Epigallocatechin Gallate (EGCG) alleviates vascular dysfunction in angiotensin II-infused hypertensive mice by modulating oxidative stress and eNOS. Sci Rep 12(1):17633. https://doi.org/10.1038/s41598-022-21107-5

Sachdeva R, Jia M, Wang S et al (2020) Vascular-cognitive impairment following high-thoracic spinal cord injury is associated with structural and functional maladaptations in cerebrovasculature. J Neurotrauma 37(18):1963–1970. https://doi.org/10.1089/neu.2019.6913

Saha A, Sarkar C, Singh SP et al (2012) The blood-brain barrier is disrupted in a mouse model of infantile neuronal ceroid lipofuscinosis: amelioration by resveratrol. Hum Mol Genet 21(10):2233–2244. https://doi.org/10.1093/hmg/dds038

Schinkel AH (1999) P-glycoprotein, a gatekeeper in the blood-brain barrier. Adv Drug Deliv Rev 36(2–3):179–194. https://doi.org/10.1016/s0169-409x(98)00085-4

Selvakumar K, Lakshmi Prabha R, Saranya K, Bavithra S, Krishnamoorthy G, Arunakaran J (2013) Polychlorinated biphenyls impair blood-brain barrier integrity via disruption of tight junction proteins in cerebrum, cerebellum and hippocampus of female Wistar rats: neuropotential role of quercetin. Hum Exp Toxicol 32(7):706–720. https://doi.org/10.1177/0960327112464798

Sharma A, Hind K, Hume P, Singh J, Patrick Neary J (2020) Neurovascular coupling by functional near infra-red spectroscopy and sport-related concussion in retired Rugby players: the UK Rugby health project. Front Hum Neurosci 14:42. https://doi.org/10.3389/fnhum.2020.00042

Sokoloff L (1996) Circulation in the central nervous system. In: Comprehensive human physiology: from cellular mechanisms to integration. Springer

Sorond FA, Lipsitz LA, Hollenberg NK, Fisher NDL (2008) Cerebral blood flow response to flavanol-rich cocoa in healthy elderly humans. Neuropsychiatr Dis Treat 4(2):433–440

Spronck B, Martens EGHJ, Gommer ED, van de Vosse FN (2012) A lumped parameter model of cerebral blood flow control combining cerebral autoregulation and neurovascular coupling. Am J Physiol Heart Circ Physiol 303(9):H1143–H1153. https://doi.org/10.1152/ajpheart.00303.2012

Sriraksa N, Wattanathorn J, Muchimapura S, Tiamkao S, Brown K, Chaisiwamongkol K (2012) Cognitive-enhancing effect of quercetin in a rat model of Parkinson's disease induced by 6-Hydroxydopamine. Evid Based Complement Alternat Med 2012:823206. https://doi.org/10.1155/2012/823206

Stickland R, Allen M, Magazzini L, Singh KD, Wise RG, Tomassini V (2019) Neurovascular coupling during visual stimulation in multiple sclerosis: a MEG-fMRI study. Neuroscience 403:54–69. https://doi.org/10.1016/j.neuroscience.2018.03.018

Sun G, Miao Z, Ye Y et al (2020) Curcumin alleviates neuroinflammation, enhances hippocampal neurogenesis, and improves spatial memory after traumatic brain injury. Brain Res Bull 162:84–93. https://doi.org/10.1016/j.brainresbull.2020.05.009

Sun X-Y, Li L-J, Dong Q-X et al (2021) Rutin prevents tau pathology and neuroinflammation in a mouse model of Alzheimer's disease. J Neuroinflammation 18(1):131

Sweeney MD, Kisler K, Montagne A, Toga AW, Zlokovic BV (2018) The role of brain vasculature in neurodegenerative disorders. Nat Neurosci 21(10):1318–1331. https://doi.org/10.1038/s41593-018-0234-x

Tarantini S, Cam HT, Tran GR, Gordon ZU, Csiszar A (2017) Impaired neurovascular coupling in aging and Alzheimer's disease: contribution of astrocyte dysfunction and endothelial impairment to cognitive decline. Exp Gerontol 94:52–58. https://doi.org/10.1016/j.exger.2016.11.004

Thakore P, Alvarado MG, Ali S et al (2021) Brain endothelial cell TRPA1 channels initiate neurovascular coupling. eLife 10:e63040. https://doi.org/10.7554/eLife.63040

Thiruvengadam M, Venkidasamy B, Subramanian U et al (2021) Bioactive compounds in oxidative stress-mediated diseases: targeting the NRF2/ARE Signaling pathway and epigenetic regulation. Antioxidants (Basel, Switzerland) 10(12):1859. https://doi.org/10.3390/antiox10121859

Tian P, Teng IC, May LD et al (2010) Cortical depth-specific microvascular dilation underlies laminar differences in blood oxygenation level-dependent functional MRI signal. Proc Natl Acad Sci USA 107(34):15246–15251. https://doi.org/10.1073/pnas.1006735107

Toledo JB, Arnold SE, Raible K et al (2013) Contribution of cerebrovascular disease in autopsy confirmed neurodegenerative disease cases in the National Alzheimer's Coordinating Centre. Brain J Neurol 136(Pt 9):2697–2706. https://doi.org/10.1093/brain/awt188

Tong V, Teng XW, Chang TKH, Abbott FS (2005) Valproic acid II: effects on oxidative stress, mitochondrial membrane potential, and cytotoxicity in glutathione-depleted rat hepatocytes. Toxicol Sci Off J Soc Toxicol 86(2):436–443. https://doi.org/10.1093/toxsci/kfi185

Toth P, Tarantini S, Tucsek Z et al (2014) Resveratrol treatment rescues neurovascular coupling in aged mice: role of improved cerebromicrovascular endothelial function and downregulation of NADPH oxidase. Am J Physiol Heart Circ Physiol 306(3):H299–H308. https://doi.org/10.1152/ajpheart.00744.2013

Turner RS, Thomas RG, Craft S et al (2015) A randomized, double-blind, placebo-controlled trial of resveratrol for Alzheimer disease. Neurology 85(16):1383–1391. https://doi.org/10.1212/WNL.0000000000002035

Turoni J, Claudio RM, Karbiner M, Muntaner J, Proto V, Peral M, de Bruno. (2011) Oxidative stress in hypertensive patients induces an increased contractility in vein grafts independent of endothelial function. Int J Hypertens 2011:902129. https://doi.org/10.4061/2011/902129

Wan W, Cao L, Liu L et al (2015) Aβ(1-42) oligomer-induced leakage in an in vitro blood-brain barrier model is associated with up-regulation of RAGE and metalloproteinases, and down-regulation of tight junction scaffold proteins. J Neurochem 134(2):382–393. https://doi.org/10.1111/jnc.13122

Wang L, Li C, Guo H, Kern TS, Huang K, Zheng L (2011) Curcumin inhibits neuronal and vascular degeneration in retina after ischemia and reperfusion injury. PLoS One 6(8):e23194. https://doi.org/10.1371/journal.pone.0023194

Wąsik A, Antkiewicz-Michaluk L (2017) The mechanism of neuroprotective action of natural compounds. Pharmacol Rep 69(5):851–860. https://doi.org/10.1016/j.pharep.2017.03.018

Wiedenhoeft T, Tarantini S, Nyúl-Tóth Á et al (2019) Fusogenic liposomes effectively deliver resveratrol to the cerebral microcirculation and improve endothelium-dependent neurovascular coupling responses in aged mice. GeroScience 41(6):711–725. https://doi.org/10.1007/s11357-019-00102-1

Wu X, Zheng X, Tang H et al (2022) A network pharmacology approach to identify the mechanisms and molecular targets of curcumin against Alzheimer disease. Medicine 101(34):e30194

Xia L, Jiang Z-L, Wang G-H, Bao-Yin H, Ke K-F (2012) Treatment with ginseng total saponins reduces the secondary brain injury in rat after cortical impact. J Neurosci Res 90(7):1424–1436. https://doi.org/10.1002/jnr.22811

Yamazaki Y, Kanekiyo T (2017) Blood-brain barrier dysfunction and the pathogenesis of Alzheimer's disease. Int J Mol Sci 18(9):1965. https://doi.org/10.3390/ijms18091965

Yang F, Lim GP, Begum AN et al (2005) Curcumin inhibits formation of amyloid β oligomers and fibrils, binds plaques, and reduces amyloid in vivo. J Biol Chem 280(7):5892–5901

Yang T, Sun Y, Zhengyu L, Leak RK, Zhang F (2017) The impact of cerebrovascular aging on vascular cognitive impairment and dementia. Ageing Res Rev 34:15–29. https://doi.org/10.1016/j.arr.2016.09.007

Yang Y-L, Cheng X, Li W-H, Liu M, Wang Y-H, Du MS (2019) Kaempferol attenuates LPS-induced striatum injury in mice involving anti-neuroinflammation, maintaining BBB integrity, and down-regulating the HMGB1/TLR4 pathway. Int J Mol Sci 20(3):491

Yang R, Shen Y-J, Chen M, Zhao J-Y, Chen S-H, Zhang W et al (2022) Quercetin attenuates ischemia reperfusion injury by protecting the blood-brain barrier through Sirt1 in MCAO rats. Asian Nat Prod Res 24(3):278–289

Ye R, Yang Q, Kong X et al (2011) Ginsenoside Rd attenuates early oxidative damage and sequential inflammatory response after transient focal ischemia in rats. Neurochem Int 58(3):391–398. https://doi.org/10.1016/j.neuint.2010.12.015

Yu X, Ji C, Shao A (2020a) Neurovascular unit dysfunction and neurodegenerative disorders. Front Neurosci 14:334. https://doi.org/10.3389/fnins.2020.00334

Yu X, Li Y, Xiaohua M (2020b) Effect of quercetin on PC12 Alzheimer's disease cell model induced by Aβ25-35 and its mechanism based on Sirtuin1/Nrf2/HO-1 pathway. Biomed Res Int 2020:8210578. https://doi.org/10.1155/2020/8210578

Zaplatic E, Bule M, Shah SZA, Uddin MS, Niaz K (2019) Molecular mechanisms underlying protective role of quercetin in attenuating Alzheimer's disease. Life Sci 224:109–119. https://doi.org/10.1016/j.lfs.2019.03.055

Zeli C, Lombardo M, Storz MA, Ottaviani M, Rizzo G (2022) Chocolate and cocoa-derived biomolecules for brain cognition during ageing. Antioxidants 11(7):1353. https://doi.org/10.3390/antiox11071353

Zenaro E, Piacentino G, Constantin G (2017) The blood-brain barrier in Alzheimer's disease. Neurobiol Dis 107:41–56. https://doi.org/10.1016/j.nbd.2016.07.007

Zhang C, Browne A, Child D, Tanzi RE (2010) Curcumin decreases amyloid-beta peptide levels by attenuating the maturation of amyloid-beta precursor protein. J Biol Chem 285(37):28472–28480. https://doi.org/10.1074/jbc.M110.133520

Zhang J, Wang Y, Dong X, Liu J (2018) Crocetin attenuates inflammation and amyloid-β accumulation in APPsw transgenic mice. Immun Ageing 15:24. https://doi.org/10.1186/s12979-018-0132-9

Zhao P, Lu Y, Wang Z (2023) Naringenin attenuates cerebral ischemia/reperfusion injury by inhibiting oxidative stress and inflammatory response via the activation of SIRT1/FOXO1 signaling pathway in vitro. Acta Cir Bras 38:e380823

Zhong Z, Deane R, Ali Z et al (2008) ALS-causing SOD1 mutants generate vascular changes prior to motor neuron degeneration. Nat Neurosci 11(4):420–422. https://doi.org/10.1038/nn2073

Zhu CW, Grossman H, Neugroschl J et al (2018) A randomized, double-blind, placebo-controlled trial of resveratrol with glucose and malate (RGM) to slow the progression of Alzheimer's disease: a pilot study. Alzheimer's Dementia (New York, NY) 4:609–616. https://doi.org/10.1016/j.trci.2018.09.009

Zonta M, Angulo MC, Gobbo S et al (2003) Neuron-to-astrocyte signaling is central to the dynamic control of brain microcirculation. Nat Neurosci 6(1):43–50. https://doi.org/10.1038/nn980

MIX
Papier aus verantwortungsvollen Quellen
Paper from responsible sources
FSC® C105338

If you have any concerns about our products,
you can contact us on
ProductSafety@springernature.com

In case Publisher is established outside the EU,
the EU authorized representative is:
Springer Nature Customer Service Center GmbH
Europaplatz 3, 69115 Heidelberg, Germany

Printed by Libri Plureos GmbH
in Hamburg, Germany